International Conference on Information Technology and Engineering Management (ITEM 2019)

IOP Conference Series: Materials Science and Engineering
Volume 648

Belitung, Indonesia
27 - 29 June 2019

ISBN: 978-1-7138-0909-8
ISSN: 1757-8981

Printed from e-media with permission by:

Curran Associates, Inc.
57 Morehouse Lane
Red Hook, NY 12571

proceedings.com

Some format issues inherent in the e-media version may also appear in this print version.

Institute of Physics
Dirac House, Temple Back
Bristol BS1 6BE UK

Phone: 44 1 17 929 7481
Fax: 44 1 17 920 0979

techtracking@iop.org

Additional copies of this publication are available from:

Curran Associates, Inc.
57 Morehouse Lane
Red Hook, NY 12571 USA
Phone: 845-758-0400
Fax: 845-758-2633
Email: curran@proceedings.com
Web: www.proceedings.com

TABLE OF CONTENTS

IOP Conf. Series: Materials Science and Engineering **648** (2019) 011001 doi:10.1088/1757-899X/648/1/011001

Dear Participants,

Welcome to ***BELITONG***! Welcome to participate in 2019 The 2019 International Conference on Information Technology and Engineering Management (ITEM), supported by **IOP Conference Series, Material Science and Engineering** and other organizations.

The 2019 International Conference on Information Technology and Engineering Management (ITEM), is going to focus on all the major aspects in the fields of information technology and engineering management. ITEM conference links researchers, scientists, engineers, students and practitioners from different branches of information technology and engineering management from around the world. ITEM provides opportunities from both academic as well as industry to discuss, share, exchange and to extend knowledge, view and experience in respective fields. ITEM will be instrumental in generating ideas and disseminating information to a greater extent than is possible through more conventional forums.

We would like to thank all the committees for their great support on organizing the conference. We also would like to thank all the reviewers for their great effort on reviewing the papers submitted to ITEM 2019. Special thanks to all the researchers and students who with their work and participate in the conference.

We hope you enjoy the conference, the food, the hospitality, and the beautiful and charming city of Belitong.

Chairman Committee,

Ikhsan Siregar

Peer review statement

All papers published in this volume of *IOP Conference Series: Materials Science and Engineering* have been peer reviewed through processes administered by the proceedings Editors. Reviews were conducted by expert referees to the professional and scientific standards expected of a proceedings journal published by IOP Publishing.

IOP Conf. Series: Materials Science and Engineering **648** (2019) 012001 doi:10.1088/1757-899X/648/1/012001

Analysis of Corporate Governance Effect and Characteristics of Companies on the Existence of Risk Management Committee

Abdillah Arif Nasution[1]

[1]Department of Accounting, Universitas Sumatera Utara, Medan-Indonesia

*Email :badinst@usu.ac.id

Abstract This study aims to analyze the influence of corporate governance and company characteristics on the existence of the Risk Management Committee (RMC) and type of RMC, whether separated and combined with the audit committee. Variables are broken down into independent commissioners, auditor reputation, market risk, leverage and company size. The statistical method for testing hypotheses is logistic regression analysis. Samples collected using random sampling are included in eighty manufacturing companies listed on the Stock Exchange for 2012-2017. This study uses agency theory, company legitimacy, and signal theory to explain the relationship between variables. This study shows that several independent variables have a positive effect on the existence of RMC, namely the size of the company. as well as the independent variables that positively influence the existence of a separate RMC are the size of the company.

1. Introduction

Risk is the potential for the occurrence of an event that can cause losses [1]. If the risk is not managed properly it will cause losses for the company and even bankruptcy experienced by a number of companies. The Company began the initiative to improve corporate governance with a significant emphasis on the role of risk management [2]. The complexity of risk management makes the quality of internal control higher when the risk management committee is compared to the absence of a risk management committee [2]. However, according to KPMG [3] it was found that the risk management committee still existed which was integrated with the audit committee.

As a result, the broad role and responsibility of a large audit committee increases criticism and doubts about its ability to function effectively [2]. Therefore, internal control of risk management is expected to be higher when the risk management committee is independent than when integrated with the audit committee.

This study aims to analyze and find empirical evidence the influence of board of commissioners independence, auditor reputation, market risk, and company size on the existence of a risk management committee.

2. Literature Review

The risk management committee (RMC) is a subcommittee that has a very important function in the company. The presence of the RMC is expected to assist the board of commissioners in carrying out the supervisory function as an effort to protect stakeholders and achieve the company's goals. The function of supervision through the RMC is mainly on risk, risk management and internal control. However, the presence of RMC is still voluntary and there are still many who are members of the audit committee. Though the task of a complex audit committee casts doubts on its accountability and credibility in monitoring corporate risk management. Therefore, this study wants to examine the actual factors that have a positive effect on the presence of RMC and type of RMC.

2.1 Agency Theory
In agency theory, both the principal and agent are assumed to be rational economic people and are solely motivated by their individual interests. From this situation arises a conflict of interest between the principal and the agent. To reduce the actions of agents that are not in accordance with their interests, the principal has two ways, namely [4]

- Supervise agent behavior by adopting audit functions and other corporate governance mechanisms that can align the interests of the agent with the interests of the principal.
- Providing attractive staff incentives to agents and establishing reward structures that can persuade agents to act in accordance with the principal's best interests

Agency theory is often used as a basis in previous researches on corporate governance, especially regarding the existence of committees. This is due to the importance of the aspect of monitoring (monitoring) for the realization of good corporate governance. When viewed from the agency perspective, there are two general management oversight mechanisms, namely internal supervision and external supervision. The internal control mechanism is the board of commissioners and committees [5; 6; 7], while the external monitoring mechanism is an external auditor [2].

2.2 Signaling Theory
One theory that can be the background of the problem of information asymmetry in markets is the signaling theory. When used in the practice of corporate disclosure, signaling theory is generally beneficial for companies to express good corporate governance practices, to create good corporate quality in the market [2]. One form of signal about the quality of the company is the formation of a committee, which provides information that the company is better in terms of supervision compared to other companies.

According to the signaling theory, although there are no regulations that mandate the establishment of RMC as committees that specifically play a role in risk monitoring, companies can still form RMC in their commitment to the practice of good corporate governance. Harrison [8] states that it may be difficult to observe what work is actually carried out by the committee, but it is possible that the supervisory committee was formed to create a favorable company appearance for outsiders.

2.3 Efficient Market Hypothesis (EMH)
In this theory, efficient terminology refers to information that is informationally efficient. This theory states that if the market is efficient, the price reflects all the information available. Regarding EMH, the Indonesian capital market itself is included in the semi-strong form [9]. Securities prices in the Indonesian capital market react to information published by the issuer, such as financial statements, dividend distribution announcements, and so on.

2.4 Risk Management Committee
RMC has become popular as an important risk monitoring mechanism for companies [2]. This is further reinforced by a survey by KPMG [3] on Australian companies, which stated that more than half of respondents (54%) had RMC, of which 70% were joined by audit committees. In general the area of duty and authority of the RMC is [2]:

- Consider organizational risk management strategies;
- Evaluate the organization's risk management operations;

- Estimating organizational financial reporting; and
- Ensure that the organization in practice complies with applicable laws and regulations.

In its formation, RMC can be incorporated into an audit or can also be a separate and independent committee. Separate committees that specifically focus on risk issues (RMC), are an effective mechanism in supporting the board of commissioners to fulfill their responsibilities in the task of risk monitoring and internal control management [2]. A separate RMC from the audit will be more able to devote more time and effort to combining the various risks faced by the company broadly and evaluating the related controls as a whole [2]. In addition, a separate RMC from the audit also allows the board of commissioners to understand the company's risk profile more deeply [10].

3. Research Method

3.1 Research Variable

In this study, the concept of the existence of RMC was measured by dummy numbers. Code 1 for companies that hold RMC and code 0 for companies that do not hold RMC [2]. The existence of the RMC type is measured by dummy numbers. Code 1 for companies that hold separate RMC and code 0 for RMC companies that are members of the audit committee. The independence of the board of commissioners is measured by the number of independent commissioners owned by the company. Auditor's reputation is measured by dummy numbers, code 1 for Big four members (Deloitte, KMPG, PWC, Ernst and Young) and vice versa is coded 0. Market risk is measured by calculating beta values with regression techniques. Size is measured from total assets company [2].

3.2 Population and Sample

The population to be studied are all manufacturing companies listed on the IDX (Indonesia Stock Exchange). The study period was conducted in 2012-2017 with the reason that there were enough statistics and observations to be obtained statistically. The sampling technique used in this study was Random Sampling. The sample in this study was randomly selected, namely as many as 80 companies listing on the IDX in 2012-2017. Retrieving several samples with sufficient reason and fulfilling statistically qualified and considered to represent the research population.

3.3 Data Analysis Method

The analytical method used to test the hypothesis in this study is logistic regression. Logistic regression does not require a normality, heteroscedastic, and classical assumption test on the dependent variable [11]. Logistic regression was chosen because this study has a dichotomous dependent variable [2] and an independent variable that is a combination of metric and nonmetric (nominal). There are two logistic regression models used to test the hypothesis, namely:

The simple linear regression general equation used in this study is:

$$\text{Logit (RMC)} = \alpha+\beta 1+\beta 2+\beta 3+\beta 4+e. \quad (1)$$

$$\text{logit(SRMC)} = \alpha+\beta 1+\beta 2+\beta 3+\beta 4+e. \quad (2)$$

4. Result and discussion

Based on the tests that have been conducted, the presence of RMC is positively influenced size (company size). Whereas, the existence of separate RMC is positively influenced by the size. The discussion of each variable will be explained as follows:

Table 1. Model I hypothesis test results

Variable	Significance value (5%)
Independent commissioner	0,988
Auditor reputation	0,550
Market risk	0,835
Size	0,012*

Note : *) significant

Source : Output SPSS

Table 2. Model II hypothesis test results

Variable	Significance value (5%)
Independent commissioner	0,036
Auditor reputation	0,075
Market risk	0,525
Size	0,000*

Note : *) significant Source : Output SPSS

The results of testing hypothesis 1a show that the independence variable of the board of commissioners has a significant probability value of 0.988 and a positive coefficient of 0.002. Whereas, the results of hypothesis 1b show a probability value of significance of 0.036 and a negative coefficient of 0.519. This shows that statistically, the independence variable of the board of commissioners does not have a positive effect on the existence of RMC or RMC separately. The absence of positive influence, because the quality and educational background of the members of the board of commissioners more determines the quality of the board's oversight function than the composition and level of independence. Another possible reason is that there are only independent commissioners to fulfill the rules and are not intended to implement corporate governance in the company. Coupled with the minimum provisions of independent commissioners in the board only 30% so that the policy for decision making to organize RMC is not yet dominant.

The results of hypothesis 2a testing indicate that the auditor reputation variable has a significance probability value of 0.550 and a positive coefficient of 0.195. Whereas, the results of the 2b hypothesis test show a probability value of 0.075 significance and a positive coefficient of 0.749. This shows that statistically, the auditor reputation variable does not have a positive effect on the existence of RMC or RMC separately. The reason that might underlie is that companies tend to use Big Four auditors just to increase their reputation. Big four auditors are also considered able to provide recommendations in corporate governance practices. However, these recommendations have not covered aspects of overall risk monitoring.

The results of testing hypothesis 3a indicate that the market risk variable has a significant probability value of 0.835 and a positive coefficient of 0.017. Whereas, the results of hypothesis testing 3b show a probability value of significance 0.525 and a positive coefficient of 0.129. This shows that statistically, the market risk variable does not have a positive effect on the existence of a separate RMC or RMC. Companies with high market risk tend not to hold RMC. When the market risk is high, the company is in a bad condition. This condition causes companies to act more carefully in activities that increase company costs. On the other hand, the company considers the principle of benefit costs. The principle of benefit costs is a comparison between the benefits obtained with the value to be received. If the costs incurred are greater than the benefits obtained, the company is better not to hold an RMC.

The test results of the size variable in model I have a significance probability value of 0.012 and a positive coefficient of 0.305. Whereas, the results in model II show a significance probability value of 0,000 and a positive coefficient of 1,188. This shows that statistically, the variable size of the company has a positive effect on the existence of RMC and RMC separately. This is consistent with the results of Andarini's (2010) study that the size of the company has a significant and positive effect on the existence of separate RMC and RMC. This is contrary to the research of Subramaniam, et al. [2] that company size does not have a positive effect on the existence of separate RMC.

5. Conclusion

The variable independence of the board of commissioners, auditor reputation, and market risk does not have a positive effect on the existence of RMC and RMC separately.

This study has several limitations that might be used as a reference in improving future research. The limitations of this study are that the sample of this study is limited to manufacturing companies listed on the Stock Exchange during 2012-2017. This research has only analyzed several corporate governance factors while there are still many corporate governance factors, company characteristics or other variables that might be indicated to influence the existence of RMC.

Based on the limitations above, further research is expected to increase the number of companies or samples so that the results of the research obtained can illustrate better research results. Second, developing corporate governance variables, company characteristics or other variables that have not been accommodated in this study such as the number of audit committees, audit committee meetings, market share, total risk, and so on.

References

[1] Minister of finance regulation Number 142/PMK.010/2009 about risk

[2] Subramaniam, Nava; McManus, Lisa; and Zhang, Jiani, "Corporate Governance, Company Characteristics and Establishment of a Risk Management Committee in Australia Companies" Managerial Auditing Journal, vol. 24, Issue 4, 2009, pp. 316-339, doi:10.1108/02686900910948170.

[3] KPMG, Strategic Risk Management Survey, Accessed November, 29, 2009, www.kpmg.com.au.

[4] Jensen, M C and Meckling W H 1976 Theory of The Firm: Managerial Behaviour, Agency Costs, and Ownership Structure. *Journal of Financial Economics*, 3(4), 1976, pp. 305-360.

[5] Ruigrok, W, Peck S, Tacheva S, Greve P, and Yan H, The Determinants and Effects of Board Nomination Committees *Journal of Management Governance*, 10, 2006, pp. 119-148.

[6] Firth M and Rui O M Voluntary Audit Committee Formation and Agency Costs" *International Journal of Accounting, Auditing and Performance Evaluation*, 4 (2), 2007, pp. 142-160.

[7] Chen,L., Kilgore A, and Radich R, Audit Committees: Voluntary Formation by ASX Non-Top 500 *Managerial Auditing Journal*, 24(5), 2009, pp. 475-493.

[8] Harrison, J.R, The Strategic Use of Corporate Board Committees *California Management Review*, 30(1), 1987, pp.109-125.

[9] Mahendra, T Analisis Efisiensi Pasar Modal Indonesia Periode 2003-2005 (Studi Pada PT. Bursa Efek Jakarta), Thesis, 2008 Universitas Muhammadiyah.

[10] Bates, E W, and Leclerc R J, Boards of Directors and Risk Committees *The Corporate Governance Advisor*, 17(6), 2009, pp. 16-18.

[11] Ghozali, I. Aplikasi Analisis Multivariate Dengan Program SPSS, Semarang 2005, Badan Penerbit Universitas Diponegoro.

Acknowledgments

Authors thank to Allah, lord of the universe who always bestows his grace along with guidance, so that the authors can finish the research.

IOP Conf. Series: Materials Science and Engineering **648** (2019) 012002 doi:10.1088/1757-899X/648/1/012002

Application of Fuel Oil Distribution Route Using Ford Algorithm

I Siregar[1]

[1]Department of Industrial Engineering, Faculty of Engineering, Universitas Sumatera Utara, Medan, Indonesia

Email : ikhsan.siregar@usu.ac.id

Abstract. The company which is engaged in Fuel Oil distribution is located in Riau City. Distribution in the company is to distribute Fuel Oil to each distribution center (SPBU) which is the scope of the company with the amount set by the distribution center. Distribution often experiences delays because there is no distribution route. This causes the company's losses in the form of increasing working hours (overtime and adding of transportation vehicles. Determination of distribution routes to the company is a key to success in handling the problem of distributing Fuel Oil to each distribution center. This method provides alternatives for the company about the shortest distance traveled for the Fuel Oil distribution and delay can be minimized. The final result in this study is the company find out the loading time of the vehicle in the form of distance traveled by vehicle, distance time and cost distribution and the company can find out how long the distribution time needed to each distribution center, the shortest path problem method with the Ford algorithm can minimize the distance as much as 13 distribution center and minimize the average savings of distance for 56 km / hour.

1. Introduction

The rapid development of oil distributor companies in Indonesia is increasingly making competition conditions in the market share increasingly of high competition. This condition because of the oil distributor company as a company engaged in services and current needs has not been replaced and always meet the needs and desires of the community. An oil distributor company is a company engaged in distributing oil to each distribution center with the amount set by the gas station. Therefore, the company must ensure that oil distribution arrives at the distribution center on time [1, 2].

Oil distributor companies are the main needs of society by presenting efficient and quality services. Distribution refers to the relationship between the production point and the last customer at the right time, the right quality specifications and cheap costs. The goal is to achieve the level of consumer needed based on cost level. This satisfaction can be a benchmark for the quality of the company and as an evaluation material. This causes companies required to provide on time services in accordance with established standards and reach all distribution centers. The business scope consists of the energy business in the upstream sector and the downstream sector. Timeliness in delivery is at the main core of good production system. In addition, the company also ensure the quality of oil arriving at the distribution center [3].

Oil distributor companies have many distribution centers are served by depots. The company distributes Fuel Oil every day according to working hours and demands from the distribution center. All of vehicle delivers FUEL OIL has a different route of delivery destination. The number of vehicles that deliver Fuel Oil with different destination routes, the company expects Fuel Oil delivery arrives at the right time and the company reach all demands from the distribution center without increase the hours of overtime work. From the distribution center also expects the delivery of Fuel Oil to be on time, if there is a delay in effect on the image of the distribution center and potentially lose customers and the distribution center suffers from financial and customer confidence. This is the main projection of the company. Based on the real condition in one period (one year) there are many delays in the distribution of Fuel Oil from Central Supply Facilities to the Distribution Center. This is caused by various factors such as the absence of a delivery schedule, the absence of route determination, processing time, and limited transportation. The delay in distributing fuel cause several losses including increasing overtime hours, transportation rent (tank cars), and losing customer trust. Avoiding and minimizing company losses, then one way can be taken is to determine the distribution route. The distribution routes currently used are still random, where the company does not know which routes are passed by transporters. The lack of clarity in choosing a route lead to inappropriate distribution. This is what the company is facing.

One method that can be used to analyze the distribution problem is using Ford algorithm. Ford algorithm is a method used in distance-vector routing protocols to saving network resource, routing information protocol and based on two metric routing problem [4]. Ford Algorithm can be used to reduce the network system for distribution problem [5]. Ford algorithm is defined by Abishek and Kalpana as shortest path from one node to another node in applied on negative weight [6]. Bellman ford algorithm can be used to solve the single source shortest path problem in which edges of a given diagraph can have negative weight as long as the graph contains no negative cycle. The computational time of his algorithm increases exponentially with the increase in the number of nodes. This algorithm returns a Boolean value TRUE if the given diagraph contains no negative cycles that are reachable from source vertex otherwise it returns Boolean value FALSE [7].

Many previous studies have been done in handling of distribution problem using Ford algorithm. Uzar & Catay (2012) do research applied *ford algorithm* on distribution planning problem of bulk lubrication at BP Turkey. Their objective is to minimize total transportation related cost. They model the problem with the first approach being a linear programming-based ford algorithm. Using a real-data of BP Lubes Logistics, they found that both variant of the rolling horizon threshold heuristic are able to provide good result very vast [8]. Other studies have been done by Onut (2013) applied ford algorithm on distribution problem in Turkey. This study works on heterogeneous fleet vehicle routing model for solving liquefied petroleum gas. Using real world data obtained from LPG Operating Company, their optimal solution aided in scheduling and assigning vehicle route daily. Zhan end Noon (1998), has conducted a research on the related matter regarding the most suitable shortest path network method with the road structure in the United State of America. They found that the ford algorithm is the most suitable method with this network. This exploratory research has been conducted in order to calculate the shortest path running time utilizing real road data of the State of Johor and Malacca. There are a few considerate matters in this research as assumption which is the tested data did not take into account the negative path in the network structure. Other than that, the network has a directed path from a node to all nodes in the network range. Thus, the network has two ways directed path. Lastly, obstructions in the network such as traffic light, traffic jam and damaged road factor has not been considered. [9, 10]. The application of Ford algorithm to improve distribution problems is still rarely applied to oil distributor companies.

2. Methodology

The research was conducted at an oil distributor in the Riau city where the object under study was a distribution system influenced and supported distribution route planning. The research begins with conducting direct observation in the company from the central supply facilities and distribution center. Activities carried out at this stage are measuring distance time from the central supply facilities to each distribution center and between distribution centers. After observation, the topic and purpose of the study are determined according to the conditions in the company. After that, data collection is needed to improve distribution service quality, namely distance. After the data has been collected, the analysis is then carried out using the Ford algorithm.. This algorithm has advantages such as cost minimization, maximizes the performance and allows splitting traffic [11]. Ford algorithm is formed by measuring distance. The stages in using the ford algorithm method are by distance time measurements. After measurements are carried out, the theoretical time of oil fuel loading is calculated by calculating cycle time, normal time and standard time. The route for distribution determination is done by ford algorithm to obtain the minimum distribution center distance. Finally, calculating the distance time of the shortest route, then the theoretical time needed to calculate the unexpected time.

3. Result and Discussion

3.1. Calculation of Theoretical in Oil Fuel Loading Time

This stage is to calculate the theoretical time when loading is done by looking for cycle time, normal time and standard time. Calculation of cycle times by paying attention to how long the workers completes their work from the beginning of the process to the final process. The normal time for each product with each conveyance capacity per product uses an adjustment factor according to Westinghouse. Standard time is done using the allowance table. Recapitulation of cycle time, normal time and standard time can be seen in table 1.

Table 1. Recapitulation of Cycle Time, Normal Time and Standard Time

Premium	16 KL	18 KL	21 KL	24 KL
WS	1:04:38	1:11:05	1:14:34	1:15:55
Wn	1:19:29	1:27:26	1:31:43	1:33:23
Wb	1:06:40	1:13:20	1:16:55	1:18:19
Biosolar	16 KL	18 KL	21 KL	24 KL
Ws	0:56:20	1:03:31	1:08:23	1:16:13
Wn	1:09:17	1:18:08	1:24:07	1:35:45
Wb	0:58:06	1:05:32	1:10:33	1:18:37

3.2. Determination of Distribution Route

Based on the result of the distribution route determination using Ford algorithm there are several distances that can be minimized. The results of the determination of distance minimization can be seen in table 2.

Table 2. Minimization of Distribution Distance

No	Distribution Center Code	Distance Before Route Determination (km)	Distance After Route Determination (km)	Difference
1	14.214226	208	120	88
2	14.214281	234	225	9
3	14.214284	217	129	88
4	14.214255	254	91	163
5	14.214225	253	165	88
6	14.227350	272	263	9
7	14.214230	312	303	9
8	14.214223	315	227	88
9	14.214219	300	291	9
10	14.214299	307	298	9
11	14.212278	362	353	9
12	14.285673	370	248	122
13	14.284617	189	157	32

IOP Conf. Series: Materials Science and Engineering **648** (2019) 012002 doi:10.1088/1757-899X/648/1/012002

From table 2 it can be seen that there are 13 minimized distribution distances from a total of 70 distribution distances.

3.3. Determination of Total Time

Travel time is obtained from the results of interviews with the relevant agencies. The agencies interviewed were drivers, employees, workers and depot heads. The results of the interview were the speed of driving a vehicle of 40 km / hour and unexpected estimation things on the trip given the 30-minute allowance. Total time can be seen in table 3.

Table 3. Total Time of Distribution Center

No	SPBU Code	Distance (km)	Total Time	
1	14.2886101	43	1:04:30	1:34:30
2	14.284602	47	1:10:30	1:40:30
3	14.288614	4	0:06:00	0:36:00
4	14.288652	13	0:19:30	0:49:30
5	14.2896120	90	2:15:00	2:45:00
6	14.287695	35	0:52:30	1:22:30
7	14.288659	9	0:13:30	0:43:30
8	14.288619	67	1:40:30	2:10:30
9	14.288626	60	1:30:00	2:00:00
10	14.287634	62	1:33:00	2:03:00
11	19.708.288	13	0:19:30	0:49:30
12	14.288619	21	0:31:30	1:01:30
13	14.287632	64	1:36:00	2:06:00

3.4. Determination of Distribution Cost

The distribution cost is Rp. 650, - / KM / KL. Distribution costs include maintenance costs and depreciation costs. Maintenance costs and depreciation costs should be recalculated and distribution costs are more presentative for company expenses. This is a usual problem because the distribution costs determined by the government. The distribution costs are set by the government usually happens and affected to disadvantage of company and losses due to the distance total time too far, to reduce losses the company create a distribution route with the shortest distance and the distribution costs can be minimized.

4. Conclusion

Oil distributor company have 70 distribution centers at first. The distance between central supply facilities to distribution center and between distribution center to distribution center does not have a clear distribution route. The distribution route based on random. The distribution route is more organized and not random and the distance total time by the transport vehicle is shorter or not too far away. The average of total time can be minimized is 56 km. The minimum distribution distance is 13 gas stations. The total time becomes shorter than before can make the distribution not take a long time and can save 1 hour of travel time so that the distribution costs can decrease.

References

[1] Reyza Theis 2013 *Jurnal EMBA*, **1** (3) 821-828
[2] Siregar I, Tarigan U, Siregar D A, Sari R M, Tambunan M M, Ginting E, and Nasution T H Determination of optimal production planning in palm oil company (Case study). In *2017 4th International Conference on Industrial Engineering and Applications (ICIEA)* (pp. 55-59). IEEE.
[3] Ivan 2012 *Optimasi Sistem Industri*, **11** (2) 243-252

[4] Shivani S., Leena J., and Bharti K 2013 *International Journal of Application or Innovation in Engineering and Management*, **2** (7) 316-320
[5] Ahmad A.Maidamisa., Noel Eckson 2017 *International Journal of Computer Applications,* **158** (1)
[6] Ms. Kalpana., Mr. Abishek Tyagi 2017 *International Research Journal of Engineering and Technology,* **4** (4)
[7] Kiruthika R and Umarani R 2012 *International Journal of Management, IT and Engineering,* **2** (4)
[8] Liangfang L., Akyene Tetteh., Qi Xu 2015 *British Journal of Economics Management & Trade*, **10** (2) 1-10
[9] Zhan, F.B. and C.E. Noon 1998 *International Journal of Engineering*, **32** (1)
[10] Rohaizan R., Shamsuddin A, Faiz S 2012 *Australian Journal of Basic and Applied Sciences* **6** (10) 278-281
[11] Jitendra B S and Tripathi R C 2018 *International Journal of engineering Development and Research*, **6** (1)

IOP Conf. Series: Materials Science and Engineering **648** (2019) 012003 doi:10.1088/1757-899X/648/1/012003

Evaluation of The Leading Work Culture 5S in Industry

I Rizkya[1]*, N Hidayati[1], R M Sari[1], U Tarigan[1]

[1]Departement of Industrial Enginnering, Faculty of Engineering, Universitas Sumatera Utara

*E-mail: indahrizkya@usu.ac.id

Abstract. At the moment the changes in the industrial world are getting faster, work demands increase based on the company wants. To support work it can be done easier and more comfortably, a effective and efficient work culture must be built. The effect of not succeeding in implementing an effective and efficient work culture is a work environment that is not clean and not neatly arranged. This paper will illustrate efforts to evaluate the application of the 5S (Five-S) work culture in the pine resin industry. 5S is applied to get a safe and comfortable working atmosphere. 5S stands for Seiri (sort), Seiton (Straighten), Seiso (Shine), Seiketsu (Standardize) and Shitsuke (sustain). The result of 5S evaluation is carried out in the pine gum industry produced a total score of 47, which 5S implementation was below the average, and a review was needed to increase the implementation of very weak of 5S component. The values of each 5S component are Seiri, Seiton, Seiso, Seiketsu and Shitsuke, respectively 17, 11, 7, 7, and 5. So the improvements are focused on Seiton, Seiso, Seiketsu and Shitsuke.

1. Introduction

Company performance cannot be separated from human factors can be observed, researched, analysed and corrected [1]. This condition done as an effort to get alternative ways of working are good, effective, and efficient. Effectively related to the right way of working to provide a short time for work completion. Efficiency is related to efforts to minimize the costs of work completion. Work effectiveness and efficiency must not ignore the quality of the products produced [2].

A good work culture within a company is also much needed for the company development in the future. In order to create a good work culture, a lot of effort is needed to achieve it. 5S is known as an effort to create a good work culture in an organization [3,4]. 5S is needed to implement Kaizen (Continuous Improvement) having a high commitment to every job, the implementation and work completion on time and on schedule, working without causing waste [5]. 5S (Seiri, Seiton, Seiso, Seiketsu and Shitsuke) is a "Digest for Kaizen" [6]. Kaizen is a Japanese term for the concept of Continuous Incremental Improvement. Kai means change and Zen means good. Kaizen means continuous improvement that involves everyone [7]. Kaizen is basically a comprehensive and integrated view that aims to carry out continuous improvement [8].

The 5S concept is the concept of utilizing the workplaces including equipment, materials, documents, and others to create a neat work environment and improve work discipline [9]. 5S implementation is able to overcome the problem of unwanted material, improper space utilization, work environment, more time to search documents, equipment, and stationary [10]. 5-S is a determination to organize, clean, maintain a stable condition and maintain the habits needed to do a good job. 5S is a process and method for creating and maintaining a safe, orderly, clean and high-performance workplace. Studies on the implementation of 5S have been carried out by many previous researchers, both in the manufacturing and service industries. The 5S manufacturing industry is capable to reduce waste and performance improvement [11,12]. 5S also be implemented in small and medium scale industries [13].

This paper describes the assessment of 5S implementation and design improvements for 5S implementation. This research is important to provide an evaluation overview of 5S implementation. Evaluation of 5S implementation is important to ensure the achievement of work culture improvement in 5S companies are not monitored and evaluated will cause failure in an effective and efficient work culture. Besides that, no specific study has been found regarding 5S implementation in industries that process pine gum.

2. Method

This research was conducted at one of the companies producing gondorukem and turpentine from pine gum. This research was conducted by following 5 steps. The first step is observation to see the current conditions of the application of 5S. The second step is evaluating the application of 5S in the pine resin industry by comparing the actual conditions and ideal conditions of 5S application. Measurements were made by distributing questionnaires to several workers and management to assess 5S implementation in their work area. The third step is to draw conclusions based on the evaluation that has been done. The fourth step is to provide improvement suggestions for improvements in the framework of implementing 5S to build a kaizen work culture.

5S measurements are carried out at each constituent components, namely seiri, seiton, seiso, seiketsu, shitsuke. Seiri is the first step in carrying out 5S culture, which is throwing / sorting / getting rid of items, files are not used anymore to the disposal site [14]. All items are on the job site, only items are really needed for work activities. Seiri aims to make the workplace look neater and not messy.

Seiton is the process of sorting all items or files are not used anymore, making sure everything must be placed according to the position specified, and it is always ready for use when needed. Seitons need to be ascertained whether each item has a place, every place has a name, make it organized and systematic, gives a name to each storage place is easy to remember, ensures that it is easily identified when needed, and there is no need to waste time searching for it [15].

Seiso is the process of cleaning the workplace, work space, equipment and work environment. Grow the thought that cleanliness is an important thing in life. Cleaning is carried out on a daily basis at work, and cleaning and hygiene maintenance are carried out [14,15]. *Seiketsu* is the process of maintaining the three stages have been carried out routinely. This stage also called the stage of care, is the standardization and consistency of each individual to carry out the previous stages. Make standards and all individuals must follow with these standards [14].

Shitsuke is discipline maintenance and it becomes a habit to run the 5S program. Recognizing by implementing practical and concise 5S has an impact on increasing efficiency, good service, working security productivity and profit [14].

3. Result and Discussion

3.1. Existing Condition and Problems Identification

The Pine Gum Industry has implemented 5S in its work area. However, the implementation is still not optimal because there is still waste, there is no need for equipment stored in the work area. The conditions of the work area in the pine gum industry are as follows:

The problem in the pine gum industry is the creation of ineffective and inefficient work culture. Meanwhile, 5S has been implemented but has never been evaluated. The efforts made in creating good work does not get a maximum benefit. Although it has implemented 5S, but there are still unclean and untidy work areas, put the equipment at random, and employees littering, equipment has no storage space, and etc.

Table 1. Existing Condition in Work Area of Pine Gum Industry

Condition	Explanation	Condition	Explanation
	Damaged drums are still in the work area and unarranged		There are items are not needed such as used wood
	Put the pedicabs are not in the right place		There is a pile of non-similar items

3.2. Assessment of 5S

There is a checklist sheet contain questions are arranged based on things must be considered in the 5S concept and adjustments to the conditions of the pine gum industry. The assessment score of the 5S score used 0 is not in accordance with the criteria, 1 to be less in accordance with the criteria, 2 to sufficiently fit the criteria, 3 to fit the criteria, 4 very appropriate with the criteria [8].

Table 2. Sheet Scoring in Existing Condition

Category	Criteria	Score					Findings Note
		0	1	2	3	4	
Seiri	There are items are not needed					√	In some areas there are items are not needed such as used wood.
	There is waste cattered on the production floor				√		In the work area there is still waste.
	There are items have been damaged are uncontrol					√	In the area of the damaged drum is uncontrol
	All unused items are easily identified				√		Unused items are of medium size and they are easily identified.
	There are written procedures for elimination or disposal of items are not needed				√		There are already a written procedure for removing items are not needed
Sub Total		17					

Category	Criteria	Score 0	1	2	3	4	Findings Note
Seiton	Items have been stored in place according to the classification				√		Some equipment / items are stored in the right place
	Drum storage has been determined and has been given a clear boundary		√				Gondorukem drum storage has no clear boundary area
	There are objects that do not have a storage area					√	There are still objects that do not have a storage area.
	There is no material handling area						Material Handling does not have its place.
	There is a neat bulletin board		√				Notification boards neatly arranged
	Sub Total Seiton			11			
Seiso	Cleaning the work area has been done regularly and scheduled according to the provisions					√	Collaborating on the cleaning of the work area has been carried out routinely
	All items are not used are always clean and shiny		√				The equipment / machine still looks unclean after it has been used
	Cleaning is always done based on the activity checklist		√				There is no checklist activity in cleaning
	Cleaning facilities / tools are available.			√			Unavailable of cleaning equipment
	There is a rotation of cleaning responsibilities within the specified work area	√					There is no rotation of cleaning responsibilities in the work area
	Sub Total Seiso			7			
Seiketsu	Staff understood the 5S procedure		√				There are still many staff who do not understand 5S procedures.
	There is a visual display about 5S				√		Visual displays about 5S already in the work area.
	Cleanliness and neatness are carried out by each worker daily.			√			Cleanliness and neatness are carried out by workers only when needed.
	There are regular audits, checks or assessments of 5S application	√					There is no periodic inspection of 5S implementation.
	Staff understood the previous 3S		√				Staff only know the previous 3S
	Sub Total Seiketsu			7			
Shitsuke	Everyone is involved in activities improvement		√				Employees have not been involved in 5S implementation
	Written procedures are implemented and communicated by each employee		√				There is no communication within the employees regarding 5S implementation
	Written rules and procedures for 5S are recognized and followed by all employees	√					There is no written procedure about 5S that is followed by employees
	Staffs have discipline in previous 4S implementation			√			Staff are disciplined in the previous 4S implementation
	Motivation from the company on the 5S implementation.		√				The company don't give the motivation to 5S implementation.
	Sub Total Shitsuke			5			
	Total 5S Score			47			It is in the range of 31-50, which means the application of 5S below the average, must review the checklist with the lowest value

If the assessment results of the existing condition are in the range of 0-30, which means they are not satisfactory, you must return to step 1, which is checklist fill. If it is in the range of 31-50 means below the average, you should review the checklist with the lowest score. In the range 51-70, it means that on average, reinforcement needs to be done on the part is still weak. In the 71-90 range it means above average, set a higher target and in the range of 91-100 which means very satisfying, must be maintained.

3.3. Recommendation

Based on 5S assessment, we found 4S components from 5S that didn't work well. There are seitons, seiso, seiketsu and shitsuke. The assessment results can be presented in a spider web diagram to facilitate the analysis. 5S assessment result in spider web diagram can be presented in figure 1.

If we seen from the total score in table 2 shows the number 52, which means that on average, it is necessary to improve weaker parts, namely seitons, seiso, seiketsu, and shitsuke. To improve the implementation of 5S, a serious handler is needed by the company. The condition of an unorganized work area causes irregular work areas that result in a time waste, energy and material.

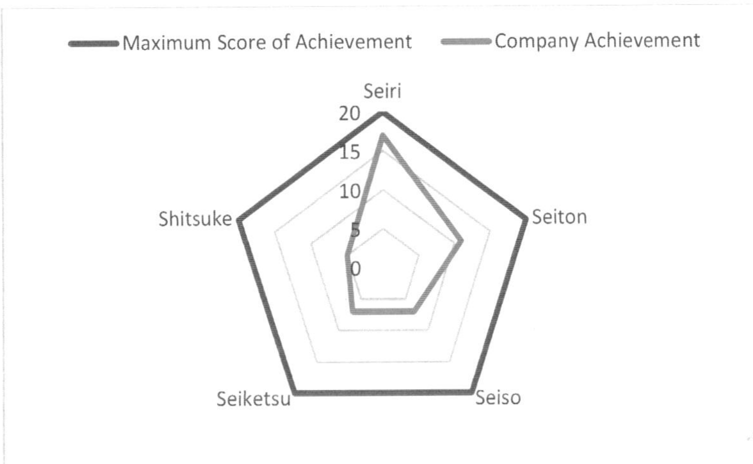

Figure 1. The Evaluation of 5S Achievement

a. *Seiton* (Straighten)
   Seiton is an activity that aims to make the item neatly arranged, and it is easy to find or use. The 5S application has not been implemented well, this is can be seen from the score of the 5S average of 11. In addition, there is no place arrangement for the running of humans and material handling. This resulted in unable to operate material handling effectively which is would cause a delay in the transfer of goods. Recommendations can be given are (1) make sure that items are needed has a storage area in the department, (2) goods and equipment storage in accordance with the place provided, and (3) provide clear boundaries on material handling and equipment in the right place of canning department and does not disturb the production process.

b. *Seiso* (Shine)
   Seiso is an activity aims to maintain the cleanliness of the work environment, both the workplace and the goods or materials contained in it. The 5S application has not been implemented well, this can be seen from the seiso score of 7. The recommendations that can be given are (1) clean the equipment after it is used, (2) determine responsibilities of cleaning for each area, (3) determine the list of cleaning equipment needed based on the number

c. *Seiketsu* (Standardize)
   Seiketsu is an activity that has a purpose, namely the activities of sorting, arranging, and cleaning have been carried out are still carried out continuously. One step taken is to make standardization.

The application on 5S is not well implemented, this can be seen from the seiketsu score of 7. The recommendations given in this process are (1) make clear rules about things might and might not done in the work environment. This regulation can be in the form of visual controls such as the prohibition of littering, not smoking, and the obligation to use personal protective equipment while in the work environment, and (2) Implement reward and punishment to maintain clean and neat conditions

d. *Shitsuke* (Sustain)
Shitsuke aims to familiarize 5S culture as an effort to create a better work environment. Shitsuke score of 5 means it has not been implemented properly. The improvement recommendation that can be given is to carry out regular audits. Audit can be conducted once a year or according to company policy. In carrying out audits, auditors must have predetermined assessment criteria and it will facilitate the audit process. Audit can be done by going directly to each department and making an assessment. Another effort was carried out, namely socialization of 5S material to employees. This socialization contains self-awareness of work ethics, such as discipline of standards, mutual respect, shame in committing violations, and others. Because in the end the most important thing remains the commitment of the workers.

4. Conclusion

Based on 5S assessment of the canning department, the values for each category are seiri, seiton, seiso, seiketsu and shitsuke, which are 17, 11, 7, 7, and 5 with a total score of 47 which is in the range of 31-50 and means 5S below the average, must review the checklist with the lowest score. Improvements are made to parts are still weak, namely seiton, seiso, seiketsu, and shitsuke.

References

[1] Cunneen P 2006 People Management **12** 1, pp. 42-43.
[2] Ephrem G, Birhanu B, Daniel K 2014) International Journal for Quality Research **8**(1) 47–60
[3] Azwan H M, et.all 5S Concept and Usage Conference: Conference on Language, Education, Engineering and Technology (COLEET2015)
[4] Asaad M N M, Saad R, and Yusoff, R Z 2012 Journal of Technology and Operations Management, **7** 1
[5] Shaman G, Sanjiv K J, 2014, Int. J. Lean Enterprise Research **1** 1
[6] Mohd N, Mohd A, Rohaizah S, and Rushami ZY 2015 International Academic Research Journal of Business and Technology **1** 2 pp. 214-226
[7] Toni L D, Eileen M V A, Jennifer A F, June M W, and Jeremy H 2008 Journal of Productivity and Performance Management, **57** 8 637.
[8] Saima A, Farasat R, Yasmin, Md. Ariful F 2015 American Academic & Scholarly Research Journal **7** 3 pp. 229-243
[9] Dossenbach T 2006 Wood & Wood Products, **111** 5, pp. 33– 38.
[10] Wazed M A and Ahmed S 2009 Australian Journal of Basic and Applied Sciences, **3** 4 pp. 3825-3835
[11] Vaidya, et. al 2017 *IJARIIE 3 2.*
[12] Arash G and Norzima Z 2013 International Journal of Business and Management Invention 2 3, pp.43-49
[13] I Rizkya, et. all 2019 IOP conference: Material Science and Engineering
[14] Peterson J and Smith R 2001 *The 5S Pocket Guide, Quality Resources* (New York)
[15] Dudek and Burlikowska M 2006 Journal of Achievements in Materials and Manufacturing Engineering **18**1-2 pp.435-438.

Acknowledgments

This paper published by funding support from the Universitas Sumatera Utara. The author thanks the Universitas Sumatera Utara and all participant those who helped in the publication of this paper.

IOP Conf. Series: Materials Science and Engineering **648** (2019) 012004 doi:10.1088/1757-899X/648/1/012004

Inventory Planning at Painting Section in the Indonesian Automotive Industry

Muhammad Haikal Sitepu[1*], Dian Pratama Situmorang[1], Meilita Tryana Sembiring[1]

[1]Industrial Engineering Department, Faculty of Engineering, Universitas Sumatera Utara, Medan, Indonesia, 20155

[*]Email: mhd_haikalkarana@usu.ac.id

Abstract. An increase of competition in the Indonesian automotive industry due to an emergence of new products and the presence of new players has pushed the companies in this industry to have proper planning for inventory. Many models are available in the literature to support good planning for inventory, However, not all models are appropriate to be used in inventory planning for a specific department in the Automotive industry. Economic impacts such as cost can be used as consideration in selecting the appropriate model. This paper aims to compare models for inventory planning at painting section in one of the Indonesian automotive industry based on inventory cost. Economic Order Quantity (EOQ), Kanban and Economic Order Interval (EOI) are models compared in this study. The results of this study indicated that EOQ and EOI are the inventory model that produces the lowest inventory cost.

1. Introduction

Competition between companies in the Indonesian Automotive industry is increasing due to the emergence of new products and the presence of new players. With increasing competition between companies, it is necessary for those companies to have an appropriate model for managing activities in producing a product. Managing inventory is one important activity to ensure the product can be produced on schedule. To ensure raw materials are available when it is requested by the production line, it is important to determine the quantity of stock for each item requested by production line. Several factors such as demand of products, delivery time for each item to arrive in the warehouse and order cost are needed to consider in calculating the quantity of stock.

Several models are available in the literature to support the calculation of stock quantity. First is a traditional economic order quantity model (EOQ) that has been used for more than 100 years [1]. In this model, the quantity of stock is a function of product demand, order cost, and inventory holding cost. Economic order quantity model is good to be used if there are continues and constant demand, the cost for ordering is fixed and holding cost is calculated based on the number of stock [2]. Second is the Economic order interval (EOI). This model focuses on determining an optimal period between two stock replenishments [3]. In this model, the optimal period is influenced by ordering cost, holding cost and demand. Third is Kanban. This is a subsystem of the Toyota Production System that focuses to control materials flow (quantity and time) in the production line and between Departments (e.g. Procurement department and production department). Cards are used to manage these materials flow [4]. However, each model has a limitation, such as economic order quantity model lacks in consideration of interactions among the items [1], economy order interval model is difficult to use in stochastic environment [3], Kanban model is not adequate to be used while the demand is volatile, the operation is not standardized and the supply of raw materials is uncertain [4].

Suzuki Indomobil motor is one of the leading company in the Indonesian automotive industry. This company produces two and four-wheel vehicles. Currently, this company uses the Assembly to order production system where the production is started after an order from a customer is received. Following this system, the Kanban system is used as an inventory model. Discussion with production persons in this company indicated that there is a gap for improvement in inventory planning. This is due to materials overstock in some sections in the production line, particularly at the painting section. This section is a part of the production department that focuses on the finishing process. This paper aims to compare three inventory models (EOQ, EOI, and Kanban) based on inventory cost to support inventory planning in the painting section.

2. Methodology

A case study is used as a research methodology. Case study is good to use when the research focuses to investigate the phenomenon in real life context [5]. In this research, inventory planning in the painting section is a phenomenon that is observed. Material overstocks have occurred for several times becomes a reason to select this section as case study.

2.1. Research Process

To achieve the aim of the research, this research has been divided into three stages. Stages 1 focuses on collecting data and information required to develop inventory models. Secondary Data such as type of parts, demand for each part, order cost and variable cost for each part are collected and reviewed at this stage. In stage 2, the focus of the research is to define the quantity of stock and total inventory cost using economic order quantity, economy order interval and Kanban. Then, this is followed by stage 3 that is focused on comparing the inventory models based on the result from stage 2. Figure 1 displays the research process used in this research.

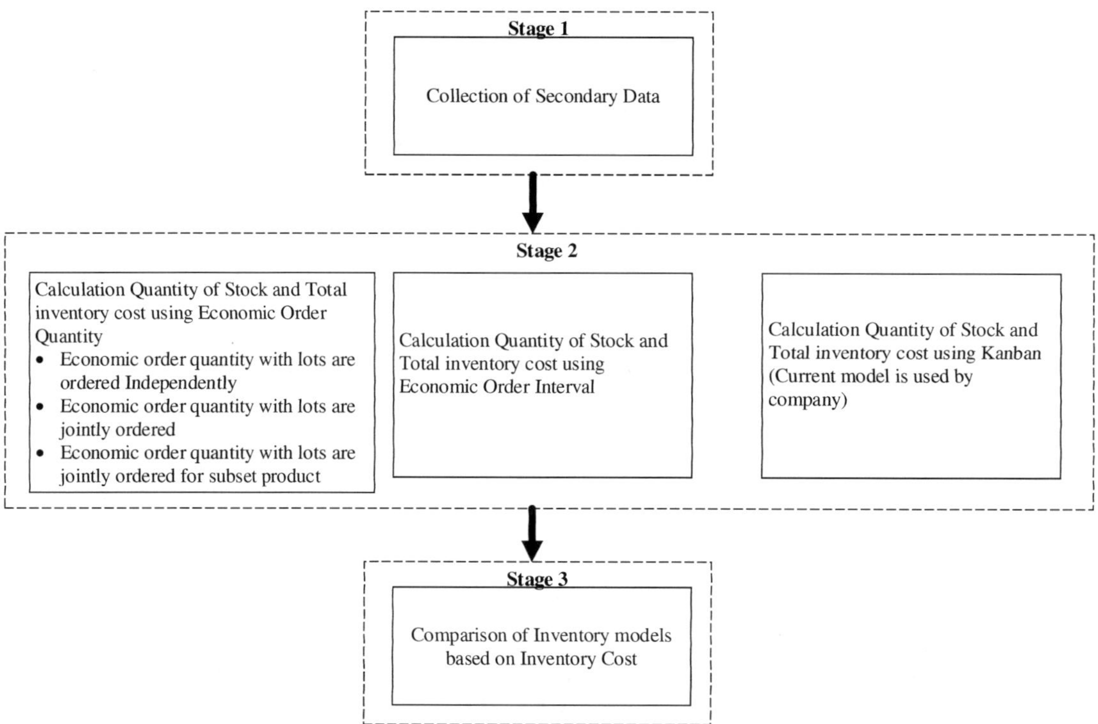

Figure 1. Research Process

2.2. Inventory models

This section focuses to introduce the equations for each inventory model compared in this paper. Each inventory model has different formulation due to different factors that are considered in developing the models. Table 1 shows those equations for each inventory model.

Table 1. Equations for EOQ and EOI Models

Inventory Models	Equations	Number of Equations
Economy order quantity for Lots ordered independently	$\mathrm{Q} = \sqrt{\frac{\mathrm{DhC}}{2\mathrm{S}}}$	(1)
	$\mathrm{N} = D/Q$	(2)
	$\mathrm{TC} = (\frac{\mathrm{D}}{\mathrm{Q}})\mathrm{S} + (\frac{Q}{2})\mathrm{hC_i} + +\mathrm{C_i}\ \mathrm{D_i}$	(3)
Economy order quantity for lots jointly ordered	$\mathrm{S} *= \mathrm{S} + \Sigma_{\mathrm{i}=1}^{\mathrm{k}} \mathrm{s_i}$	(4)
	$\mathrm{N} = \sqrt{\frac{\Sigma_{\mathrm{i}=1}^{\mathrm{k}} \mathrm{D_i h_i}}{2\mathrm{S} *}}$	(5)
	$\mathrm{TC} = (\Sigma_{\mathrm{i}=1}^{\mathrm{k}} \mathrm{D_i C_i}) + (\frac{\mathrm{D}}{2\mathrm{N}}\ hC_i) + (\mathrm{N\ S}^*)$	(6)
Economy order quantity for lots jointly ordered for subsequent product	$\bar{\mathrm{n}}_\mathrm{I} = \sqrt{\frac{\mathrm{hC_i D_i}}{2(\mathrm{S} + \mathrm{s_i})}}$	(7)
	$\bar{\bar{n}} = \sqrt{\frac{\mathrm{hC_i D_i}}{2\mathrm{s_i}}}$	(8)
	$\mathrm{m_i} = [\bar{\boldsymbol{n}} / \bar{\bar{\boldsymbol{n}}}]$	(9)
	$N = \sqrt{\frac{\Sigma_{1=1}^{\mathrm{i}} \mathrm{hC_i} m_\mathrm{i} \mathrm{D_i}}{2(\mathrm{S} + \Sigma_{1=1}^{\mathrm{i}} \mathrm{s_i}/m_\mathrm{i}}}$	(10)
	$\mathrm{TC} = \mathrm{NS} + \Sigma_1^{\mathrm{i}} n_\mathrm{i} s_\mathrm{i} + \Sigma_1^{\mathrm{i}} \frac{D_i}{2n_i}\ h\ C_i$	(11)
Economy order interval	$\mathrm{T} *= \sqrt{\frac{2(\mathrm{S})}{\Sigma_{\mathrm{i}=1}^{\mathrm{k}} \mathrm{D_i h_i}}}$	(12)
	$\mathrm{Q_i} = \mathrm{D_i T}^*$	(13)
	$\mathrm{TC} = (\Sigma_{\mathrm{i}=1}^{\mathrm{k}} \mathrm{D_i\ C_i}) + (\Sigma_{\mathrm{i}=1}^{\mathrm{k}} \frac{\mathrm{Di\ hi}}{2}) + (\frac{S*}{\mathrm{T}*})$	(14)
Kanban	$N = \frac{\mathrm{d}\ (\mathrm{c} + \mathrm{Wp} + \alpha)}{\mathrm{k}}$	(15)
	$\mathrm{TQ} = \mathrm{N\ k}$	(16)
	SS = (Service Level) * (Std)* Lead time	(17)
	$\mathrm{I} = (\mathrm{Q} + \mathrm{SS}) / 2$	(18)
	$\mathrm{TC} = (\frac{\mathrm{D}}{\mathrm{Q}})\mathrm{S} + \mathrm{I\ h\ C_i} + +\mathrm{C_i\ D_i}$	(19)

Where:
TC is the total inventory cost, Q is the quantity of stock for part (i), Di is demand for part (i), Ci is unit price for part (i), N is the frequency of order for part (i), S is the total order cost for joint parts, Si is ordered cost for part (i), hi is holding cost for part (i), T^* is interval of order, $\overline{n_1}$ is to determine part with most order frequency, $\bar{\bar{n}}$ is order frequency for all parts, m is to evaluate order frequency part i with part with most order frequency, d is the daily demand, c is order cycle, Wp is order time, k is capacity of container, TQ is total stock in one period, SS is safety stock, Std is sthe tandard deviation of demand and I is average inventory.

3. Result and Discussion

This section presents results from the implementation of three inventory models (EOQ, EOI, and Kanban) in determining the quantity of stock for each part used in the painting section. This section is divided into three subsections: Kanban result, economic order quantity result, and economic order interval result.

3.1. Kanban result

Inventory under Kanban system is designed based on a pull system where the stock of part is replenished whenever that part is taken by production line [6]. Kanban system focuses to supply the right quantity of parts in the right time for right production lines in order to minimize inventory level. The calculation of inventory is started by determining the number of Kanban card (order frequency) using Equation 15. Several factors such as daily demand, order cycle, order time, and capacity of containers affect the number of Kanban card. This is followed by calculating total stock in one year, safety stock and average inventory. Table 2 displays the total quantity of stock for 10 parts under Kanban system for August 2017- July 2018 at painting section.

Table 2. Quantity and Order Frequency of Parts Using Kanban System

Materials	Demand	Quantity (k)	Order Frequency (Kanban Cards)	Average Inventory (I)
AUTONET DISC 3" MIRKA # 320	5350	53,5	105	3910
AUTONET DISC 5" MIRKA # 400	29350	294	107	18980
PRIMER K98 ZF1 DARK GREY	3420	8	463	1991
NEO A'LAC DEEP BLACK	79660	166	537	48874
M'CRON 7100 CLEAR HG	19000	33	618	10588
M'CRON B/C THINNER	25560	47	579	14464
AMILAC I/M THINNER	20840	50	448	11947
MAG. 5500 RADIANT RED ZKB	1678	4	463	976
NEO AMILAC SUPER.WHITE	65120	116	619	38409
AMILAC KP KAI THINNER	7880	14	613	4438

3.2. Economic order quantity result

Quantity of stock for each part can be calculated using economy order quantity under three conditions [7]. The first condition is a situation where each part is ordered independently. In the second condition, all part are joined in each order. The third condition is a situation wherein each order, there are multiple parts to be joined but not all parts. Equations 1 to 3 is used for the first condition while

Equations 4 to 6 is used for the second condition. The third condition is calculated using Equations 7 to 11. Table 3 shows the quantity of stock and order frequency under these three conditions.

Table 3. Quantity and Order Frequency of Parts Using EOQ

Materials	Demand	Parts are ordered independently (First condition)		Parts are jointly ordered (Second condition)		Only certain part are jointly ordered (Third condition)	
		Quantity (Q)	Order Frequency (N)	Quantity (Q)	Order Frequency (N)	Quantity (Q)	Order Frequency (N)
AUTONET DISC 3" MIRKA # 320	5350	2182	2	92	58	206	26
AUTONET DISC 5" MIRKA # 400	29350	4502	7	506	58	554	53
PRIMER K98 ZF1 DARK GREY	3420	353	10	59	58	132	26
NEO A'LAC DEEP BLACK	79660	2012	40	1372	58	1503	53
M'CRON 7100 CLEAR HG	19000	1014	19	327	58	359	53
M'CRON B/C THINNER	25560	1854	14	440	58	483	53
AMILAC I/M THINNER	20840	1651	13	359	58	394	53
MAG. 5500 RADIANT RED ZKB	1678	147	11	29	58	65	26
NEO AMILAC SUPER.WHITE	65120	1769	37	1122	58	1229	53
AMILAC KP KAI THINNER	7880	1011	8	136	58	303	26

It can be seen from table 3, a total of order frequency is higher for the first condition. This occurs because of only one type of material in each order. As a result, the order cost increases significantly following the number of order. An investment for buying the stock of part is lowest in the first condition. This is due to the low quantity of parts must be bought in each order comparing to other condition. In the second condition, all ten parts are combined in each order with different number of units for each part. As a result, a big quantity of parts should be bought in each order. This condition requires 58 orders to fulfill the quantity of all parts. The lowest order frequency can be observed in the third condition with only 53 orders. Four parts (AUTONET DISC 3", PRIMER K98 ZF1 DARK GREY, MAG. 5500 RADIANT RED ZKB, and AMILAC KP KAI THINNER) are combined and ordered 26 times in one year.

3.3. Economic order interval result

Economic order interval model is started by calculating interval times between order using equation 10 (see table 2). Then this is followed by determining quantity for each part by using equation 11. From the calculation, it is found that the interval time between orders is 0,017 year or 6.2 days. This means that the company should orders all parts every 6 days. In total, to fulfill the demand of parts, 59 orders are required. Table 4 displays the quantity and order frequency for each part calculated using EOI.

It can be seen from table 4, order frequency calculated using EOI is similar with order frequency calculated using EOQ under the second condition. This is because the order cost for combining parts (S^*) is not too big compared to the order cost for individual part (see Equation 5 in Table 2). Moreover, both models require combining all parts in each order. Hence, a big investment is required to buy total quantity of stock for all parts.

Table 4. Quantity and Order Frequency of Parts Using EOI

Materials	Quantity	Order Frequency
AUTONET DISC 3" MIRKA # 320	91	59
AUTONET DISC 5" MIRKA # 400	499	59
PRIMER K98 ZF1 DARK GREY	58	59
NEO A'LAC DEEP BLACK	1354	59
M'CRON 7100 CLEAR HG	323	59
M'CRON B/C THINNER	435	59
AMILAC I/M THINNER	354	59
MAG. 5500 RADIANT RED ZKB	29	59
NEO AMILAC SUPER.WHITE	1107	59
AMILAC KP KAI THINNER	134	59

4. Comparison between inventory models based on inventory cost

Inventory cost consists of three types of cost including the cost of buying the parts, order cost and holding cost [7]. However, order frequency can be different for each inventory model. For example, order frequency for EOQ with lots ordered independently is influenced by the quantity of stock (Q) while order frequency for EOI is affected by the interval between orders. As a result, the equations for calculating inventory cost can be varied (see Table 2). Table 5 shows inventory cost for each inventory model.

Table 5. Comparison of Inventory Cost for Each Inventory Model

Method	Total Cost (IDR)
EOQ Independent	16,226,190,176
EOQ with lots jointly ordered	**16,174,770,877**
EOQ with lots selected jointly ordered	16,186,107,008
EOI (*Economic Order Interval)*	16,174,952,195
Kanban	17,121,778,890

The calculation of inventory cost in Table 5 shows that Kanban has the highest order cost compare to other models. This is due to the highest number of order frequency (Kanban cards) in this model. For purchasing cost, EOQ with lots jointly ordered and EOI has the highest purchasing cost due to the highest quantity of parts in each order. EOQ with lots selected jointly ordered has the lowest order cost due to low order frequency. Based on the calculation of the inventory cost above, the EOQ model (Economic Order Quantity) with lots jointly ordered and EOI (Economic order interval) is suggested to adopt by the company due to lowest cost produced by these models.

5. Conclusion

Based on the results, some conclusions are obtained as follows.

- Inventory cost can be used to select appropriate inventory models to support inventory planning in business.
- The implementations of several inventory models could give new insight for inventory planner in managing inventory for their businesses.
- Inventory model based on Kanban system and EOQ with lots ordered independently might enhance the order cost due to high order frequency while EOQ with lots jointly ordered and EOI might increase the purchasing cost due to the high quantity of parts in each order.
- Based on the implementations of EOQ, EOI, and Kanban system in calculating the quantity of stock and order frequency, the EOQ method (Economic Order Quantity) with lots jointly ordered and EOI is suggested to use due to lowest inventory cost.

References

[1] Holmbom M and Segerstedt A 2014 Economic Order Quantities in production: From Harris to Economic Lot Scheduling Problems *Int. J. Prod. Econ.* **155** 82–90

[2] Godichaud M and Amodeo L 2018 Economic order quantity for multistage disassembly systems *Int. J. Prod. Econ.* **199** 16–25

[3] Montanari R, Bottani E, Shekarian E and Kazemi N 2017 A model for the analysis of procurement strategies in the economic order interval environment *Math. Comput. Simul.* **134** 79–98

[4] Lage Junior M and Godinho Filho M 2010 Variations of the kanban system: Literature review and classification *Int. J. Prod. Econ.* **125** 13–21

[5] Yin R 2013 *Case study research: Design and methods*

[6] Naufal A, Jaffar A, Yusoff N and Hayati N 2012 Development of Kanban System at Local Manufacturing Company in Malaysia-Case Study *Procedia Eng.* **41** 1721–6

[7] Chopra S and Meindl P 2007 *Supply chain management. Strategy, planning & operation* (Springer)

Delivery Lot Size Optimization and Schedule of Shipment in Distribution Centre

I Rizkya[1]*, J Hidayati[1], M M Tambunan[1], J Utaminingrum[1]

[1]Departement of Industrial Enginnering, Faculty of Engineering, Universitas Sumatera Utara, Medan, Indonesia

*E-mail: indahrizkya@usu.ac.id

Abstract. This paper describes the optimal delivery lot sizes and plan delivery schedules from distribution center to manufacturer with a Distribution Requirement Planning (DRP). A company that produces nails has a problem that the planning and scheduling of goods distribution is not coordinated well. the demand for each product is out of control resulting in shortages or excesses as well as delays in distribution at each Distribution Center (DC). The purpose of this research is to analyze the distribution needs by using the Distribution Resources Planning (DRP) method to avoid delays in distributing products to each DC. Distribution analysis using the DRP method is used to get when and how many products will be sent for each DC. This method starts with calculating demand forecasting for the next 12 months. After obtaining the request results for the next 12 months, then the EOQ is calculated and the safety stock profit of each DC. With DRP method the optimal delivery quantity in DC Aceh 51.180 kgs, DC Padang 65.830 kgs, DC Pekanbaru 54.100 kgs, DC Palembang 67.225 kgs, and DC Medan 42.950 kgs. Using the optimal quantity delivery, frequency of shipment decreased until 51,5% and inventory cost decreased until 20,4%.

1. Introduction

The development of the business world has experienced intense competition and increased service for customers. The distribution system is one of the pillars that is important in the continuity and success of the company. Distribution is an effort to deliver products from the final production line to consumers with the goal on target, on time and at reasonable prices [1]. Distributors are required to deliver products well to prevent stock vacancies. Therefore, good management is needed to regulate the distribution system to work properly. A good distribution system performance can be seen from the availability of goods at each distribution center. Distribution is also a whole logistics process that involves sending the final product or service to the right place, at the right time and at the lowest cost [2].

DRP was suggested as an integrated approach to scheduling delivery and controlling inventory for a logistics system [3]. Distribution Requirement Planning is a distribution planning method to calculate product requirements at each distribution point. The DRP system is a complex information and control system, combining sales forecasting, inventory levels, and delivery problems, to schedule how much is needed, where and when [4,5]. DRP is able to resolve the problem of quantity, time, and location of delivery [6]. In multilevel distribution networks, distribution process can continue for various levels of warehouses such as the main warehouse, factory warehouse, distribution center, agent, etc [7,8]. The distribution process can be tiered at several levels of costumers ranging from distributors, wholesellers, retailers, to products to end users [9].

Several studies have been conducted on the application of DRP in several manufacturing industries. The first DRP implementation was carried out in the pharmaceutical industry to improve inventory turnover, improve service levels, and reduce distribution costs [10]. DRP method is widely used in some research before, one of which is the research conducted by Chrwan-jyh Ho (2007) concern to generalized system for delivery scheduling in a multi-sourcing logistics system [11]. A DRP system for increasing delivery flexibility and enhancing responsiveness to customer demand in the distribution network. This modified DRP is demonstrated to deal with an operating environment in which multi sourcing is a preferred distribution strategy. This system also allows a channel member to evaluate the performance of its suppliers using the criterion of on-time delivery. Finally, related research issues are discussed. The approach to DRP in the area of logistics. In addition, there are some of the following studies describing problem solving / solutions using the Distribution Resource Planning method. Distribution Resource Planning method with the addition of forecasting methods shows that the application of DRP can reduce distribution costs, make delivery scheduling to ensure delivery the right product, the right quantities, the right place, and the right time [4,5].

Companies engaged in production of building materials have experienced several problems in the planning and scheduling of distribution. Products do not arrive at the distribution center on time. Because the company does not have good distribution planning and scheduling to anticipate the problems. Distribution Resources Planning (DRP) method is known to be able to schedule deliveries well so that inventory costs and delivery costs are minimum [6]. Many studies determine order lot size in the manufacturing industry, this time it is done at the distribution center.

2. Methodology

The study was carried out on one of the industrial factories in the city of Medan that produces building materials. The object examined in this study is the product distribution activity at the factory against each Distribution Center (DC), namely Aceh, Padang, Palembang, Pekanbaru, and Medan. The method used in this study is DRP. Distribution requirements planning (DRP) has been suggested as a scheduling method in the physical distribution system [1]. The data used in this study are data on the number of requests in 2017, lead time, project on hand (POH), inventory costs and delivery costs. The data that has been collected is then processed and analyzed with a predetermined procedure. DRP Calculation in 5 steps. There are:

a. Determine each DC demand forecast

   At this initial stage, the demand data for each DC is predicted using the forecasting method.

b. Calculate the optimum delivery lot size.

   The optimum delivery quantity calculation is done using the Economic Order Quantity (EOQ) method which is used to determine the most economical delivery quantity. The purpose of EOQ to determine delivery quantity that will minimize the amount of inventory costs per period and delivery costs. EOQ calculation is done using the following formula [12]:

$$EOQ = \sqrt{\frac{2xDxk}{h}} \quad (1)$$

   Where D is the demand distribution center in time horizone planning, k is the delivery cost, and h is the inventory cost per period

c. Calculate the amount of safety stock for each DC

   In order for the stock status to remain safe, a safety supply for each distribution center (DC) must be made according to the average demand lead variance. Calculation of safety stock is done using the following formula [12]:

$$\textit{Safety Stock} = s \times Z \times \sqrt{L} \quad (2)$$

   Where s is a standard deviation of requests at the distribution center, Z is the value under the normal curve which is determined by the level of service the company's service level to the consumer is 95% and the Z value is 1.65. L is the lead time value.

IOP Conf. Series: Materials Science and Engineering **648** (2019) 012005 doi:10.1088/1757-899X/648/1/012005

d. Create DRP worksheets
After an optimal delivery quantity is obtained, the monthly product delivery plan determined using the DRP method. DRP worksheet [11] can be seen in Table 1.

Table 1. DRP Worksheet

On hand balance	:	Lead time	:	
Safety stock	:	Delivery Lot Size	:	
			Period	
	Past Due	1	2	......n
Demand				
In Transit				
Projected on Hand				
Net Requirement				
Planned Shipment - Receipt Date				
Planned Shipment - Ship Date				

e. Create pegging information every DC
The source of information used to see how many requests for each DC at a certain time. This source of information is a list of requests that indicate where the request came from, both in terms of time and amount.

3. Result and Discussion

3.1. Demand Forecast in Every Distribution Centers

Forecasting is an important tool in effective and efficient planning [13]. To forecast the number of product requests for the next 12 months, it is done by looking at the historical demand in 2018. With ARIMA model (Autoregressive Integrated Moving Average} obtained the results of demand forecasting for each distribution center. The result of Demand forecasting for each distribution center 2019 can be seen in figure 1.

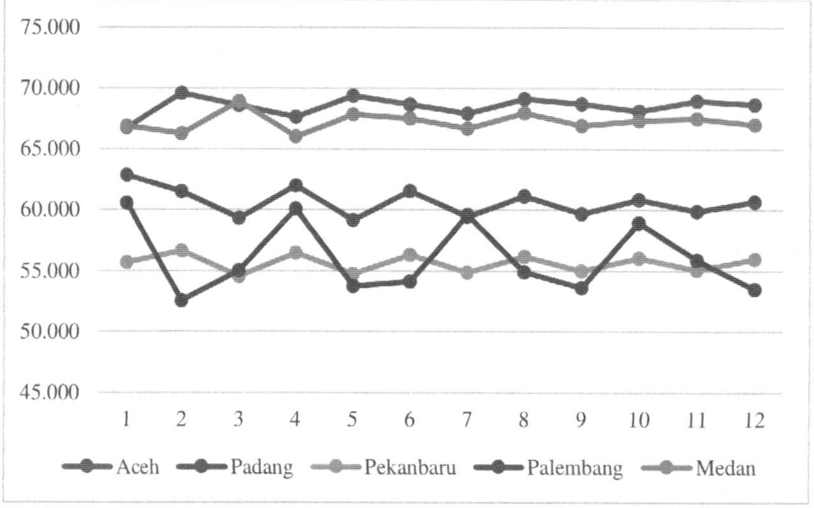

Figure 1. Demand Forecasting Results for Each Distribution Center

Based on the picture above, it can be seen that the demand for nail products has fluctuated every period. The number of requests for each DC product also varies. The highest average number of DC requests for each month is in DC Aceh. These results indicate that there is a fluctuating product demand for each period so that planning and scheduling of distribution activities is needed for each period to be fulfilled.

3.2. Optimization Delivery Lot Size

The optimum delivery quantity calculation for each distribution center is calculated using the Economic Order Quantity (EOQ) method. The delivery quantity calculation is also used to schedule distribution activities. Delivery quantity in this study uses the economic order quantity (EOQ) method that takes into account the distribution cost, holding cost, and product demand factors for each distribution center. Calculation of delivery quantity to each distribution center can be seen in Table 2.

Table 2. Optimum Delivery Quantity each Distribution Center

Distribution Center	Optimum Delivery Quantity (Kgs)
Aceh	51.180
Padang	65.830
Pekanbaru	54.100
Palembang	67.225
Medan	42.950

Based on table 2, the delivery quantity is quite high due to the high number of requests and delivery costs. By implementing delivery quantity based on EOQ calculations, it is expected that the company will not experience out of stock conditions due to uncertain shipments.

3.3. Calculation of safety stock

Safety stock is a stock that is reserved for the needs while waiting for the goods to arrive [8]. Safety stock in DRP is used to anticipate uncertainty of demand relative to the predictions made. The result of safety stock in DC Aceh, Padang, Pekanbaru, Palembang, and Medan namely 320 kgs, 645 kgs, 357 kgs, 870 kgs, and 256 kgs.

3.4. Distribution Requirement Planning

To make a product delivery schedule with DRP, it requires a bill of distribution. Bill of distribution has information about the distribution network chain and lead time delivery from the building material industry to the distribution center.

Planning of distribution schedule by DRP in weekly (52 weeks). Besides bill of distribution, it is needed the status on hand inventory and consumer demand prediction to perform DRP. Next step is calculating net requirements to know the product quantity needed to meet consumer at each distribution center. Net requirements are obtained by demand minus on hand inventory and in transit product plus safety stock. The Planned shipment receipt date is the time when product must be delivered in each DC to fulfil consumer demand. The planned shipment release date is the time when each DC must be ordered the product according to the optimal lot size that has been determined. Schedule of planned shipment release date is a pegging information.

IOP Conf. Series: Materials Science and Engineering **648** (2019) 012005 doi:10.1088/1757-899X/648/1/012005

3.5. Pegging information
Pegging Information is the source of information used to see how many requests for each DC at a certain time. This distribution scheduling is summarized in pegging information. Pegging information for each DC can be seen in table 3.

Table 3. Pegging Information each Distribution Center

Weeks	*Distribution Center* (kgs)				
	Aceh	Padang	Pekanbaru	Palembang	Medan
1	51.180	65.830	54.100	67.225	42.950
2					
3					
4	51.180				42.950
5			54.100		
6		65.830		67.225	
7					42.950
8	51.180				
9					
10			54.100		42.950
11		65.830			
12	51.180			67.225	
13					42.950
14			54.100		
15	51.180				
16		65.830			42.950
17					
18				67.225	
19	51.180		54.100		42.950
20					
21		65.830			
22	51.180				42.950
23			54.100	67.225	
24					42.950
25	51.180	65.830			
26					
27			54.100		42.950
28	51.180			67.225	
29					
30					42.950
31		65.830			
32	51.180		54.100		
33					42.950
34				67.225	
35	51.180				
36		65.830	54.100		42.950
37					
38	51.180				
39				67.225	42.950
40		65.830	54.100		
41					42.950

Table 3. Pegging Information each Distribution Center (continue)

Weeks	*Distribution Center* (kg)				
	Aceh	Padang	Pekanbaru	Palembang	Medan
42	51.180				
43					
44				67.225	42.950
45	51.180	65.830	54.100		
46					
47					
48					42.950
49	51.180				
50		65.830	54.100	67.225	42.950
51					
52	51.180				

With the DRP worksheet and pegging information, the company will know when and how many products will be sent to each distribution centers. Table 3 shows the delivery lot size that will be ship to distribution center in Aceh, Padang, Pekanbaru, Palembang and Medan. Based on Table 3 also shows the frequency of delivery from building material industry to each distribution center in a year. Table 3 shows the delivery frequency of nail products that compare with historical data in company.

Table 4. Comparison of Delivery Frequency Actual by DRP Method

Distribution Center	Actual Frequency of delivery (times)	Delivery Frequency with DRP (times)
Aceh	26	16
Padang	26	11
Pekanbaru	26	12
Palembang	26	10
Medan	26	18

Actual product delivery is done every 2 months, so that there are 26 times shipments in a year. Actual delivery frequency to distribution center about 130 times in a year and decreased by 51,5% using DRP method. Using DRP method inventory cost decreased around 20,4% in distribution center. This result clarifies some previous research that DRP able to reduce inventory and frequency of shipment, delivery cost, inventory cost [7,11].

4. Conclusion

Planning distribution activities in the form of DRP Worksheet has been clearly projected that ordering requirements in accordance with the delivery quantity calculation to meet the demand so there is no delay in fulfillment in the future. With the design of the distribution system with the DRP method, the company is expected to maintain the smooth and effective and efficient distribution activities so that the company can continue to have sustainable profits and savings in a market that fluctuates up and down. Using DRP method frequency of shipment decreased until 51,5% and inventory cost decreased until 20,4%.

References

[1] Fogarty D W, Blackstone J H and Hoffmann T R, 1991 *Production and Inventory Management.* (South-Western: 2nd Revised edition)
[2] Weiss H J and Gershon M E, 2002 *Production and Operation Management*
[3] Stenger A J and Cavinato J L,1979 Production and Inventory Management **20** 4 pp. 1-14
[4] Vollman T E, Berry W L and Whybark D C, 1984 *Manufacturing Planning and Control System, Homewood* (IL: Irwin)
[5] Sekhar J and Balasubramanian V, 2012 *International Journal of Computer Trends and Technology* **3**
[6] I Rizkya et al 2018 *IOP Conf. Ser.: Mater. Sci. Eng.* **434** 012243
[7] Franks S IET JOURNALS & MAGAZINES 38-40
[8] Alessandri A, Gaggero M and Tonelli F, 2011 *Transactions on Control Systems Technology* **1075** 1089
[9] Salmela H, Toivonen S and Scholliers J 2010 *Transactions on Intelligent Transport Systems* **307** 317
[10] Martin A J 1983 *Distribution Research Planning, Essex Junction* (VT: Oliver Weight Publications Inc)
[11] Ho C J 1990 *International Journal of Physical Distribution & Logistics Management* **20** 2 pp.3-8,
[12] Zipkin P H 2000 *Foundations of Inventory Management* (New York: McGraw-Hill)
[13] Rahmat R F, Nurmawan, Sembiring S, Syahputra M F, Fadli 2018 Adaptive neuro-fuzzy inference system for forecasting rubber milk production *IOP Conference Series : Materials Science and Engineering* **308**(1), 012014

Acknowledgement

Universitas Sumatera Utara has provided financial support to publish this paper. The author is grateful for the help and support so that this paper has been completed and published.

Identification of Fertilizer Quality Factors from Distributors, Retailers and Farmers Perspective

Muhammad Haikal Sitepu[1*], Natasya Agnes[1], MeilitaTryana Sembiring[1]

[1]Industrial Engineering Department, Faculty of Engineering, Universitas Sumatera Utara, Medan 20155

*Email: mhd_haikalkarana@usu.ac.id

Abstract. Indonesia is one country in the world with a developing agricultural industry. Around 50% of Indonesian citizen is involved in the agricultural sector. Fertilizer is one important material for this industry to control the productivity of plantation or farming areas. In Indonesia, Fertilizer is produced by the fertilizer industry and is distributed to the farmer through two channels: distributor and retailer. With increasing competition in the Indonesian fertilizer industry due to the presence of fertilizer products from outside the country, it is necessary for this industry to maintain the quality of its product. Identification of product quality factors becomes a critical step in managing the quality of fertilizer products. Moreover, discussion with people in the Indonesian fertilizer industry, it is substantial to know the perception of customers about quality factors for fertilizer products. This research aims to identify quality factors for fertilizer products from distributors, retailers and farmers perspective. It was found that distributors and retailers emphasized the accuracy of weight and composition from fertilizer products as prominent quality factors while farmers stressed fertilizer price and fertilizer composition as key quality factors.

1. Introduction

Indonesia is a developing agricultural country with more than 50% of its citizen are involved in the agricultural sector. The Indonesian economy is supported by exporting commodities from this industry such as palm oil, natural rubber, coffee, and chocolate [1]. The fertilizer industry is one supporting industry for agricultural sector by producing fertilizer to maintain land nutrient which in turn affecting the productivity of plantation or farming areas. Indonesia has significant fertilizer enterprises to support Indonesian agricultural industry. However, Indonesian fertilizer industry faces the increase of competition due to the appearance of fertilizer products from outside the country. Hence, it is important to maintain the quality of their products in order to keep these companies competitive [2,3].

Fertilizer is one key factor to manage agricultural productivity which becomes the reason for the high consumption of fertilizer [4]. Fertilizer plays an important role in improving food security in several ways such as sustainable intensification [5]. Although, fertilizer is a source of emission, a significant increase in fertilizer demand in the future is expected due to the increase of agricultural areas [6]. To fulfill this demand, the number of fertilizer enterprises increases in some agricultural countries such as Indonesia, China, Malaysia, and Thailand. However, the use of fertilizer in some regions is influenced by farmer decisions [7], agricultural practices [5] and type of fertilizers [4]. Hence it is important to identify fertilizer quality factors from customers' perspective since it influences fertilizer use.

Pupuk Iskandar Muda is a leading fertilizer enterprise in Indonesia. This company is a state-owned company that focuses to produce and to distribute fertilizer for farmers. Initial discussion with people in this company indicated the desire of the company to make its products meet customer expectation. Moreover, this company has a target to distribute the right type of fertilizer at the right time, the right amount, the right place and the right price. This company distributes its fertilizer to farmers through distributor and retailers which scattered across districts and sub-districts. This paper aims to identify fertilizer quality factors from the perspective of distributors, retailers, and farmers.

2. Research Methods

A case study is used as a research method. To investigate the phenomenon in real-context, the case study is a good approach. This is because case study can provide a depth insight into the problem from a real case. Pupuk Iskandar Muda North Sumatera supply region is selected as the case study. This supply region distributes fertilizers for plantation areas in North Sumatera Province. In this research, product quality in perspective of distributors, retailers, and farmers is a phenomenon to be investigated.

2.1. Research Process

To achieve the aim of the research, the research process is divided into two stages: a qualitative stage and a quantitative stage. In the qualitative stage, the research focuses to investigate fertilizer quality factors from the perspective of distributors, retailers, and farmers. A depth investigation using interview was used to identify these factors. Then, it is followed by the quantitative stage that focuses to determine the significance of fertilizer quality factors. A survey using a questionnaire was used to define significant fertilizer quality factors. Figure 1 displays the research process used in this research.

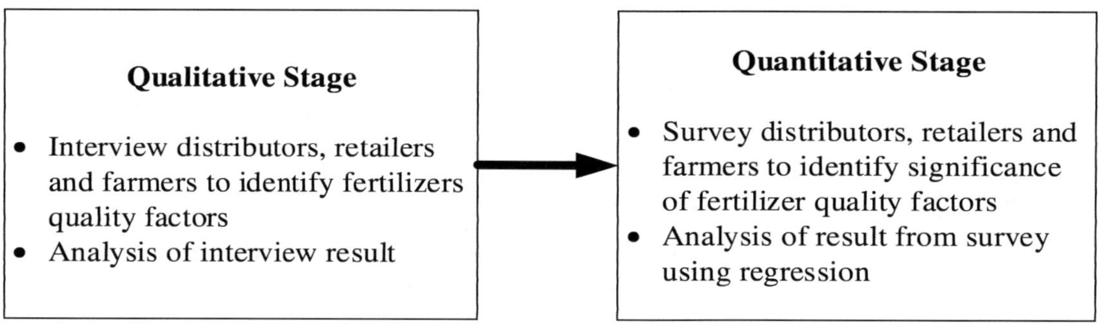

Figure 1. Research Process

2.2. Analysis Methods

An interview result is analyzed using thematic analysis to define fertilizer quality factors stated by distributors, retailers, and farmers. Multiple linear regression analysis is used to determine whether there is an influence of independent variables on the dependent variable. In this case, fertilizer quality is the dependent variable. SPSS software is used to apply regression analysis to result from the survey. The regression equation can be formulated as follows [8].

$$Y=a + b_1X_1 + b_2X_2 + b_3X_3 + b_4X_4+e \quad (1)$$

Where:

Y = Fertilizer quality (Distributor, Retailer or Farmer)

IOP Conf. Series: Materials Science and Engineering **648** (2019) 012006 doi:10.1088/1757-899X/648/1/012006

a	= Constant
b1, b2, b3, b4	= coefficient of determination
X_1	= Fertilizer quality factor 1
X_2	= Fertilizer quality factor 2
X_3	= Fertilizer quality factor 3
e	= Error

3. Result and Discussion

3.1. Qualitative Result

From an interview with distributors, retailers and farmers, some fertilizer quality factors have been identified. Table 1 shows these factors.

Table 1. Fertilizer Quality Factors from Interview

Fertilizer Quality Factors from the Perspective of Distributors	**Fertilizer Quality Factors from the Perspective of Retailers**	**Fertilizer Quality Factors from the Perspective of Farmers**
Quality of fertilizer granules (XA1).	Quality of fertilizer granules (XB1).	Quality of fertilizer granules (XC1).
The quality of fertilizer Packaging (XA2).	The quality of fertilizer Packaging (XB2).	The quality of fertilizer Packaging (XC2).
Accuracy of fertilizer weight (XA3).	Accuracy of fertilizer weight (XB3).	Accuracy of fertilizer weight (XC3).
Suitability of the amount of fertilizer received with the amount ordered (XA4).	Suitability of the amount of fertilizer received with the amount ordered (XB4).	Suitability of the amount of fertilizer received with the amount ordered (XC4).
Suitability of the type of fertilizer received with requested (XA5).	Suitability of the type of fertilizer received with requested (XB5).	Suitability of the type of fertilizer received with requested (XC5).
Conformity between subsidized prices and the company selling price (XA6).	Conformity between subsidized price and distributor selling price (XB6).	Conformity between subsidized price and retailer selling price (XB6).
Accessibility of the company warehouse (XA7).	Accessibility of distributor warehouse (XA7).	Comparison with competitor price (XC7).
Adequate company warehouse facilities (XA8).	Adequate distributor warehouse facilities (XB8).	Accessibility of retailer store (XC8).
Availability of fertilizer in the manufacturer's warehouse at the time of purchase (XA9).	Availability of fertilizers at the distributor at the time of purchase (XB9).	Availability of fertilizers at the retailers' store when needed (XC9).
		Impact of fertilizer on land yields (XC10).
		Ease of use (XC11).

From the appearance of the product, three quality factors have been identified from distributors, retailers, and farmers as quality factors. The first factor is a fertilizer granule. According to them, fertilizer with good quality must have fine granules with similar size and shape. The second factor is fertilizer packaging. Fertilizer is distributed from the factory to farmers through long trips that need strong and good packaging. From distributors and retailers perspective, good packaging makes the product easier to distribute and sell while in farmers' perspective good packing is required to ensure that the product is easy to handle in the farming area. The third factor is fertilizer weight. Distributors, retailers, and farmers agreed that fertilizer with good quality must have real weight same with the weight written on the packaging.

From the distribution process, five quality factors have been identified. First is the suitability amount of fertilizer delivered with the amount of fertilizer ordered. For farmers, if the amount of fertilizer delivered less than the amount requested, it might cause some of their lands will not get adequate fertilizers. For distributors and retailers, if the amount of fertilizer received less than the amount requested, it might cause farmers in their network will not get sufficient fertilizers. The second factor is the type of fertilizer. For farmers, a different type of fertilizers might not suitable for their plant or might cause different impacts on their plant. Hence, if the type of fertilizer delivered different from the type of fertilizer requested, that fertilizer might not be bought. The third factor is the fertilizer price. There are two issues in fertilizer price: the different with the subsidized price and the different with competitor price. In Indonesia, the Central Government has a program to subsidize fertilizer price with a view, more farmers can buy the fertilizer. Hence, the subsidized price becomes a consideration for the farmers to select the fertilizer. If the difference between fertilizer price with the subsidized price is too high, farmers might not select this fertilizer. The similar situation occurred while the fertilizer price is compared with competitor price.

The fourth factor is the accessibility of the warehouse for distributors and retailers. It is necessary for distributors and retailers to access the warehouse. If the company warehouse or distributor warehouse is easy to reach, the distribution process will be easier and faster. For farmers, the accessibility of the retail store is the fourth factors. The location of the retail store can influence farmer decision to buy fertilizer from that store. The fifth factor for distributors and retailers is a facility in the warehouse. If the warehouse is supported by adequate facilities, the distribution process is more convenient. The sixth factor for distributors, retailers, and farmers is the availability of fertilizer. This is related to the stock of fertilizer at the company warehouse, distributor warehouse, and retailer store. Good quality fertilizer is always available when it is needed.

For farmers, there are two additional quality factors identified including impacts of fertilizer to land's yield and easiness to use. For farmers, the productivity of land is essential factors because it can influence their income. In their view, giving fertilizer could maintain the productivity of their land. Furthermore, for farmers, good quality fertilizer is easy to use. This means that it will not require complex procedures to apply fertilizer to their land.

3.2. Quantitative Result

This stage has the purposes to determine significant quality factors and to develop fertilizer quality models. To achieve these objectives, quality factors identified from the interview with distributors, retailers, and farmers are used to develop a questionnaire. Significant quality factors and fertilizer quality models from the perspective of distributors, retailers, and farmers are defined through an analysis of questionnaire answer. This section is divided into three subsections: Significant quality factors from the perspective of distributors, significant quality factors from the perspective of retailers and significant quality factors from the perspective of farmers.

3.2.1. Significant Fertilizer Quality Factors from Distributors Perspective

Distributors play a role to distribute fertilizer from fertilizer factory to fertilizer retailers. Five distributors have been surveyed related to fertilizer quality factors. Figure 1 shows significant quality factors from the distributor's perspective.

Coefficients[a]

Model	Unstandardized Coefficients		Standardized Coefficients	t	Sig.	Collinearity Statistics	
	B	Std. Error	Beta			Tolerance	VIF
2 (Constant)	6.250	.714		8.748	.013		
XA5	5.708	.321	.828	17.761	.003	.415	2.408
XA3	1.458	.321	.212	4.537	.045	.415	2.408

a. Dependent Variable: Y

Figure 2. Results of Multiple Linear Regression Analysis on Distributors

It can be seen from figure 2, Distributors emphasized the accuracy of weight and suitability of fertilizer's type as significant quality factors. Both have positive b value that means the quality of fertilizer is likely to increase if the company could maintain the weight of fertilizer so it is same with the weight written on the packaging and could manage the type of fertilizer sent to distributor same with the type of fertilizer requested. Based on Figure 2, it can be obtained the regression equation as follows:

$$Y = 6{,}250 + 1{,}458\ XA3 + 5{,}708\ XA5 + e \quad (2)$$

3.2.2. Significant Fertilizer Quality Factors from Retailers Perspective

Retailers have a role to distribute fertilizers from distributors to farmers. Retailers sell fertilizers in their store. Ten retailers have been surveyed related to fertilizers quality factors. It can be observed from Figure 3, the accuracy of fertilizer weight and suitability of fertilizer's type are statically significant quality factors from the perspective of retailers. Both factors have positive b value that means the quality of fertilizer is likely to increase if the distributor could maintain the weight of fertilizer and the type of fertilizer to be sent. Figure 3 displays significant quality factors from the retailer's perspective.

Coefficients[a]

Model	Unstandardized Coefficients		Standardized Coefficients	t	Sig.	Collinearity Statistics	
	B	Std. Error	Beta			Tolerance	VIF
2 (Constant)	6.725	3.916		1.717	.130		
XB3	3.627	1.030	.552	3.521	.010	.796	1.256
XB5	3.539	1.030	.539	3.436	.011	.796	1.256

a. Dependent Variable: Y

Figure 3. Results of Multiple Linear Regression Analysis on Retailers

Based on figure 3, the regression model for fertilizer quality is displayed by Equation 3

$$Y = 6{,}725 + 3{,}627XB3 + 3{,}539XB5 + e \quad (3)$$

3.2.3. Significant Quality Factors from Farmers Perspective

Farmers is the final customer that use fertilizer for their lands. Thirty farmers have been surveyed related to fertilizer quality factors. All ten quality factors identified in the qualitative stage are stated to be significant quality factors (see Table 2). Furthermore, all quality factors have positive b value that means the increase in these factors will be likely to increase the quality of fertilizer. Figure 4 shows significant quality factors from the farmers perspective.

Coefficients[a]

Model		Unstandardized Coefficients		Standardized Coefficients	t	Sig.	Collinearity Statistics	
		B	Std. Error	Beta			Tolerance	VIF
10	(Constant)	1.398	1.373		1.018	.322		
	XC6	1.376	.267	.261	5.150	.000	.401	2.492
	XC5	1.258	.183	.261	6.881	.000	.719	1.391
	XC3	1.022	.164	.215	6.243	.000	.874	1.144
	XC4	1.266	.168	.263	7.522	.000	.844	1.184
	XC11	1.037	.196	.178	5.294	.000	.910	1.099
	XC10	.793	.199	.157	3.995	.001	.665	1.503
	XC1	1.264	.243	.231	5.204	.000	.525	1.905
	XC8	1.058	.190	.194	5.581	.000	.854	1.171
	XC7	1.013	.180	.207	5.622	.000	.763	1.310
	XC2	.588	.191	.110	3.085	.006	.814	1.229

Figure 4. Results of Multiple Linear Regression Analysis on Farmers

Based on Figure 4, we can obtain the regression formula as follows:

$$Y = 1{,}398 + 1{,}264XC1 + 0{,}588XC2 + 1{,}022XC3 + 1{,}266XC4 + 1{,}258XC5 + 1{,}376XC6 + 1{,}013XC7 + 1{,}058XC8 + 0{,}793XC10 + 1{,}037XC10 + 1.037XC11 + e \quad (4)$$

4. Conclusion

Based on the results, some conclusions are obtained as follows.

- Fertilizer Quality Factors could be different between distributors, retailers, and farmers.
- From the distributor's perspective, two fertilizer quality factors are found to be statistically significant: the accuracy of fertilizer weight and suitability of fertilizer types.
- From the retailer's perspective, there are two statistically significant fertilizer quality factors: the accuracy of fertilizer weight and the suitability of fertilizer types.
- From the farmers perspective, there are ten statistically significant fertilizer quality factors including fertilizer granules, fertilizer packaging, Accuracy of fertilizer weight, suitability of fertilizer amount, suitability of fertilizer type, Different of fertilizer price with subsidize price, Different of fertilizer price with competitor price, Accessibility of retailers' store, Effect of fertilizers on yields and ease of processing fertilizers.
- Fertilizer enterprises might consider these quality factors in maintaining customer satisfaction.

References

[1] Sitepu M H, McKay A and Holt R J 2016 Towards a Framework for Sustainable Development Planning in the Indonesian Natural Rubber Industry Supply Network *Procedia CIRP* **48** 164–9

[2] Teh P-L, Adebanjo D and Ahmed P K 2014 Factors affecting product quality and reliability: A comparison of developed and developing countries *2014 IEEE International Conference on Industrial Engineering and Engineering Management* (IEEE) pp 1481–5

[3] Lombard R, van Waveren C C and Chan K-Y 2014 Factors affecting quality in a manufacturing environment for a non-repairable product *2014 IEEE International Conference on Industrial Engineering and Engineering Management* (IEEE) pp 137–42

[4] Sun Y, Hu R and Zhang C 2019 Does the adoption of complex fertilizers contribute to fertilizer overuse? Evidence from rice production in China *J. Clean. Prod.* **219** 677–85
[5] Holden S T 2018 Fertilizer and sustainable intensification in Sub-Saharan Africa *Glob. Food Sec.* **18** 20–6
[6] Jawjit W, Kroeze C and Rattanapan S 2010 Greenhouse gas emissions from rubber industry in Thailand *J. Clean. Prod.* **18** 403–11
[7] Haider H, Smale M and Theriault V 2018 Intensification and intrahousehold decisions: Fertilizer adoption in Burkina Faso *World Dev.* **105** 310–20
[8] Regression P 2013 Chapter 5 Multiple Linear Regression *Handb. Regres. Anal.*

IOP Conf. Series: Materials Science and Engineering **648** (2019) 012008 doi:10.1088/1757-899X/648/1/012008

Determination of Priority Supply Chain Distribution Using DEMATEL Method in Instant Noodle Company

I Siregar[1,2], J.B. Saedon[2], Mohd Shahriman Adenan[2]

[1]Faculty of Engineering, Universitas Sumatera Utara, Medan, Indonesia
[2]Faculty of Mechanical Engineering, Universiti Teknologi MARA, Shah Alam, Selangor, Malaysia.

Email : ikhsan.siregar@usu.ac.id

Abstract. The company has a supply chain aims to improve company performance. The supply chain has a distribution center matches the company goals. Companies produce instant noodles have many branches spread in various regions. In order to achieve this goal, the company must determine the priority of the distribution center in the supply chain and analyze the relationship between each factor affects the distribution center to increase the flow of materials and information and the productivity of the company increase. Determination of company supply chain distribution priorities is done by using the DEMATEL method in knowing the criteria influences the selection of supply chain distribution centers. The criteria used are the four Balance Scorecard criteria, namely Financial, Internal, Customer and Learning & Growth. Then distributing priorities are obtained and the relationship of influence between criteria and distribution center. Based on the analysis of the distribution center in the company (Jambi, Medan, Aceh, Riau and Padang), it is obtained that the weighting value for prioritizing the distribution in sequence is 1,0035; 1,5649; 1,3789; 1,2605 and 1,2688. Then it is find out the level of importance sequentially, namely Medan, Aceh, Padang, Riau and Jambi.

1. Introduction

In this era of globalization with very high business competition, every company is required to develop a business strategy and it can compete with other companies. Companies must begin to develop the process of producing goods or services effectively and efficiently to be distributed to consumers. For most companies, generally using supply chain management to deliver of product sales to the consumers. Supply chain aims to deliver products to arrive at the final consumer. In this chain there are several organizations works in different scopes.

The problem often occurs is the criteria that influences the selection of the distribution center are unknown, which is distribution is the priority and the relationship between each criterion exists and its relationship with the distribution center. The company has several distribution centers spread across several regions. From the distribution center, the company does not yet know what factors can affect the performance of the distribution center and has not yet analyzed and determined the priorities of the existing distribution. This condition causes the priority of the distribution center and the criteria influences the distribution of products at the distribution center are unknown. This affects the productivity of the company. The DEMATEL method is used to identify priority distribution centers as well as criteria affects the company. Therefore, it is necessary to determine distribution priorities and factors affects the performance of the distribution center.

The importance of Supply Chain starting from the initial buying operations till providing the services and products to the customers has been increasingly growing in today's fast market. The future and the challenges of Supply Chain Management (SCM) have been taken the interest of scholars and companies [1]. Supply chain is seen as the internal and external integration of enterprise processes with the customers and suppliers to create value for the customers [2]. The supply chain management has been touted as one of the ways to reduce volatility and improve outcomes for all involved with the supply chain [3]. Supply chain management examines the efficiency and effectiveness of the flow of goods, information and money flow that occurs continuously by involving related parties. In practice, the SCM can integrate manufactures', suppliers, retailers, and sellers efficiently. So, they can produce and distribute the goods in precise quantity and minimal cost. One of applications of supply chain management is the logistic efficiency system that is reliable. values of the supply chain by optimizing the flow of goods, information, and money in the supply chain so that the products which reach the consumer can provide satisfaction in terms of timeliness of delivery, goods quality, and an affordable price. In turn, it will provide the maximum benefit to all members involved in the supply chain. One of the most important things in supply chain management is the sharing of information, therefore, the material flow, cash flow and information flow are elements in the overall supply chain which need to be integrated [4] [5]. Over the past two decades, supply chain management has proven to be an unavoidable solution to meet these contradictory constraints. As a result, several researchers and practitioners focus on integrating the various supply chain functions to increase flexibility, improve cycle times, and reduce costs. In supply chain management, both functions, production and distribution operations, can be decoupled if there is a sufficient buffer between them. [6].

One of the methods used for supplier selection and criteria priority determination is DEMATEL method. DEMATEL method can be used to analyze and can help companies accurately predict where suitable suppliers by focusing on important factors were found. DEMATEL method can also be used to obtain a causal relationship between each factor can provide a basis for determining the price and also provides the most important reference data for decision makers [7]. DEMATEL method was originally developed by Batelle Memorial Institute of Geneva Research Center. DEMATEL can help managers measure the importance and casual relationship of system components through assessing their direct and indirect relations and constructing a map [8]. Different criteria have different internal relationship with other criteria. Criterion with high prominence should play a more important role in the process of computing comprehensive weight. Therefore, DEMATEL method as an important tool to criteria and supplier selection.

The problem of suppliers and criteria determination has been resolved in previous studies. The research was conducted at one of the fast food industries which is has suppliers and branches spread throughout the world. The study was conducted using the supply chain by Kshitiz Sharma in priorities determination and they could serve consumers well. Low prices and customer satisfaction are the company's motto and the selected suppliers must have a good criteria [9]. The proposed research model analyzing the relationship among SCM practices and operational performance and its validation using the Vietnam garment industry provided valuable insights both from theoretical and practical perspectives. The textile supply chain is complex structure consist of cotton or other raw material supplier, manufacturer and wholesale distributor, retailer and consumer. It requires some complex processes to supply product to customer. But, irrespective of its complexity, the supply chain has separated links in which communication is carried out as needed with specific methods [10].

Based on the research has been done before, this study uses DEMATEL. The company has a distribution center spread in the country, namely Aceh, North Sumatra, Padang Riau and Jambi. From the distribution center, the company does not yet know what factors can affect the performance of the distribution center and has not yet analyzed and determined the priorities of the existing distribution. The DEMATEL method is used in distribution centers identification and influencing criteria in the supply chain run by the company.

IOP Conf. Series: Materials Science and Engineering **648** (2019) 012008 doi:10.1088/1757-899X/648/1/012008

2. Methodology

This study was conducted at one of the companies engaged in the manufacture of food namely instant noodles in the Medan city where the object examined in this study was to find out the criteria needed to determine the priority of the distribution center. The company has many branches and companies need to determine the distribution center selection. The study begins with observations to observe and see the state of the company directly. From the results of observations, it is determined the formulation of the problem in accordance with the conditions that occur in the company and the research objectives can be applied. The research objectives determined are a solution to existing problems. Furthermore, the data collection is the input in conducting this research. The data needed is in the form of SWOT and DEMATEL data from the results of questionnaires. With these data, data processing is carried out with several stages, namely data validity test, data reliability test and variables test between variables and distribution centers. Data validity test is needed to see the relationship between each variable and can be responsible for. Reliability test is needed and the information obtained can be trusted as a data collection tool and is able to reveal actual information in the factory.

Identification determination of criteria needed for the selection of distribution centres spread in the country, namely Aceh, North Sumatra, Padang Riau and Jambi using the DEMATEL method. DEMATEL method can be used to analyze and can help companies accurately predict where suitable suppliers by focusing on important factors were found. DEMATEL method can also be used to obtain a causal relationship between each - each factor can provide a basis for determining the price and also provides the most important reference data for decision makers. Due to its advantages and capabilities, the approach of DEMATEL has received a great deal of attention in the past decade and many researchers have applied it for solving complicated system problems in various areas. In addition, the DEMATEL has been extended for better decision making under different environments since many real-world systems include imprecise and uncertain information. The DEMATEL can confirm interdependence among factors and aid in the development of a map to reflect relative relationships within them and can be used for investigating and solving complicated and intertwined problems. This method not only converts the interdependency relationships into a cause and effect group via matrixes but also finds the critical factors of a complex structure system with the help of an impact relation diagram. As stated earlier, the DEMATEL technique can convert the interrelations between factors into an intelligible structural model of the system and divide them into a cause group and an effect group. Hence, it is an applicable and useful tool to analyze the interdependent relationships among factors in a complex system and rank them for long-term strategic decision making and indicating improvement scopes. The formulating steps of classical DEMATEL are generate the best result for criteria and supplier selection [11].

3. Result and Discussion

3.1. Validity Data Test

Validity data test is done to see the relationship between each variable. Test validity is based on perspective parameters, namely the perspective section of Balance Scorecard, Financial perspective section, Internal perspective section, Customer perspective section, Learning & Growth perspective section, BSC of the distribution center (Financial), BSC of the distribution center (internal), BSC section of the distribution center (Costumer), BSC of the distribution center (Learning & Growth) and the distribution center for the Financial perspective. Balance Scorecard perspective section has 12 relationships, Financial perspective section has 3 relationships, Internal perspective section has 4 relationships, Costumer perspective section has 4 relationships, Learning & Growth perspective section has 4 relationships, BSC section of the distribution center (Financial) has 5 relationships, BSC of the distribution center (Internal) has 5 relationships, BSC of the distribution center (Costumer) has 5 relationships, BSC of the distribution center (Learning & Growth) has 5 relationships and the distribution center for the Financial perspective has 5 relationships. The results obtained indicate the overall parameters are valid.

3.2. Reliability Data Test

Reliability data test is needed, and the information obtained can be trusted as a data collection tool and is capable to reveal actual information in the factory. The results recapitulation of the calculation of reliability data test can be seen in table 1.

Table 1. Recapitulation of Reliability Data Test

Parameter	Category
Balance Scorecard Perspective	Reliable
Financial Perspective	Reliable
Internal Perspective	Reliable
Costumer Perspective	Reliable
Learning & Growth Perspective	Reliable
Financial of Distribution Centre Perspective	Reliable
Internal of Distribution Centre Perspective	Reliable
Costumer of Distribution Centre Perspective	Reliable
Learning & Growth of Distribution Centre Perspective	Reliable
Distribution Centre of Financial Perspective	Reliable

3.3. Calculation of Distribution Centre

The Distribution Center consists of Jambi, Medan, Aceh, Riau and Padang. From the results of calculation, it is known that the D + R value from a balanced scorecard perspective is 1,0631; 1,5516; 1,3994; 1,3090 and 1,3133. The results of calculation with D-R are obtained -0,3141; -0,8934; -0,5300; -0,6386 and -0,7327. Then it can be seen in order of importance, namely Medan, Aceh, Padang, Riau and Jambi.

4. Conclusion

The problem of distribution priorities determination often occurs in large companies because they do not know the appropriate criteria in supplier selection. Validity and reliability tests show the results of questionnaires are valid and reliable. The DEMATEL method is one method can be used in overcoming the problem of distribution centers and selection criteria determination. The results obtained by this method shows that the criteria influences the selection of the companies distribution center in the supply chain are based on the balance scorecard perspective consisting of Financial, Internal, Customer and Learning & Growth perspective. By using the DEMATEL method, it can be seen the priority distribution center is Medan followed by Aceh, Padang, Riau and Jambi.

Acknowledgement

Authors thank to Kartika Widya Astuti, Omaar Al Faridzi Siregar, and Alisha Rizki Siregar for the support provided during this research and Nidia Juneika who actualize the research.

References

[1] Faruk Hasanspabic 2013 *Global Business and Economic Research Journal* **2** (6) 26-37.
[2] Phuoc Van Nguyen 2018 *European Journal of Economic Management* **7** (1) 11-18.
[3] Julio Kyosen Nakatani, Marcos Fava Neves 2018 *Journal of Business Administration* **4** (30) 1-11.
[4] Natelda R. Timisela, Ester D. Leatemia, Febby J. Polnaya, Rachel Breemer 2018 *Journal of Applied Management* **15** (1) 135-145.
[5] Dunne A.J. 2001 *Australian Agrobusiness Perspectives*.

IOP Conf. Series: Materials Science and Engineering **648** (2019) 012008 doi:10.1088/1757-899X/648/1/012008

[6] Achraf Touil, Abdelwahed Echchatbi, Abdelkabir Charkaovi 2019 *International Journal of Supply and Operations Management* **6** (1) 30-50.
[7] Fu Xiaoyong,dkk 2010 *A Grey-Dematel Methodology For Green Supplier Development Program Evaluation* (Guangzhou: GPMI Working).
[8] Tianyu Liu, Yong Deng, Felix Chan 2017 *International Journal of Fuzzy Systems*.
[9] Kshitiz 2013 *Asia Pacific Journal of Marketing and Management Review* **2** (1) 112-120.
[10] Phouc Van Nguyen 2017 *Department of International Business Administration*
[11] Sheng-Li Si, Xiao Yue You, Hu Chen Liu, Ping Zhang 2018 *Mathematical Problems in Engineering* 1-33

IOP Conf. Series: Materials Science and Engineering **648** (2019) 012009 doi:10.1088/1757-899X/648/1/012009

Minimization of Makespan Using FCFS and Gupta Method Comparison in Machine Spare Part Industry

I Siregar[1]

[1]Faculty of Engineering, Universitas Sumatera Utara, Medan, Indonesia

Email : ikhsan.siregar@usu.ac.id

Abstract. Companies that produce spare parts for manufacturing industry machines are experiencing rapid development. The company produces production activities based on job orders where the process uses FCFS (First Come First Served) rules. Company scheduling often experiences delays especially on products have the same flow. The study was conducted to find the optimal sequence of scheduling with makespan minimization criteria using the Gupta Method and FCFS Method. The FCFS method used by the company produces makespan of 12.160,21 minutes = 202,67 hours. The recommendation method has a better level of performance compared to the company shown in efficiency index (EI) and relative error (RE). The EI value of the Gupta method is 1.14 which is indicates that the Gupta method has a good performance compared to the FCFS method. While the Gupta method RE value is 13,76%. The conclusion is Gupta method produces the minimum makespan value with a reduction of makespan of 178,15 hours. The Gupta method is a better method than FCFS method, which is shorter makespan, tardiness and flow time.

1. Introduction

A company's performance will not be optimal if the company is only trying to maximize the use of existing resources without looking at other factors. Companies must make a systematic work plan and the company can provide good service to consumers by sending products on time and completeness of production in accordance with the due. A systematic work plan can be carried out by scheduling measurable production. Good scheduling will have a good impact on increasing the effectiveness and efficiency of the company's production system so that it can reduce production costs.

The problem often occurs is that the production process activities on the production floor often experience delays in activities through the same machine. Delays occur due to the use of the same machine for working on different products. In addition, the type of product variation and the stage of the work process with the use of the same machine also causes delays due to the bottleneck. Therefore, rescheduling is needed to minimize the total completion time in order to meet customer orders.Scheduling can be defined as the allocation of resources over a period of time to perform a collection of tasks [1]. Makespan minimization in tasks scheduling is important of infrastructure [2]. The theme of scheduling in distributed clusters and systems is one of the important themes in utilizing available resources and reducing response time. Job scheduling software are some important software in groups that are often used. The job scheduling function involves allocating the right processor and further maximizing the use of system resources and reducing the average response time created from existing conditions [3]. Regarding scheduling in parallel has been very widely used, scheduling is very useful in the application of knowledge. Work in parallel consists of several processors that perform simultaneously all of which do certain calculations [4]. Dissimilar job and process scheduling algorithms are often used for very long periods of time in various domains such as microprocessor-based systems, software production management techniques [5]. Production scheduling is one of the significant factors related to consumer demand. This is interrelated with the optimization of several indicators, including performance indicators, all of which are directed towards the functioning of the system and customer satisfaction, such as: efficient use of resources, delivery of products with a given deadline, and reduction of production costs [6].

By far the modest scheduling algorithm is the first-come, first served scheduling (FCFS) algorithm. With this system the process that requirements the object first is allocated the object first. The implementation of the FCFS policy is simply succeeded with a FIFO queue [7]. NEH (Nawaz, Enscore, Ham) Algorithm is an efficient algorithm that works by minimizing the makespan. There are algorithm issues in job scheduling where are theoretical and applied research can both contribute a lot of solutions. First come first served scheduling a modest job scheduling method which can frequently be found in actual systems but is regularly measured insufficient by numerous researchers. Many workloads good performance can be expected from an FCFS schedule [8]. Gupta anticipates heuristic techniques to reach the makespan to the minimum. In the Gupta heuristic algorithm, all work is divided into two parts by linking the allowance time of the first machine and the last machine in each job. For each group, analyze the quantity of processing time from two end-to-end tasks in a job and find the minimum processing time, and then schedule the work in sequential order according to the minimum processing time added. [9, 10].

Production forecasting problems [11] and the problem of production scheduling has been resolved in previous studies. The study was conducted at one of the juice companies have 5 different production flows. The study was conducted using the Gupta method by Soheil Sadi-Nezhad and Samira Borhani Darian and managers were easier to make decisions and an effective way of making production scheduling models [12]. The new tie-breaking mechanism is based on the estimated idle time of various work sequences in selecting jobs then the one with the lowest estimated value. Some computational combinations carried out by them show that this mechanism outperforms preexisting, both for FCFS and iterated greedy algorithms for different CPU times [13].

Based on the research has been done before, this study uses the FCFS method and the Gupta method. The Gupta method is used in analyzing production scheduling which is has a series of machines with a number of jobs (n-job m-machine). The FCFS method is used by companies to obtain the minimum makespan value and it can reduce the level of delay occurs on the spare parts company. The study was conducted by comparing the performance of the FCFS method with the Gupta method to get a better method in minimizing of lead time.

2. Methodology

The study was held in industries engaged in spare parts machine in Medan city where the objects studied in this research is doing production sequencing processes and the production scheduling system for each spare part machine. The study begins with observations to observe and see the state of the company directly. From the results of observations, it is determined the formulation of the problem in accordance with the conditions occurs in the company and the research objectives can be applied. The research objectives determined are a solution to real problems. Furthermore, the data collection is the input in conducting this study. Data needed in the form of cycle time, standard time, number of demands and the number of machines used. Cycle time is obtained by measuring time using a stopwatch time study. The stop time watching study technique is to use the most basic and easiest time measurement method, and this method is usually used rather than other time measurement methods [14]. With these data, data processing is carried out with several stages, namely conducting data uniformity test, data adequacy test, standard time calculation and calculation of lead time (processing time), and production scheduling. The cycle time has been obtained from the measurement results is then given allowances in the westinghouse method and normal time can be input into the standard time calculation. Calculation of standard time is done for each type of door leaf in the entire work center with the formula.

$$Ws = Wn \times \frac{100\%}{100\%\text{-}\%Allowance} \quad (1)$$

With the formula above, a standard time will be obtained which is will be used to calculate the lead time. Calculation of lead time is measured using the following formula:

$$\text{Lead Time } (t_{ij}) = \text{Setup time} + \left(\frac{\text{standard time x demand}}{\text{product capacity(unit/process)}}\right) \quad (2)$$

Then production scheduling is carried out using the First Come First Served (FCFS) and Gupta method. Production scheduling using the FCFS method is done to make makespan minimum with several steps. The first steps are the amount of the processing times of each job and do the sequence of jobs affording to the amount of processes opening from the largest to the smallest and the results of this classification are called the list of sorting jobs. Then take the job of ranks primary and another on the list of job sequences then calculate the number of partial makespan and mean flow time to determine the selected partial sequence with the smallest makespan value, then do the similar steps as in the previous stage until the overall sequence of jobs is obtained. Purpose of job arrangements is done by observing at the smallest makespan rate for each job. Scheduling with the Gupta method is done by adding up the lead time of the closest machines to obtain the minimum value of the minimum time to determine the value of EI and SI. Furthermore, the calculation results are sorted from jobs have the largest to the smallest value. Then the best method is chosen to produce the minimum makespan between the Gupta method and the FCFS method. Gupta and FCFS in this research are combine method.

3. Result and Discussion

3.1. Uniformity Data Assessment and Adequacy Data Assessment

The uniformity data test is carried out to see whether the data collection done undeviating observation (in the controller limits) or not. When the data competence test is done to understand which is data has been gotten in sufficient for data dealing out. The quantity of observational data needed for a confidence level of 95% and accuracy level of 5%. The recapitulation of results in uniformity data test and data adequacy test shown in Table 1.

Table 1. Recap of Uniformity Data and Adequacy Data *Assessment* at Each Job

Job Type	Uniformity Data	Adequacy Data
Wheel Scales	Uniform	Sufficient
Pulley	Uniform	Sufficient
Sprocket 14T	Uniform	Sufficient
Gear	Uniform	Sufficient
Sprocket 12T	Uniform	Sufficient
Impeller	Uniform	Sufficient

Based on the table above, it can be seen that the data collection during the observation was uniform and sufficient and it can be used for the next stage.

3.2. Calculation of Standard Time

Standard time is planned after normal time and allowance are known where the proportion of allowance is allowance for rest given to each worker. The result gets from formula 1 and 2 above. The results recapitulation of standard time calculations shown in Table 2.

Table 2. Recap of Standard Time

Job Type	Standard Time (Minute)				
	M 1	M 2	M 3	M 4	M 5
Wheel Scales	15,12	23,34	38,38	63,70	15,17
Pulley	18,18	27,19	44,55	106,66	15,22
Sprocket 14T	7,67	19,61	35,31	41,21	5,63
Gear	12,60	41,76	47,83	153,41	27,44
Sprocket 12T	6,51	16,82	31,96	36,07	5,50
Impeller	13,82	28,55	44,26	60,88	10,42

3.3. Calculation of Lead Time

Calculation of lead time is influenced by the quantity of machineries owned by the company, production volume or machine and standard time to make a product. Recapitulation of lead time at each job in all work stations can be seen in Table 3.

Table 3. Recapitulation of Lead Time

Job Type	Lead Time (Minute)				
	M 1	M 2	M 3	M 4	M 5
Wheel Scales	640,24	985,42	1228,89	2685,55	502,84
Pulley	308,04	458,12	576,84	1787,73	215,21
Sprocket 14T	209,64	527,92	726,10	1108,97	137,68
Gear	193,97	152,51	558,03	2311,22	333,73
Sprocket 12T	152,51	386,21	563,36	827,65	118,50
Impeller	115,54	233,44	285,58	497,01	87,53

3.4. Actual Scheduling using First Come First Serve (FCFS)Method

Based on the sequence of jobs according to the company method in the form of First Come First Serve (FCFS) where the first order arrives will be done or completed first, then the work order obtained is job 1 - job 2 - job 3 - job 4 - job 5 - job 6 with the resulting makespan value of 202.67 hours.

3.5. Scheduling using Gupta Method

Scheduling with the Gupta method is done by adding up the lead time of closest machines to obtain the minimum value of the minimum time to determine the value of EI and SI. Furthermore, the calculation results are sequenced from jobs have the largest to the smallest value. Recapitulation of the sequence of all jobs with the Gupta method can be seen in Table 4.

Table 4. Scheduling Sequences using Gupta Method

Job Type	Time (Minute)				
	M 1	M 2	M 3	M 4	M 5
Gear	193,97	631,47	558,03	2311,22	333,73
Wheel Scales	640,24	985,42	1228,89	2685,55	502,84
Pulley	308,04	458,12	576,84	1787,73	215,21
Sprocket 14T	209,64	527,92	726,10	1108,97	137,68
Sprocket 12T	152,51	386,21	563,36	827,65	118,50
Impeller	115,54	233,44	285,58	497,01	87,53

The sequence of jobs obtained results in a makespan value of 10.689,13 minutes or equal to 178,15 hours. Gupta method makespan value is shorter when associated to the company authentic method, FCFS method with the order of job 4 - job 1 - job 2 - job 3 - job 5 - job 6 where job 1 is gear.

4. Conclusion

Production scheduling limitations often occur due to delays in products sent to customers due to the long process of product processing. Gupta method is one method can be used to overcome production scheduling problems. The results facile by this method shows that the time needed to complete the product is shorter than the actual method currently applied by the company, namely the First Come First Serve method with a makespan value of 202,67 hours with the order in which is the product 1-job 2-job 3-job 4-job 5-job 6 when makespan with Gupta method is 178,15 hours with the order of job 4 - job 1 - job 2 - job 3 - job 5 - job 6. Performance with the Gupta method also proved to be good than the actual method of the company with an efficiency index value of 1,17.

References

[1] Siregar I, Nasution A A, and Sari R M Effect of Total Quality Management on the Quality and Productivity of Human Resources *IOP Conference Series: Materials Science and Engineering* (Vol. 180, No. 1, p. 012116). IOP Publishing.

[2] Abdulhamid S M, Abdul Latif M S, Idris I 2014 *Journal of Engineering and Applied Science* **9** (12) 2528-2533.

[3] Hafshejani Z M, Mirtaheri S L, Khanegah E M 2013 *The International Journal of Soft Computing and Software Engineering* **3** (3) 422-428.

[4] Zhang Y, Franke H, Moreira J E, Sivasubramaniam A 2000 *Procceedings 14th International.*

[5] Sanyal J 2016 *Journal of Operation Research* **4** (3) 177-179.

[6] H Y Fuchigami, S Rangel 2017 *Journal of Computational Science*, **17**.

[7] Ashok K, Harish R, Arya S 2013 *International Journal of Advanced Research in Computer Science and Software* **3** (5) 1158-1162.

[8] Schwiegelshohn U, Yahyapour R 2013 *Jurnal Universitas Gajah Mada.*

[9] Gupta J N D 1972 *Heuristic Algorithm for Multistage Flow Shop Problem* (Skotland: AIIE Transaction).

[10] Kumar S, Jadon P 2014 *International Journal of Computer Science and Information Technologies* **5** (4) 5057-5061.

[11] Rahmat R F, Nurmawan, Sembiring S, Syahputra M F, Fadli 2018 Adaptive neuro-fuzzy inference system for forecasting rubber milk production *IOP Conference Series : Materials Science and Engineering* **308**(1), 012014

[12] Sadi-Nezhad S, Darian S B 2010 *International Journal of Engineering and Technology* **2** (6) 410-418.

[13] Viagas F, Framinan J M 2014 *Computers & Operations Research*, **45** 60-67.

[14] Sutalaksana I Z, et. al 2006 *Perancangan Sistem Kerja Edisi Kedua* (Bandung: Institut Teknologi Bandung).

The Success Model of TQM on Managerial Performance

Fadli[1]
[1]Department of Management, Universitas Sumatera Utara, Medan, Indonesia

*Email : fadlifauzi28@gmail.com

Abstract The objectives of this study were to analyze technology of information, measurement of performance systems and systems of prize for performance of managerial of total management quality as a moderating variable. currently the managerial performance of PT. Pegadaian Kanwil I Medan is good, but the management has not been satisfied with the results to date. Therefore, this research is conducted to find out factors that can further improve management performance. The population in this study were employees at managerial level at PT. Pegadaian Kanwil I Medan (Persero). The method of determining the sample used in this a study is using criteria/purposive sampling method, primary data is the data used in this study, while multiple regression analysis method used on the data processing and interaction testing. This study showed that technology of information variables and measurement of performance systems partially had a positive significant effect on performance of managerial. TQM variables have a significant influence, this means that moderation is able to reinforce the influence of technology of information, measurement of performance systems and systems of reward for performance of managerial with positive coefficients.

1. Introduction

Today's world continues to grow rapidly and there is a lot of competition. Progress in all fields of technology, information, knowledge and transportation requires all parties, including the business world to further improve quality and services in the form of customer service improvements, meeting market demand, improving the quality of goods or service, in order to attract as many customers as possible.

Ways that have been widely used by companies to improve products, services and exercise through the exercise of their employees. either quality management instrument it can be outworn is Total Quality Management (TQM). TQM is a technique commonly used by good corporation engaged in services and manufacturing to help increase customer satisfaction, employee satisfaction, and productivity [1].

Strong competition among the same types of companies, requires every company to try their best to improve their respective performance not only that, the organization must be able to overcome all obstacles that can hinder the development of the company and also support activities of its human resources. Improving and maintaining morale is not easy, because every employee who works for a company has a different goal that is sometimes not the same as the company's goals. If this happens there needs to be a motivational action against employees to have the same goals as the company, so that the survival of the company can be guaranteed.

This is very influential on economic actors including PT. Pegadaian Persero, Medan Regional Office I, as one of the State-Owned Enterprises (BUMN) in Indonesia which is engaged in the field of credit distribution servicesto the public or mortgage law. The company competes closely with similar companies, so it demands management to manage its companies and business units efficiently through improved performance, in addition to improving the quality of services to meet satisfaction for its customers, the demands faced by the service industry today are creating a service system that provides convenience in exchanging information on resilience to disruptions, stability, loyalty to customer authority and high performance.

Various pressures in the business world force changes in business practices themselves and continue to make improvements in everything, improve skills and work skills of employees, this of course will increase the motivation of employees in carrying out their work to produce the best performance. Consumer satisfaction is very important for the survival of the company, if the consumer is not satisfied it will cause the company and will become a customer of a competing company. This will cause a decrease in income and in turn will reduce profits and even suffer losses [2].

TQM is a technique that is often used bycompanies to help increase customer satisfaction, employee satisfaction, and productivity [3]. In companies with low performance, it is usually caused by their dependence on the company's Accounting Management System which fails in determining the right targets, performance measurement, and reward system.

2. Literature Review

Company performance dimensions can be divided into two dimensions [4], namely; performance of operational and performance of organizational. Performance of operational describe the corporation internal performance of operating about costs and reduced trash, repairing product quality, developing fresh products, improving performance of delivery, and increasing productivity. These indicators and variables are considered as the main factors because they follow directly from the actions taken in the company's operations. Whereas organizational performance is measured by financial measures such as income growth, net income, profit ratio with income and profit on assets, and non-financial measures such as investments in R&D, and the company's capacity to develop competitive profiles.

According to Zehir and Esin [5], measurements business performance can be done through 2 dimensions namely: innovation performance and employee performance. Performance innovation is measured through product innovation rather than competitors in the market, the number of new products being marketed the last 5 years, and the speed of introduction new products/services on the market. Whereas performance employees are measured through 3 indicators, namely level employee satisfaction, attendance, and employee morale.

Empirical studies that examine the relationship between TQM practices with a lot of company performance found in the operations management literature. For example Demirbag et al [6] conducted a study empirical to identify important factors for the success of TQM in SMEs in Turkey. They concludes that there are seven critical successes factors (CSF) TQM practices, namely data and reporting quality, top management role, employee relations, supplier quality management, training, quality policy and process management. Factors the practice of TQM can improve performance company. Prayogo and Hong [7] did research on 130 manufacturing industry R&D units in Korea. The results of his research found that implementation of increasingly effective TQM practices has a significant influence on performance company.

3. Methodology

In this study the type of data used is a type of qualitative data. Qualitative data (is the type of research data in the form of opinions, attitudes, experiences or characteristics of a person or group of people who are the subject of research or respondents). Whereas, for data sources used in this study are primary data, namely the source of data obtained directly from the questionnaire. Distribution of questionnaires was conducted to obtain respondents' personal data and assessment of Information Technology, Performance Measurement System and Reward System on Managerial Performance with Total Quality Management as a moderating variable.

Population in this study is defined as employees at managerial level which included staff, assistantsManagers, Senior Assistant Managers, Bureau Heads, Managers, Senior Managers for all departments. In the sampling criteria and numbers have been determined, for the number of samples as many as 36 employees, and for the criteria are as follows:

1. Has worked in a managerial level of at least 1 (one) year at PT. Pegadaian Persero, Regional Office I Medan
2. Employees who work at Medan Head Office.

The data collected in this study is primary data. Primary data is obtained by using the Survey method, namely through a questionnaire. by visiting one by one prospective respondent, asking whether the candidate meets the requirements as a prospective respondent to fill out the questionnaire.

The Hypothesis Testing Model used is Multiple Regression Analysis, which aims to test the effect of one variable on other variables. Multiple linear regression involves more than one independent variable. By using two or more independent variables (independent) in making a regression equation, it is expected to better explain the characteristics of the dependent variable and the value of the coefficient of determination is expected to be greater and the standard error value is smaller so that the resulting regression equation is better. Multiple linear regression models are said to be a good model if the model meets the assumptions of normality of data and is free from classical statistical assumptions, both multicollinearities, autocreation and heteroscedasticity

So, to test the first, second and third hypotheses, multiple regression analysis will be used, the aim is to see the effect of the independent variable with the dependent variable, using the following formula:

$$Y=a + b1X1 + b2X2 + b3X3 + e \quad (1)$$

Information :

Y	: Managerial Performance
a	: constant
X1	: Information Technology
X2	: Performance measurement system
X3	: Reward system
b1	: Information Technology regression coefficient
b2	: Performance Measurement System regression coefficient
b3	: Reward System regression coefficient
e	: error

Testing the fourth, fifth and sixth hypothesis, namely, the independent variable, the dependent variable and the moderating variable, with the Interaction Test method, while the regression equation is:

$$Y = a + b1X1 + b2X2 + b3.X3 + b4X4 + b5X1X4 + b6X2X4 + b7X3X4 + e \quad (2)$$

4. Result and Discussion

4.1. Multiple Linier Regression Model

Multiple regression testing aims to determine the coefficients of each independent variable. So that the magnitude of the influence of each independent variable is simultaneous and partial. The multiple regression equations used in this study are as follows:

$$Y = 1.279 + 0{,}023X1 + 0{,}253X2 + 0{,}335X3 + e \quad (3)$$

Table 1. t-test

Coefficients[a]

Model		Unstandardized Coefficients		Standardized Coefficients		
		B	Std. Error	Beta	T	Sig.
1	(Constant)	1.279	.414		3.089	.004
	LNX1	.023	.128	.030	.179	.859
	LNX2	.253	.150	.296	1.684	.102
	LNX3	.335	.145	.417	2.313	.027

a. Dependent Variable: LNY

Source : SPSS Software

4.2. Moderate Regression Analysis (MRA)

The multiple regression equation used in the hypothesisfourth, fifth and sixth are as follows:

$$Y = 1.088 + 0{,}016X1 + 0{,}205X2 + 0{,}298X3 + 0{,}149X4 + 0{,}599X1.X4 + 0{,}526X2.X4 + 0{,}576X3.X4 + e \quad (4)$$

Table 2. t-test Moderating

Coefficients[a]

Model	Unstandardized Coefficients		Standardized Coefficients		
	B	Std. Error	Beta	T	Sig.
(Constant)	1.088	.433		2.514	.017
LNX1	.016	.127	.021	.123	.903
LNX2	.205	.153	.240	1.346	.188
LNX3	.298	.146	.371	2.046	.049
LNX4	.149	.110	.202	1.349	.187

a. Dependent Variable: LNY

Source : SPSS Software

The results of the first and second hypotheses show that all independent variables have a positive and not significant effect on Managerial Performance, namely Information Technology and Performance Measurement Systems, with each having a positive influence of 0.023, and 0.253 with significant levels of 0.859 and 0.102 (Sig. 5 0.05) and the fourth hypothesis shows that all independent variables have a positive and significant influence on Managerial Performance, namely the Reward System has a positive effect of 0.335 with a significant level of 0.027 (Sig. 5 0.05). While Information Technology, Performance Measurement Systems and Reward Systems simultaneously affect Managerial Performance, which is a positive effect of 8,350 and a significant level of 0,000 (Sig. 5 0.05). This research is in line with the research conducted [8] which states that Information Technology, Performance Measurement Systems and Award Systems partially and simultaneously affect Managerial Performance.

The results of the research on the fourth, fifth and sixth hypotheses indicate that Total Quality Management as a moderating variable can strengthen the influence of Information Technology on Managerial Performance, Performance Measurement Systems for Managerial Performance and Reward Systems for Managerial Performance. Can be seen from the results of the study of the interaction effect on the fourth hypothesis, namely the interaction between Information Technology and Total Quality Management has a positive coefficient of 0.599. And the study results of the interaction effect on the fifth hypothesis, namely the interaction between the Performance Measurement System and Total Quality Management has a positive coefficient of 0.526. And the results of the study of the interaction relationship on the sixth hypothesis, namely the interaction between the Prize System and Total Quality Management has a positive coefficient of 0.576. Then it can be concluded that Total Quality Management is able to strengthen the influence of Information Technology, Performance Measurement Systems and Gift Systems on Managerial Performance.

5. Conclusion

Based on the results of this study, some conclusions can be drawn as follows:

a. Partially only the reward system has a positive and positive effectsignificant to managerial performance, while information technology anda partial performance measurement system does not have a positive effect not significant to managerial performance at PT Pegadaian Kanwil I(Persero).

b. Information technology, performance measurement systems and reward systemssimultaneous influence on managerial performance at PT Pegadaian Kanwil I (Persero)

c. Total quality management is able to strengthen the influence of technology information, performance measurement systems and reward systems for performance management at PT Pegadaian Kanwil I Medan (Persero).

References

[1] Ibrahim, 2000, Total Quality Management : Guide to Facing Global Competition, Jakarta,

[2] Suwastiko, 2011, Pengaruh Total Quality Management (TQM) Terhadap Kinerja Manajerial Dengan Sistem Pengukuran Kinerja Dan Sistem Reward Sebagai Variabel Moderating Pada Rumah Sakit Haji Di Surabaya, Skripsi, Surabaya,

[3] Sim and Killough, 1998. The Performance Effects of Complementarities Between Manufacturing Practice and Manufacturing Practice and Management Accounting Systems. *Journal of Management Accounting Research, Volume 10,PP::325-346.*

[4] Brah, s. And Lim, H. 2006. The effects of technology and TQM on the performance of logistics companies", International Journal of Physical Distributions & Logistics Management, Vol. 36, No. 3, pp. 192-209.

[5] Zehir, Cemal and Esin Sadikoglu. 2009. The relationship between total quality management (TQM) practices and organizational performance; An empirical investigation, *International Journal of Production Economics*, Vol. 11, No. 3, pp. 140-156

[6] Demirbag, M, Tatoglu, E. Tekinkus, M and Zaim. S. 2006. An Analysis of the relationship between TQM implementation and organizational performance; evidence from Turkish SMEs, *Journal of Manufacturing Technology Management*, Vol. 17, No. 6, pp. 829-847

[7] Prajogo, Daniel I. And Soon W. Hong, 2008. The effect of TQM on performance in R&D environments: *A Perspective from South Korean firms, Technovation* 28, pp. 855-863.

[8] Noviyanti, 2013. Pengaruh Teknologi Informasi, Sistem Pengukuran Kinerja Dan Sistem Reward Terhadap Kinerja Manajerial Dengan Total Quality Management Sebagai Variabel Moderating Pada PT. Pelabuhan Indonesia- I Medan.

Acknowledgments

Author is grateful to God, the ruler of the universe who always brings his grace together with guidance, so the writer can complete the research.

IOP Conf. Series: Materials Science and Engineering **648** (2019) 012011 doi:10.1088/1757-899X/648/1/012011

Automatic Number Plate Recognition for Indonesian License Plate by Using K-Nearest Neighbor Algorithm

D Gunawan, W Rohimah, R F Rahmat

Department of Information Technology, Universitas Sumatera Utara, Medan, Indonesia

E-mail: danigunawan@usu.ac.id

Abstract. Transportation surveillance requires many human resources to cover all the road in a city. The human resources might be reduced by utilizing automation. For example, a task to recognize the license plate can be automated by utilizing CCTVs or surveillance cameras that usually installed in the crossroads. The automatic number plate recognition is required to record the license plate of the vehicle that violates the traffic rules. This number can be processed for future references. This research utilizes K-Nearest Neighbor (KNN) to recognize the license plate. The research is conducted to 125 license plate images which are divided into 100 training images and 25 testing images. The success rate of this research is 92.86%. The condition of the license plate and light intensity influence the recognition result.

1. Introduction

Automation in public transportation surveillance such as speeding violation or traffic light violation requires automatic number plate recognition. A surveillance camera has an important role in intelligent transportation system [1]. However, to observe the traffic in a city requires many human resources. The automatic number plate recognition might reduce the human resources to observe and record each vehicles license plate that runs in the road. A surveillance camera can provide the inputs for the automatic number plate recognition system. Previous research had implemented optical character recognition to recognize a license plate (or a number plate) [2]. In addition, the system was simulated by using Matlab. The violation is logged and sent to the user via SMS. Another research also uses optical character recognition to recognize a license plate [3]. However, instead of sending the result via SMS, this research compares the recognized license plate with the existing database. The research conducted by Bhavin et al [4] proposed an improvement to OCR-based number plate recognition by using a trained neural network. In 2015, a research conducted by Islam [5] proposed a method to recognize license plate by using structured elements. The method is developed based on morphological operations based on elements structure. The aim of this method is maximally excluded non-interested region and improving object area. This research yields a 92% success rate with the varying light condition. Automatic number plate recognition researches are also conducted for a specific region such as India and Sudan [6] [7]. A research to recognize number plate in Indonesia was conducted by using optical character recognition [8]. They improved the optical character recognition by construct higher resolution template, using more relevant font type, and removing available noise. Our research utilizes K-Nearest Neighbor (KNN) instead of optical character recognition to recognize license plate. Because of the difference among license plate designs in many countries, this research limits the number plate recognition to Indonesian license plate only.

This paper organization is as follow: the first section is an introduction which describes the background of the research, related works, and research aim. The second section is the elaborate description about research method used in this research. The third section is the result of the research and followed by the research summary in the last section, which is the conclusion.

2. Research Method

There are several processing stages to detect number plate in the vehicle as illustrated in figure 1. The first stage is capturing the license plate images for dataset collection. These images will be used as the training and testing data. Second, the image pre-processing stage includes cropping, resizing and grayscaling. The third stage is the segmentation process includes edge detection and thresholding. The fourth stage is feature extraction which applies contour detection and Principal Component Analysis (PCA). In the next stage, all the images from the previous stage are segmented to obtain letters and numbers images. These images will be used as the input for training and testing data. The training and testing data are processed by using K-Nearest Neighbor (KNN) algorithm.

The detail of the processing stages is described as follow:

2.1. Dataset collection

License plate images in this research are collected directly from the parking area. These images will be used as the training and testing data. All the images are taken by using a digital camera which has been set up with 1080 x 810 pixels resolution. The captured images are saved in JPG format. The captured license plate has standard form and size, which is divided into three sections. The first section is the area code or regional code, which represents the place of the registered vehicle. This section can have one or two digits of the alphabet. The second section consists of 2-4 digits of the number. And the last section consists of 1-3 digits of the alphabet.

This research used 125 images, divided into 100 training images and 25 testing images. All the images are taken in good quality. The example of testing data is shown in figure 2. Some of these images are taken indoor and outdoor. This research only takes the pictures of the license plate of the private vehicles. The private vehicles license plate has the black background and white foreground. In some cases, the owner of the vehicle modifies the license plate by adding sticker or writing.

2.2. Image pre-processing

This stage prepares the images before being the classification process is applied. The image pre-processing stage includes cropping, resizing and grayscaling. The detail of each process is explained below.

2.2.1. Cropping The original images are taken from the close distance. However, the images are still not focus on the license plate (shown in figure 3a). Therefore, the images should be cropped to the license plate size manually. The result of the cropped image is shown in figure 3b.

2.2.2. Resizing All the cropped images have different size because the image was taken within the various range of distance. However, the input of the images should be in the same size. Therefore, we should resize all the cropped images to the same size. This research resizes all the cropped license plate images to 216 x 106 pixel.

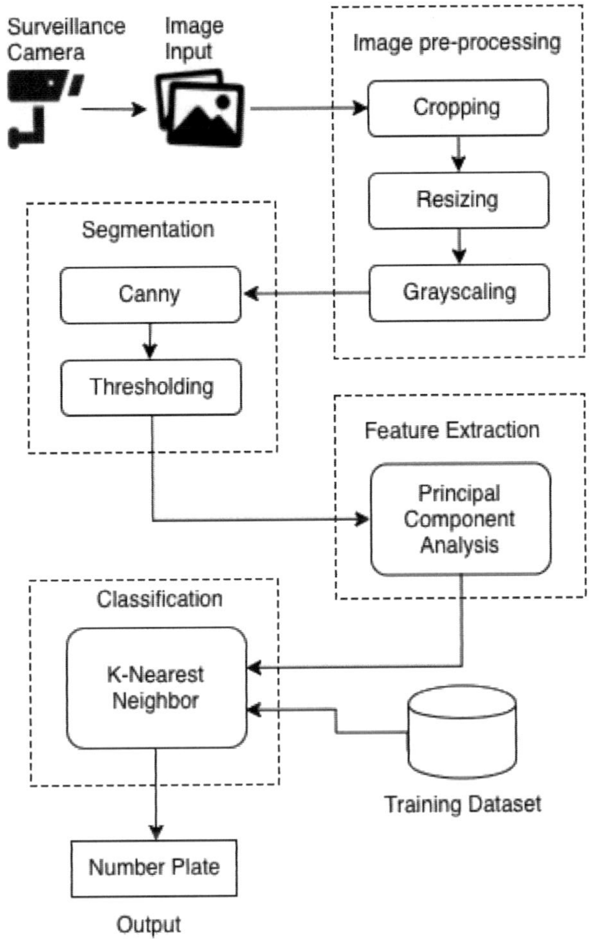

Figure 1. General Architecture

2.2.3. Grayscaling Basically, the cropped license plate images have two dominant colors, such as black and white. The background of the license plate is black. On the other hand, the foreground is white. However, some vehicle owners put things such as stickers, pin or else which are stuck to the license plate and have various colors. Therefore, in order to perform segmentation, the image should be converted to grayscale. This research utilizes OpenCV library to complete this operation.

2.3. Segmentation

The segmentation stage consists of three processes, such as edge detection, thresholding, and gaussian blur. The edge detection divides two homogenous areas which has different brightness. This research uses the canny algorithm to perform edge detection. This method is used because it has the ability to put and mark every edge according to convolutional parameter selection. In addition, it provides high flexibility to determine preferred edge thickness. Canny also supports edge detection with minimum error level. The result of the edge detection is shown in figure 4.

Moreover, thresholding is used to distinguish background and foreground (letters and numbers on license plate). This process converts gray-level images to bi-level images (binary images). The next process is removing the noise by using Gaussian. This process is required to avoid

Figure 2. Original images of license plate

(a) Image before cropping (b) Image after cropping

Figure 3. A license plate image before and after cropping

(a) (b)

Figure 4. Edge detection result

detecting noise as the object. All these three processes are implemented by using the OpenCV library.

2.4. Feature Extraction
This research utilizes the Principal Component Analysis (PCA) to extract features from the images. PCA is a method to obtain important features from a set of data by performing decomposition to the data, therefore it yields uncorrelated coefficient. Feature extraction is done to reduce the data dimension by maintaining as much information as possible. PCA maintains that information from the original dataset. Its projection data has high variants determined by eigenvectors.

2.5. Classification
Extracted features from the images will be classified by using K-Nearest Neighbor (KNN) algorithm. This process will classify the object by determining the closest object based on the training data. The KNN algorithm is divided into two different processes, namely training and testing processes. The training process is the data validation stage. As shown in figure 5, each image of license plate will be divided into nine pieces. For each piece, we should put the corresponding value to the available text field below the image. The number of training data is 60% of the total dataset images. The testing process classifies the images based on the training data.

Figure 5. Data input for training process

3. Result and Discussion
This section is discussing the implementation result of the K-Nearest Neighbor algorithm in automatic plat number recognition. This research is tested on a computer which has the specification as follow:

(i) CPU: 1.80 GHZ
(ii) Harddisk: 500 GB
(iii) RAM: 2 GB
(iv) Operating system: Microsoft Windows 8.1 Pro 64-bit (6.3, build 9000)
(v) Library: OpenCV

The testing is conducted to 25 license plate images which will be recognized based on 100 training images. The result is shown in table 1. According to the result, we obtain 92,86% of success rate.

Table 1 no. 17 shows the license plate BK6892XZ is only 87.5% recognized by the system. The license plate is not fully recognized because the plate has a wavy surface. Therefore, because of its wavy surface, the light distracts the system to read the character correctly. The final result of the license plate recognition is BW6892XZ. The system recognizes BW instead of BK.

IOP Conf. Series: Materials Science and Engineering **648** (2019) 012011 doi:10.1088/1757-899X/648/1/012011

Table 1. Testing Result

No	Image	Result	%	No	Image	Result	%
1		BK3856RAG	100	14		B45626ZAG	89
2		BK4235ADD	100	15		BB5038HI	87.5
3		KK4152AGO	89	16		BK6750MZ	100
4		BK4452AGD	100	17		B6 6892XZ	87.5
5		BK5274GG	100	18		B94442AEW	89
6		KK6808AGF	89	19		BK3597AGN	100
7		BK6130CW	100	20		B55550VAL	89
8		BK6063AEO	100	21		B02763ACB	89
9		P1905PZB	87.5	22		BK5700AGB	100
10		B6703WJF	100	23		B26889 NS	87.5
11		B15362QI	87.5	24		BZ3069 GP	87.5
12		B05966KF	62.5	25		B26545ABW NS	89
13		BK4560FF	75	-	-	-	-

Table 1 no. 13, the success rate of the automatic number plate recognition is 75%. The system yields BK4560FF instead of BK4566MF because of the leak of light and brightness in the license plate. The light is the main problem of the automatic number plate recognition.

Table 1 no. 12 has a lower accuracy of 65.25%. This license plate is an example of broken license plate. The accident causes the size of the plate changed. We suspect the change in the plate dimension influence the plate number recognition. The ratio of the plate number will be different from the standard plate number. In addition, the accident brings up some curves and cause the reflection of the sunlight in the surface of the plate. This reflection still remains and cannot be eliminated in the pre-processing. The reflection will be segmented and considered as characters. Both reasons might decrease success rate.

According to the Table 1, most of the plate number recognition mistakes occur in the recognition of regional code such as BK, BL, BB, B, an so on. The mistakes might occur if one or more condition below presents. For example:

- The plate number condition is changed because of third-party modification or accident
- The plate number recognition requires enough light
- The characters are printed closely to each other
- Some characters are similar, for example, the character 6 is similar to G, 0 is similar to O
- The license plate position

4. Conclusion

This research aims to recognize license plate images automatically to fulfill the requirement for automation in transportation surveillance. This research obtains the license plate images by taking their pictures indoor and outdoor manually. The license plate usually has two dominant colors. The background of the plate is black, and the foreground is white. This kind of plate is used for private vehicles. This research utilizes K-Nearest Neighbor in order to recognize the numbers and characters in the plate. The result shows the success rate of number plate recognition is 92.86%. There are some obstacles to recognize the number plate such as license plate modification, the characters are printed closely to each other, similar characters to each other, and the license plate position. These issues should be covered in the next research.

References

[1] Qiu D, Lu J, Sun X and Jiang J 2010 Application of graphic minimal covering algorithm in the distribution of surveillance cameras in small and medium-sized city road networks *2010 International Conference On Computer Design and Applications* vol 1 pp V1–200–V1–203

[2] Shreyas R, Kumar B V P, Adithya H B, Padmaja B and Sunil M P 2017 Dynamic traffic rule violation monitoring system using automatic number plate recognition with sms feedback *2017 2nd International Conference on Telecommunication and Networks (TEL-NET)* pp 1–5

[3] Qadri M T and Asif M 2009 Automatic number plate recognition system for vehicle identification using optical character recognition *2009 International Conference on Education Technology and Computer* pp 335–338 ISSN 2155-1812

[4] Kakani B V, Gandhi D and Jani S 2017 Improved ocr based automatic vehicle number plate recognition using features trained neural network *2017 8th International Conference on Computing, Communication and Networking Technologies (ICCCNT)* pp 1–6

[5] Islam R, Sharif K F and Biswas S 2015 Automatic vehicle number plate recognition using structured elements *2015 IEEE Conference on Systems, Process and Control (ICSPC)* pp 44–48

[6] Kulkarni P, Khatri A, Banga P and Shah K 2009 Automatic number plate recognition (anpr) system for indian conditions *2009 19th International Conference Radioelektronika* pp 111–114

[7] Mubarak H, Ibrahim A O, Elwasila A and Bushra S 2017 Sudanese license plate identification using automatic number plate recognition *2017 Joint International Conference on Information and Communication Technologies for Education and Training and International Conference on Computing in Arabic (ICCA-TICET)* pp 1–8

[8] Hidayatullah P, Syakrani N, Suhartini I and Muhlis W 2012 Optical character recognition improvement for license plate recognition in indonesia *2012 Sixth UKSim/AMSS European Symposium on Computer Modeling and Simulation* pp 249–254

IOP Conf. Series: Materials Science and Engineering **648** (2019) 012012 doi:10.1088/1757-899X/648/1/012012

Automatic regulator design for Liquified Petroleum Gas

T H Nasution[1*], R Y Nasution[1], K A Putri[1], and C F Nasution[1]

[1]Department of Electrical Engineering, Universitas Sumatera Utara, Medan, Indonesia

[*]Email: tigor.nasution@usu.ac.id

Abstract. In Indonesia, cases of explosions due to gas cylinder leakage have begun to become the attention of the government and the development company. If a leakage happens when there are no people around it this can be dangerous because no one open the regulator on the LPG tube. To handle this, we design a system to detect leakage and release regulators if a gas leakage is detected. We build the automatic regulator system, we use several equipment including Arduino Nano, MQ-2 sensor and servo motor. The MQ-2 sensor is used as an input for gas leakage detection when the servo motor is used as a control output to open the regulator knob. Arduino Nano is used as a control device to read sensor signal input and control the output in the servo motor. Based on the results of the tests have been carried out the automatic regulator work well when a gas leakage happens by opening the regulator knob and the gas does not flow continuously. From the test results obtained an average response time to open the knop after the detected gas is 1,796 seconds.

1. Introduction

Fuel is one of the most important needs for humans. The fuel divided into two, oil and natural gas. The petroleum fuel will be processed into gasoline, diesel and kerosene. Meanwhile, natural gas can be processed into LNG (Liquified Natural Gas) and LPG (Liquified Petroleum Gas). In Indonesia before, mothers use kerosene as a fuel for household needs. With the decreasing supply of natural petroleum, in 2007 the Indonesian government establishes a policy to switch LPG as a fuel for household needs[1]. However, this LPG is airborne, allowing leakage causing dangerous explosions.

In Indonesia. cases of explosions due to gas cylinder leakage began to become the concern of the government and the development company. Actually, leakage known by humans when they are near the LPG location because LPG has a special typical odour. If a leakage is known then we simply release the regulator and the LPG tube will be closed. However, if a leakage happens when there are no people around, this can be dangerous because no one will open the regulator on the LPG tube. To handle this, we propose to create an automatic regulator to release regulators in the event of a gas leakage.

This study was conducted based on the literature of several previous studies on LPG leakage. Previously there have been several studies to detect LPG using sensors[2–4]. There are also several studies to make safeguard against LPG leakage[5–7]. Based on some of these studies we designed a system to detect leakage and release the regulator if a leakage is detected.

2. Method

2.1. Material Collection

To build the automatic regulator system we use several equipment including Arduino Nano, MQ-2 sensor and servo motor. Arduino Nano is used as a control device to read sensor signal input and control the output in the servo motor. The selection of Arduino Nano is due to small size and low price in terms of the control board. Arduino widely used in building control or monitoring systems[8–11]. The Arduino Nano board can be seen in Figure 1.

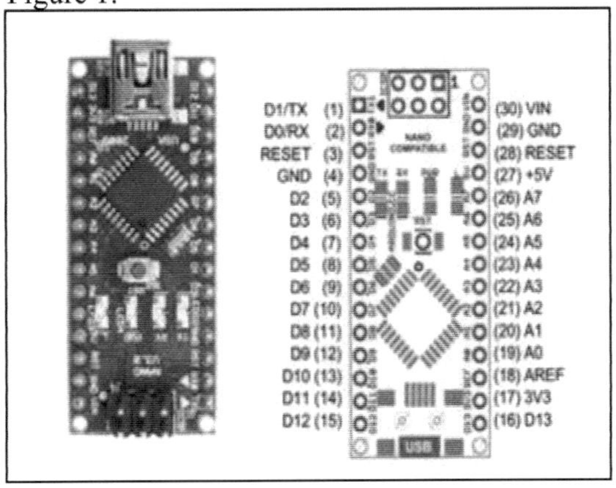

Figure 1. Arduino Nano

The MQ-2 sensor is used as an input for monitoring when the servo motor is used as a control output. The MQ-2 sensor is useful for detecting gas leakage (at home and industry) and it is suitable for detecting H2, LPG, CH4, CO, Alcohol, Smoke or Propane[12]. Because of its high sensitivity and fast response time, measurements can be made as soon as possible. Sensor sensitivity can be adjusted using a potentiometer. Sensor MQ-2 can be seen in figure 2. Servo motor is a DC motor with a close feedback system where the rotor position will be informed back to the control circuit in the servo motor[13]. This motor consists of a DC motor, a series of gears, a potentiometer, and a control circuit. Servo motors can be seen in figure 3. In addition, we also use the buzzer as a marker sound when a gas leakage happens.

Figure 2. MQ-2 Sensor

Figure 3. Motor Servo

2.2. System Design
In the system design, the system is made according to system block diagram in Figure 4.

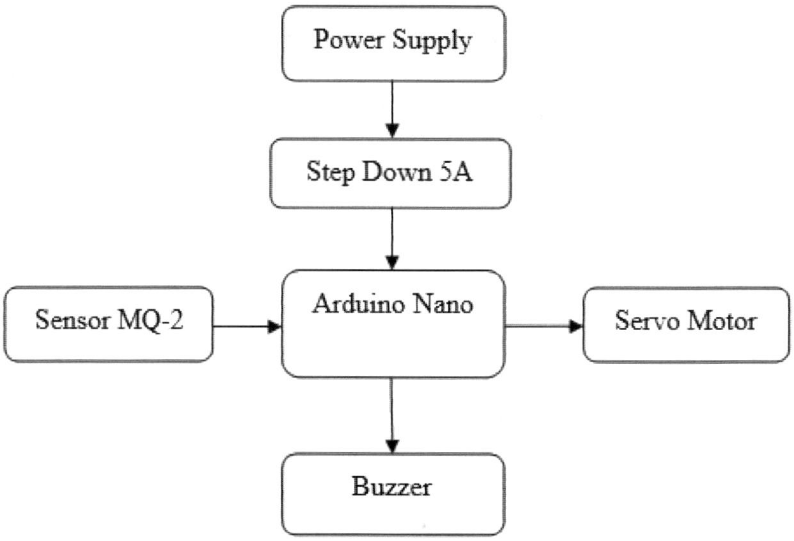

Figure 4. System Block Diagram

Based on Figure 4. the working principle of the system being built is:

1. When a gas leakage happens, the MQ-2 sensor detects a gas leakage in the tube due to changes in conductivity on the sensor.
2. The sensor sends a signal to the Arduino Nano Microcontroller.
3. Arduino Nano Microcontroller receives the signal and processing the signal.
4. The signal is sent to the buzzer and sounds an alarm.
5. After a few seconds, the Microcontroller also sends a signal to the servo motor to open the regulator knob.

3. Result and discussion

From the design results obtained an automatic regulator prototype as shown in Figure 5.

Figure 5. Prototype System

Furthermore, a functionally tested prototype test work well or not. Testing is done by calculating the response time needed to open the regulator knob when a gas leakage is detected. The results of the test can be seen in Table 1.

Table 1: Test Results of Automatic Regulator Time Response

No	Testing	Response Time (s)
1	First	3.34
2	Second	1.32
3	Third	3.47
4	Fourth	1.57
5	Fifth	1.33
6	Sixth	0.99
7	Seventh	1.63
8	Eighth	1.77
9	Ninth	1.23
10	Tenth	1.31

Based on the results of testing in Table 1. it can be calculated the average time required as follows:

$$\bar{t} = \frac{\sum_{i=1}^{n} t_i}{n}$$

$$\bar{t} = \frac{17.96s}{10} = 1.796\ s$$

4. Conclusion

Based on the results of the tests it can be seen that the automatic regulator work well when a gas leakage happens by opening the regulator knob and the gas does not flow continuously. From the test results obtained an average response time to open the knop after the detected gas is 1,796 seconds.

For further work, development will be carried out by adding a communication module and LPG leakage can be monitored remotely.

References

[1] Thoday K, Benjamin P, Gan M and Puzzolo E 2018 The Mega Conversion Program from kerosene to LPG in Indonesia: Lessons learned and recommendations for future clean cooking energy expansion *Energy Sustain. Dev.* **46** 71–81

[2] Dhawale D S, Dubal D P, Jamadade V S, Salunkhe R R, Joshi S S and Lokhande C D 2010 Room temperature LPG sensor based on n-CdS/p-polyaniline heterojunction *Sensors Actuators, B Chem.* **145** 205–10

[3] Jain A, Nyati P, Nuwal N, Ansari A, Ghoroi C and Gandhi P R R D 2014 Pre-detection of kitchen fires due to auto-ignition of cooking oil and LPG leakage in Indian kitchens *Fire Safety Science* vol 11 pp 1285–97

[4] Unnikrishnan S, Razil M, Benny J, Varghese S and Hari C V. 2018 LPG monitoring and leakage detection system *Proceedings of the 2017 International Conference on Wireless Communications, Signal Processing and Networking, WiSPNET 2017* vol 2018–Janua pp 1990–3

[5] Bhosale P P 2014 Automatic control of air pressure loop in industrial 2–5

[6] Kumar A, Kumar M and Singh B 2017 Designing and implementaion of smart LPG trolley with home safety *Proceedings on 2016 2nd International Conference on Next Generation Computing Technologies, NGCT 2016* pp 185–90

[7] Ugurlu A and Gumus M 2017 Exergetic analysis of an LPG evaporator/regulator with thermal storage *Int. J. Hydrogen Energy* **42** 17984–92

[8] Nasution T H, Muchtar M A, Siregar I, Andayani U, Christian E and Sinulingga E P 2017 Electrical appliances control prototype by using GSM module and Arduino *2017 4th International Conference on Industrial Engineering and Applications, ICIEA 2017*

[9] Nasution T H, Siregar I and Yasir M 2017 UAV telemetry communications using ZigBee protocol *J. Phys. Conf. Ser.* **914**

[10] Nasution T H, Siagian E C, Tanjung K and Soeharwinto 2018 Design of river height and speed monitoring system by using Arduino *IOP Conference Series: Materials Science and Engineering* vol 308
[11] Rahmat R F, Athmanathan, Syahputra MF, Lydia M S 2017 Real time monitoring system for water pollution in Lake Toba, *International Conference on Informatics and Computing*, pp. 383-388
[12] Data T 2016 MQ-2 Semiconductor Sensor for Combustible Gas *Pololu* 2
[13] Haidar A M A, Benachaiba C and Zahir M 2013 Software interfacing of servo motor with microcontroller *J. Electr. Syst.* **9** 84–99

Acknowledgement
Author would like to thank Universitas Sumatera Utara for funding the financial support to publish this paper and the meubel company which is willing to be the object of this research. We appreciate the efforts of all participants who have a role in conducting of this study.

IOP Conf. Series: Materials Science and Engineering **648** (2019) 012013 doi:10.1088/1757-899X/648/1/012013

FMEA Approach to Analysis Crude Palm Oil Quality Parameters

I Rizkya[1], K Syahputri[1], R M Sari[1], J M Haryono[1]

[1]Department of Industrial Engineering, Faculty of Engineering, Universitas Sumatera Utara, Medan, Indonesia

Email : indahrizkya@usu.ac.id, khalida@usu.ac.id

Abstract. Palm oil companies are an important part of development in all countries, including in Indonesia. The development of the oil palm plantation business in Indonesia has experienced significantly progress. This condition can be seen from many companies set up the Palm Oil Factory (PKS) and the increasing variety of products can be produced from palm oil so that it can increase the selling value of the product. There are many problems faced by the palm oil industry in producing a product, one of them which is a quality problem. In the Crude Palm Oil (CPO) production process, it is known that the products produced are not in CPO quality standards. Products are not in this standard result in decreasing product quality and causing major problems for the company. Therefore, it is carried out to overcome problems above, an analysis is carried out by identifying the causes of the decline in the quality of CPO. Based on the results of the study, the highest RPN value of 280 was obtained with the cause of failure, namely the quality of raw materials is not good and is a type of failure which is the top priority for immediate improvement. The recommendation improvement can be done to prevent disability to carry out careful inspection of raw materials.

1. Introduction

Indonesia is one of the largest crude palm oil (CPO) producing countries in the world from palm oil plantations. The development of the oil palm plantation business in Indonesia has experienced significantly progress. This can be seen from the many companies set up Palm Oil Factory (PKS) and the increasing variety of products can be produced from palm oil and increases the value sales of the product. The high demand for CPO has an impact on business competition among CPO producers.

One of the efforts made by the company to meet CPO demand is to produce good quality of CPO to increase production utilization and company competitiveness. The quality of the product which is satisfies the users needed may include the different features and enhances the performance of product [1]. The quality of CPO is said to be good if it meets the quality standards. CPO quality standards becomes quality parameters are Free Fatty Acid (FFA), water content and impurities.

The palm oil industry in one of the industries in Medan city has experienced problems with the quality of CPO produced. In the CPO production process, it is known that the products produced are not in CPO quality standards. Products are not in this standard result in decreasing product quality and causing major problems for the company. There are several factors affects these problems both from the engine, work methods, materials used and operators. However, these factors are not yet known specifically which is parts are affect the most product defects. The impact of this problem is the quality of CPO is not homogeneous and gives the disadvantage to the company because the company must reprocess CPO whose quality exceeds the standards and it causing of considerable waste. Therefore, to overcome the problems above, an analysis is carried out by identifying the causes of decline quality of CPO.

Many methods can be used to analyze these problems, one of them is the Failure Mode and Effect Analysis (FMEA) method. FMEA is a method to identify and analyse potential failures and consequently aimed to plan the production process is steady and can avoid the failure of the process of production and losses desired [2]. FMEA is one method that is often used to measure the reliability of a product or process. It can identify the highest risk so that actions can be determined to prevent these risks. So that it can significantly reduce industry losses in terms of money and time [3,4]. The main purpose of the FMEA method is to identify potential defects or failures in products, processes or designs. Then the next step is to eliminate or reduce the risk associated with the potential for the defect. By consistently applying the FMEA method, it is expected to continue to improve product quality, process and design [5].

Many researches have been done to overcome the causes of quality decline. Amir et.al (2014) conducted research for define point of failure, analyze the effect of implying appropriate corrective actions applying failure modes and the effects analysis method (FMEA). Through FMEA, the risk and analysis are considered as different aspects. In order to reduce and reduce risks and reduce threats, continuous analysis must be carried out within a certain period [6]. Another study has been done by Nurul *et.al* (2014) by using failure mode effect and criticality analysis method in the automotive parts industry. The purpose in this case study is to analyze and identify the characteristics of product defects based on data rejecting products received on the industrial inspection line [7]. However, this study is more significant to know the main causes of the priority of declining quality of CPO and improves can be done as soon as possible. This study aims to identify the factors that cause problems in CPO quality using Failure Mode and Effect Analysis.

2. Methodology

The research was conducted in one of the palm oil industries with the object of research in the form of the quality of crude palm oil. The research began with direct observation of the palm oil industry to see and observe the conditions of the industry. From the results of observations, it is determined the formulation of the problem in accordance with the conditions of the industry. After the formulation of the problem is determined, then the research objectives are determined for the problems occur. Then data collection is needed in the research. Data collected comes from the results of observations directly and the results of interviews with the company. Based on these data, an analysis of the problems in the quality decline of crude palm oil happened in the company was carried out.

The steps of problem analysis can be done in several steps. The first step is identifying the causes of product quality decline. In this case, the identified factor is the type of content contained in crude palm oil which the content does not meet the standard included in the category of one of the causes in product decline quality. Furthermore, analysis of the causal factors is carried out. This analysis is done using *Why-Why Analysis Table*. After the cause of the quality decline of crude palm oil is known, then determine the value of the risk priority number (RPN) using the FMEA method. FMEA is a structured procedure to identify and prevent as many as possible modes of failure (failure mode) [8]. FMEA is used to identify the sources and root causes of quality problems [9]. The use of FMEA is based on the value of the RPN obtained. The value of RPN is obtained by multiplying the rating severity, occurance, and detection [10]. RPN calculations can be done using formulas.

$$SS = S \times O \times D \quad \text{.................. (1)}$$

The largest RPN value obtained will be a reference to determine the cause of the quality decline of the products are major at risk and will be a top priority in making improvements later.

3. Result and Discussion

3.1. Identification of Causes Factor
Analysis of the causes of declining quality of crude palm oil is done by using the *Why-Why Analysis Table*. This table is useful for analyzing and finding factors that have a significant influence in determining the quality characteristics of work output. In addition, this diagram is useful for finding the real causes of a problem. The *Why-Why Analysis Table* for the three types of content causing the decline in the quality of crude palm oil can be seen in Table 1, Table 2, and Table 3.

3.2. Analysis of Causes Factor
Analysis of the causes in declining quality of crude palm oil is done by using cause and effect diagrams. This diagram is useful for analyzing and finding factors that have a significant influence in determining the characteristics of the quality in work output. In addition, this diagram is useful for finding the real causes of a problem. The causal diagram for the three types of content causing the quality decline of crude palm oil can be seen in figure 1, figure. 2, and figure 3

Table 1. Why-Why Analysis for Free Fatty Acid (FFA)

Why	*Why*	*Why*	*Why*	*Why*
Decreasing of CPO Quality	Content Variation of Free Fatty Acid (FFA)	Man	1. The operator is not careful in sorting fruit 2. No discipline in maintaining the boiling time	1. Fatigue factor 2. Supervision is not implemented
		Material	1. The quality of raw materials is unappropriate 2. High rest fruit 3. Broken seed	1. There are crude palm oil fractions 5 to 6 2. Late fruit transportation 3. Seed crushed by the truck
		Machine	1. Temperature of *Crude Oil Tank* (COT), Temperature of *Crude Stelling Tank* (CST) are unappropriate 2. Overheat engine	1. Machine unmaintenance 2. Usage of *non-stop* machine
		Method	Lack of procedures at the inspection stage	Unstructured at SOP at the Inspection Stage

Table 2. Why-Why Analysis for Water Level

Why	*Why*	*Why*	*Why*	*Why*
Decrease of CPO Quality	Variation of Water Level	Man	1. The operator is not careful in regulating temperature 2. Not discipline to operate the procedure	1. Hurriedly and careless 2. The lack of supervision

		Material	1. Quality of raw materials are unappropriate	1. Harvesting time is late 2. Fruit are piles up
		Machine	1. *Sterilizer, vacuum dryer* and *oil purifier* is not working optimlly. 2. Over heat machine	1. Failure of machine setting 2. The usage of *non-stop* machine
		Method	Lack of procedures at inspection stage	Unstructured SOP at the Inspection Stage

Table 3. Why-Why Analysis for Impurities Level

Why	*Why*	*Why*	*Why*	*Why*
Decrease of CPO Quality	Variation of Level Impurities	Man	1. Operators are careless 2. Operator do not concenterate	1.Operators are talking 2. Fatigue factor
		Material	1. Quality of Raw Material is unappropriate	1. Raw materials have thick fibers 2. Dirty storage of raw materials
		Machine	1. *Sterilizer, vacuum dryer* and *oil purifier* machines is not working optimally 2. The machine dies suddenly	1. Lack of maintenance 2. *Overheat* machine
		Method	1. Lack of procedures at inspection stage 2. Inspection not in the entire area	Unstructured SOP at the Inspection Stage

Based on the text above shows that the causes of quality decline of CPO include humans, materials, machinery and methods. Each of these causes has a different solution.

3.3. Analysis with Failure Mode and Effect Analysis

This analysis is carried out by using the failure mode and effect aalysis method to identify the main causes of the declining quality in crude palm oil products produced in the palm oil industry. The purpose of implementing FMEA is to prevent problems occurring in processes and products. Analysis using FMEA is done by calculating the risk priority number. The value of the risk priority number (RPN) obtained by determination first of the values in severity, occurance, and detection. The value of RPN is obtained by multiplying the rating severity, occurance, and detection. The results of RPN calculations can be seen in Table 4.

IOP Conf. Series: Materials Science and Engineering **648** (2019) 012013 doi:10.1088/1757-899X/648/1/012013

Table 4. Recapitulation of Calculation in Risk Priority Number at FMEA Table

Failure Mode	Failure Effect	S	Failure Causes	O	Detection Method	D	RPN
Variation of FFA Level	Causes low quality of oil quality which is damage human health	8	The operator does not carefully sort the fruit	6	Do intensive supervision	5	240
			The quality of raw materials is not suitable	7	inspection of raw materials carefully	5	280
			The temperature of COT and CST is not suitable	3	Periodic temperature checks	5	120
			Lack of procedures at the inspection stage	3	Create a new SOP	4	96
Variation of Water Level	Causes hydrolysis in the rancid taste and oil smell	6	Operators whom are not disciplined operate the procedure	5	Do supervision	4	120
			Improper quality of raw materials	7	Do inspection of harvesting time	4	168
			Sterilizer, vacuum dryer and *oil purifier* machines doesn't work optimally	5	Supervision of Engine Settings	4	120
			Lack of procedures at the inspection stage	4	Create a new SOP	6	144
Variation of Impurities Level	Activates microbes and enzymes that react and produce free fatty acids and discoloration in oil	7	The operator does not concentrate	5	Do supervision	4	140
			Raw material quality is unappropriate	7	Do maintenance of the storage area	3	147
			Sterilizer, vacuum dryer and *oil purifier* machines doesn't work optimally	4	Inspection of each machine	6	210
			Inspection is not comprehensive	4	Inspection at each process whether it is in good condition or not	5	140

Based on the table above, the highest RPN value of 280 is obtained with the cause of failure, namely the quality of raw materials are not good and is a type of failure which is the top priority for immediate improvement. The recommendation improvement can be done to prevent defects is to carry out careful inspection of raw materials.

4. Conclusion

By using FMEA, it can be seen that the cause of failure is the quality of raw materials are not good with an RPN value of 280. This failure is a type of failure mode which is the top priority for immediate improvement. The recommendation improvement can be done to prevent defects is to carry out careful inspection of raw materials

References

[1] Asma S, Abdul G, Muhammad I, Muhammad Y, and Naveed A 2015 *Global Journal of Management and Business Research: EMarketing* **15**

[2] Khawarita S and Syahrul F S 2017 *The 2nd Annual Applied Science and Engineering Conference (AASEC)* **288** (2017), pp. 1-6

[3] Ambekar S, Edlabadkar A and Shrouty V 2013 *Int. J. Of Eng. And Innovative Tech. (IJEIT)* **1**

[4] Parsana T S and Patel M T 2014 Int. J. Of Industrial Eng. and Mgt Sc **4** (3), pp. 145-152

[5] Patrycja H and Michal M 2017 *6th International Conference on Operations Research and Enterprise Systems (ICORES)*

[6] Amir A N, Danial Z, Idin Z K, and Mona T 2014 *International Journal of Poultry Science* **13** (12), pp. 718-728

[7] Nurul A A, Rosmaini A, Shaza S, and Noorhafiza M 2014 *8th MUCET* (Melaka Malaysia, 10-11 November 2014)

[8] Helia V N and Wijaya W N 2017 *IOP Conf. Series: Materials Science and Engineering* **215** (2017), pp.1-10

[9] Khalida S, Jelly L, Rahmi M S, Indah R, Ikhsan S, and Mangara M T 2018 MATEC Web of Conferences **197** (2018), pp.1-4

[10] Foster S 2010 *Managing Quality* (*Integrating The Supply Chain, 4th Edition*) (New Jersey: Pearson Education, Inc)

Acknowledgement

Author would like to thank to Universitas Sumatera Utara for funding the financial support to publish this paper and the meubel company which are willing to be the object of research. We appreciate the efforts of all those who have cooperated in conducting this study.

Design of automatization of trolley cover based on microcontroller AT Mega 8535 as a learning media at ATKP Medan

S D Novalina[1*], A Panjaitan[1], R P Simanjuntak[1], B Y Tamba[1]

[1]Department of Airport Electrical Engineering, Civil Aviation and Safety Engineering Academy of Medan, Medan, Indonesia

*Email: susidiriyanti@atkpmedan.ac.id

Abstract. As an air transportation service and the operation of flight operations, airports must be arranged in an integrated operational system and capable of providing balanced transportation services. Inside the airport there are equipment that can improve service to aircraft passengers, one of which is a trolley. For this reason, maintenance of these trolleys is necessary and even to develop into. The study used experimental methods of integrating AT Mega microcontrollers, and prototype trolley covers connected with servo motors as the main drivers of the gates. From the research is concluded that this prototype has been successfully designed and between the tools can function properly so that the trolley-writing movement is fully controllable. This automatic trolley has a part that functions to open and close the trolley door automatically. This is very important, because this trolley will work on its own or automatically arrive at the final destination, so the door to close the trolley is very necessary to avoid losing passenger goods. With this design, it is expected to be able to improve services within the airport and as a learning media in the field.

1. Introduction

Technological developments have experienced very rapid development, especially in the field of electronic instruments which are already popular in the community and are not something foreign. The progress of electronic technology is able to overcome even complex problems, with precision and very high speed and accuracy [1,2]. Various equipment can be operated or used with the help of robots. Science and technology are progressing along with the times. The airport requires reliable and adequate facilities and infrastructure for the safety, security and comfort of the passengers. Convenience for passengers at the airport is an important aspect that must be considered in the operation of airport operations.

Inside the airport there are equipment that can improve service to aircraft passengers, one of which is a trolley [3-5]. Trolleys are used by passengers to carry items to the check-in counter, within the departure terminal. Counter check-in is one of the very important stages that must be passed in the service of air passengers at airport terminals. For this reason, maintenance of these trolleys is necessary and even to develop into a trolley robot. Automation is increasingly desirable nowadays. One in their field is the carrying of goods. Usually people are required to push the trolley. However, if there is a trolley that can run automatically, the activity of carrying goods will be easier [6].

2. Methods

Research was conducted using the method of experimental approach, namely by conducting various experiments conducted repeatedly until the conclusion and the results of the previous research. Prior to research, the design was first performed using an Atmega microcontroller and a prototype trolley cover [7].

A microcontroller is a computer built on one chip and can be connected to a device (hardware) and control functions. Because the microcontroller is an electronic component in which there is a series of microprocessors, memory (RAM / ROM) and I / O, the circuit is contained in the chip level or commonly called a single chip microcomputer. On the microcontroller there are components of the microprocessor with interconnected internal buses. These components are RAM, ROM, timer, parallel and serial I / O components, and interrupt controller. There are several types of microcontrollers, namely ATMEGA8535, AT89S51.

AVR microcontrollers have quite complete features. Starting from the program memory capacity and data memory is quite large, interruptions, timers / counters, PWM, USART, TWI, analog comparator, internal EEPROM and also the internal ADC are all in ATMEGA8535.

A program is a collection of instructions that are used to set a computer to perform certain actions. Without a program, the computer really can't do anything [8]. That is why it is often said that computers cover three important aspects, in the form of hardware, software, in this case in the form of programs, and brain ware or people who play a role in computer operations and software development [9][10][11]. In other words, the program is an important part of the computer that regulates the computer in order to take action in accordance with the will of the author [12].

Software is the most core part of a microcontroller system. Because the microcontroller hardware will function if the software in the form of instructions (functions) has been given to it. In this writing the author uses programming with C language to write a list of programs that will operate the microcontroller [13]. C is a very short computer language.

DC motors are the most widely used motors in everyday life. DC motor is a machine that functions to convert direct current electricity into motion or mechanical energy [14]. The basic construction of a DC motor consists of 2 main parts, namely the rotor and stator. The rotor is a rotating part or armature, in the form of a coil where an electric current can flow. While the stator is a fixed part and produces a magnetic field from the coil. Most motors used in the game; cars are DC motors [15]. This causes the production of DC motors to be larger than other motors. A DC motor has wire coils mounted in the slots of a cylinder.

An electric motor is an electromagnetic device that converts electrical energy into mechanical energy. This mechanical energy is used for, for example, turning the pump impeller, fan or blower, moving the compressor, lifting material, etc. Electric motors are also used at home (mixers, electric drills, wind fans) and in the industry. Electric motors are sometimes called industrial "work horses" because it is estimated that motorcycles use about 70% of the total electrical load in the industry.

Voltage regulator functions as a voltage filter to suit your needs. Therefore, usually in a power supply circuit, the voltage regulator IC is always used to stabilize the voltage output. The rectifier circuit is good enough if the ripple voltage is small, but there are stability problems. If the voltage of the PLN rises / falls, the output voltage will also go up / down, if the current is getting bigger it turns out that the output dc voltage also goes [16].

2.1. Framework of thinking

The working principle of a microcontroller learning tool is how to provide convenience for the layman to try to conduct microcontroller learning such as previous research (2015) ever conducted at the University of Serang Jaya Using microcontroller as a design build learning tool.

Trolley is one of the airport facilities that serves to support the convenience of airport users or passengers. So that this tool is very necessary, in order to create a better airport service. At this time trolleys are needed by passengers, because passengers usually carry a lot of goods to travel to their destination. For this reason, trolleys are needed so that maintenance of these items must be carried out even to develop the trolley automatically. Passengers do not need to manually push the trolley, where it is expected that this trolley will work automatically until the check-in counter. Counter check-in is one of the very important stages that must be passed in the service of air.

On this automatic trolley there is a part that serves to open and close the trolley door automatically. This is very important because this trolley will work on its own or automatically arrive at the final destination, so the door to close the trolley is very necessary to avoid losing passenger goods. With this design, it is expected to improve services within the airport.

3. Result and Analysis

At this time the conditions at the airport for the purpose of lifting passenger goods, the tool used to carry goods to the check-in room is to use a trolley. Trolleys at the airport still use manual methods that are driven or driven directly by passengers. Then if the items are carried a lot, then passengers will have difficulty pushing the trolley to the check-in room.

The desired condition is when the passenger arrives at the airport directly there is an automatic trolley. First of all, the visitors who have arrived at the airport and put the items that were taken and took the ID card contained in the place of the card provided. Then the visitor enters the ID card in the ID card slot, to open the lid from the trolley, the visitor must enter security code and choose what airline is the destination. After the visitor enters the item he is carrying, there is a process of weighing the goods, after weighing the goods is completed and in accordance with the provisions of the scales, the visitor presses the close button so that the trolley door closes again and will move forward following the black line on the floor. The trolley will continue to follow the line with no barrier in front of the trolley. If the ultrasonic sensor detects an obstacle or an obstacle then the trolley will stop and sound the alarm until the barrier in front of the trolley no longer exists. After the barrier is gone and the alarm is off, the trolley goes back to automatically following the line. When arriving at the x-ray there is a deviation that makes the trolley stop and the visitor ejects the item empty and automatically the trolley goes to the front of the x-ray to receive the item back from the x-ray. After all items are entered into the x-ray, the trolley will run to the check-in room. After arriving at the check-in room, the trolley will run towards the check-in counter which airline has been chosen. After the item is removed from the trolley and the id card is removed from the slot, the trolley will return to its initial position.

In this design, the first source of power supply is a battery with a capacity of 12 v dc. And with the power supply circuit using a voltage lowering or LM 7805 regulator IC, so it will produce 5 v dc output. From the power supply then the data will be processed by the microcontroller using C language. The microcontroller processes the given data in the form of commands or programs. With these commands, the microcontroller will provide output to the servo motor to open and close the door.

Servo motors require a voltage of 5 v dc to work, and this voltage is received from vcc. Servo motors will work in accordance with the timer or modular pulse width (PWM) provided by the microcontroller. In this design the PWM is used to open the door from trolley which is at level 12 of 255 a timer counter2 while to close the door trolley is used at level 18 of 255 a timer counter2. The motor will work according to the order given at that level.

In the AT Mega 8535 microcontroller, the port used to adjust the servo motor is on port D, which is on pin 21. Pin 21 on the microcontroller is a pin that functions as the output of compare register 2 (OCR2), where OCR is a comparison of TCNT2. When TCNT2 reaches the level that is ordered, OCR2 will capture the signal, and directly input the servo motor. Thus, the servo motor will work according to the command of pin 21, whether the motor works to open the door or to close the door trolley.

From the results of the discussion and the workings of the circuit that has been designed, some conclusions can be drawn, including the opening and closing of the trolley can work if there is input given by the microcontroller on the servo motor, by making this automatic trolley easier for passengers without having to push the trolley because the trolley has run automatically until the check-in counter, with this automatic trolley will advance the development of increasingly advanced and modern technology in the airport, by making this automatic trolley, we can understand and understand the programming language namely C language, and after the whole series is successful, this tool can be used as a learning media in laboratory navigation lessons in practice so that it will arouse the spirit of cadets to apply the theoretical knowledge.

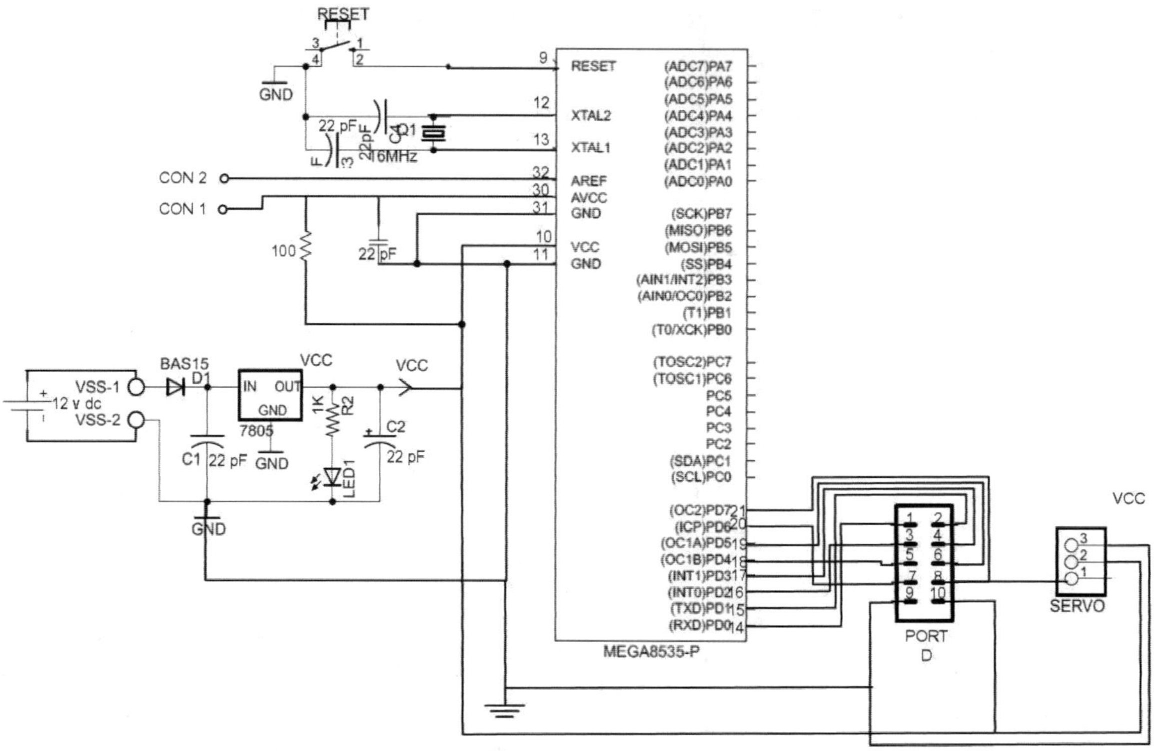

Figure 1. Image of opening circuit and trolley door closers

From the results of the discussion and the workings of the tool, the authors can provide advice including automatic trolleys using sophisticated electronic equipment so that people who use this trolley must understand and follow the procedures for using the trolley, automatic trolleys using dry battery power so that the deadline or the ability of dry batteries will be reduced when often used, therefore there must be maintenance or refill of the dry battery, with the development of technology, the cadets should be equipped with a variety of the latest science learning media so that it makes it easier for cadets to apply theory to laboratory practices regarding electrical equipment at the airport. This will also motivate cadets to be more enthusiastic to explore knowledge according to the latest developments in technology that develops, especially airport.

4. Conclusion

From the results of the discussion and how the series work has been designed then there can be several conclusions, among others:

1. The opening plan and door of the trolley can work if there is input given by microcontroller on the servo motor.
2. By making this automatic trolley can facilitate passengers without having to drive the trolley because the trolley is already running automatically until the check-in counter.
3. With the existence of this automatic trolley will advance the development of advanced and modern technology in the airport.
4. This suite of learning tools can be developed for other application development.
5. By creating this automatic trolley, we can understand and understand from the programming language of the C language.

References

[1] Abdul K H 2005, Algoritma Pemrograman, Yogyakarta, ANDI
[2] Deni A 2011, Kamus Komponen Elektronika, Jakarta, PT Kawan Pustaka

[3] Martono H K 2002, *Hukum Penerbangan*, Bandung : CV. Mandar Maju
[4] Prihono 2009 *Jago Elektronik Secara Otodidak*, Jakarta Selatan, Kawan Pustaka.
[5] Sakti A A, 2012 *Penerbangan dan Bandar Udara*, Graha Ilmu
[6] Widodo B,2010 Robotika, Jakarta, Andi Offset
[7] Dony A and Rum A K R, 2008 *Komunikasi Data,* Yogyakarta, ANDI Yogyakarta,
[8] Hartono J 2015, *Pengenalan Komputer Ed. V.* Yogyakarta: Andi
[9] Kadir A 2015, *Buku Pintar Pemrograman Arduino,* Yogyakarta: MediaKom
[10] Syahwil M 2013 *Panduan Mudah Simulasi dan Praktek Mikrokontroler Arduino.* Yogyakarta: Andi
[11] Rahmat R F, Athmanathan, Syahputra MF, Lydia M S 2017 Real time monitoring system for water pollution in Lake Toba, *International Conference on Informatics and Computing*, pp. 383-388
[12] Prihono, 2009, *Jago Elektronika Secara Otodidak*,: Kawan Pustaka, Jakarta Selatan 2009
[13] Thomas S W 2002, *Elektronika Dasar,* Salemba teknika, Jakarta,
[14] Widodo P P, and Herlawati 2011 Menggunakan UML, *Informatika* Bandung.
[15] Habi M 2013 Perancangan alat Pembelajaran Microcontroller di Universitas Serang Raya, *Sistem computer* , Universitas Serang Raya
[16] Sahrul, 2014,Pemograman Mikrokontroler AVR Bahasa Assembly dan C, *Informatika*, Bandung

IOP Conf. Series: Materials Science and Engineering **648** (2019) 012015 doi:10.1088/1757-899X/648/1/012015

Determination of Total Labor Plant 1 using Workload Analysis (WLA) Method In Compound Fertilizer's Industry

Farida Ariani[1], Khawarita Siregar[2], Tugiman[1]

[1]Department of Mechanical Engineering, Faculty of Engineering, Universitas Sumatera Utara, Medan, Indonesia
[2]Department of Industrial Engineering, Faculty of Engineering, Universitas Sumatera Utara, Medan, Indonesia

Email: arianiida@yahoo.com

Abstract. The compound fertilizer's industry did trading as a first business in compost using Mahkota brand which is a subsidiary of Wilmar Group Indonesia, which became one of the largest agribusiness companies in the world. The results of preliminary observations show that the number of activities and methods of work done by each operator varies. Thus, the calculation workload against any worker in Plant 1 using Work Load Analysis (WLA) to calculate the percentage of productivity using the Work Sampling method. Furthermore, the excellent determination rating factor value using Westinghouse. The high determination Allowance values using tables Industrial Labor Organization (ILO). The results of the workload calculation show that the workload accepted by six workers is high because the value is above 100%. Proposed recommendations for improvements presented in connection with high workload conditions is to increase the number of workers as much as two workers so that the number of workers in Plant 1 is eight workers.

1. Introduction

Humans are the main assets in the organization so that human resources (HR) must be managed and utilized fully. HR is the biggest asset of the organization. HR supports the ability of workers to support overall economic and social development, including intellectual workability and physical work. HR does not mean only helping the people and the ability of the employee. Talent, competence, ideas, energy from employees, productivity, and the performance of the organization collected were identified as HR from the company [1]. Organizations or companies that are just developing well cannot be separated from human resources within the organization/company. The role of human resources (HR) in the company is significant because HR is carrying out all activities to meet the objectives of the company. With efficiency and productivity, companies can find out how to optimize the resources used and know the targets that have been carried out by the company. With human resource planning, companies can determine the number of employees needed to run a company system efficiently [2].

This industry is one of the companies engaged in the compound fertilizer production process in Dumai, Pelintung. The company was first established since 1999 and conducts fertilizer trade as the company approves Mahkota brand which is a subsidiary of the Wilmar Group Indonesia which is one of the largest agribusiness companies in the world.

Workload is one of the key factors in working efficiency that affects work productivity. The amount of oxygen used by the body is one of the loading indicators during work [3]. The method used in this study is the Workload Analysis (WLA) method because this method is suitable for analysing the workload of workers at this industry. WLA is one way that can be used to calculate the amount of workload caused by activities carried out. This study will analyse several causes of increased work and improve improvements to reduce high workloads. Besides, the workload that must be received by workers can also be used to determine the number of workers who need to be approved by the company [4 -5].

2. Workload Analysis

Job analysis is the procedure for determining the duties and skill requirement of a job and the kind of person who should be hired for it. Job analysis produces information used for writing job specifications and job descriptions [6]. This is a systematic activity to collect, assess, and integrate information related to tasks within the organization or some information about the characteristics of a particular position. This analysis is done to learn about workloads and possible obstacles that arise during work, and function as foundations of all human resource management activities in the organization. Some activities that are based on the results of job analysis are restructuring the quality of incentive enhancements, human resource planning, design work, training, career development, and evaluation system work. Job analysis is a process, method, and technique for obtaining some work data that is treated as job information and presented to fulfill work programs and as feedback for organization and governance. Local governments must have job analysis documents so that the human resource system can function correctly. Job analysis allows one to identify job descriptions and specifications. As a result, existing human resources can be allocated appropriately to each unit, and this will ensure the type of work that is effective and efficient so that unit performance can be optimal. There are ten essential reasons why job analysis in an organization needs to be carried out. Job analysis aims to (1) evaluate challenges from the environment that can affect one's work; (2) avoid unnecessary job requirements that can lead to job discrimination; (3) expressing elements of work that can help or ignore the quality of work life; (4) planning the future needs of human resources; (5) match compatibility between job applicants and vacancies; (6) define the type of training for new and experienced employees; (7) determine plans for developing potential employees; (8) determine realistic work standards; (9) assign employees in appropriate positions where they can use their skills effectively; (10) giving fair rewards to every official.

Employee workload can also be calculated through job analysis. This workload must be by the needs of the service unit. Thus, employee management is applied every time an evaluation to improve the service unit is carried out. As a result, planning the formation of employees will be by workloads and vacancies. In general, job vacancies are the number of employees needed in public organizations to build high-quality services. Each employee must receive the correct amount of workload. Therefore, realizing maintaining the high performance of a service, local governments must implement their total employees and workload, so that it is the same between workload and performance. Preferably, a prosperous workload, which is not equivalent to the number of existing employees, will result in poor performance. On the other hand, too little workload will reduce the efficiency of a post. This will be a loss for the regional government as a service unit. This is why employee planning needs to be done in the management of local government apparatus resources. Human resource planning has a close relationship with the issue of human resource allocation which includes the number, type or qualifications of skills and expertise.

Workload analysis is a methodology to determine the time, effort and resources necessary to carry out the product department's operations, resulting in identifying the organization's actual needs of human resources both in terms of quality, and develop these resources to achieve the goals and strategies that the organization wants to achieve in the various work sites [7]. In other words, workload analysis aims to ensure the number of employees needed is by some workloads and specific responsibilities given to employees. Workload analysis is an activity carried out to identify the number and qualifications of the employees required to realize organizational goals. In line with an opinion, Minister of Home Affairs Regulation Number 12 of 2008 states that workload is a set or number of activities that must be completed by a unit or official within an organization for a specified period. Besides, workload measurements are also considered as management techniques for obtaining information about specific positions or jobs. This is solved through a systematic research and evaluation process. This work information will be useful in improving the apparatus whether it is in institutions, government or human resources. Workload analysis calculates the time and ability of employees to complete tasks. Not much attention is paid to workload so that problems that arise are often seen as a result of low work motivation and lack of incentives. Besides, the volume of work in institutions is often unstable; there is too much workload at one time and too little at a later time. Therefore, workload analysis will be important in studying the workload of each position and the lowest work unit. Finally, the right amount of work volume imposed on a unit can be known, and this will later be useful in interpreting strategic policies in the future [8].

3. Methodology

The steps in conducting this study is as follows.

1. Field Study
   Field studies conducted to obtain information on the general picture and the actual condition of the company.
2. Literature review
   Studies conducted to obtain and understand the theories related to problem-solving. Sources of literature come from books, journals, and a study of previous research with the main topic.
3. Formulation of the problem
   Based on the literature study and field studies will be known existing problems can be formulated so that the problems being studied.
4. Determining Objectives
   Determining the purpose of the research is used to describe any goal to be achieved by the holding of the study.
5. Data collection
   The data collected for this study included primary data, and secondary data include:
   a. A general overview of the data firm
   b. The data structure of the organization
   c. Data on the number of workers today
   d. The data job description of each worker
   e. Data productive and non-productive operator.
   f. Data rating factor
   g. Data allowance.
6. Data processing
   Data processing is performed in this study, namely:
   a. Calculate the percentage of productive and non-productive at work sampling method.
   b. Determine the rating factor by the method of Westing House System.
   c. Determine allowance by using tables ILO.
   d. Calculate the workload with WLA methods.
   e. Determine the number of workers.
7. Analysis and Conclusions
   Analysis and conclusions made in this study, namely:
   a. Analyze the percentage of productive and non-productive individual operators
   b. Analyze workload conditions related to the cause of high workload.
   c. Analyze related to the number of workers which will compare the number of workers that exists today with the number of workers based on workload.

d. Conclude is a final summary that can answer the formulation of the research conducted and provide research advice.

4. Result

4.1. Productive and Non Productive Activities

Definition of non-productive activity is an activity that does not generate added value in improving the quality and speed of task completion process, while the productive activity is an activity that corresponds to a predetermined job description and this activity was to create a product or service.

When calculating the workload used to calculate the number of workers needed, productive activities are used as a mere activity by the job description of each machine operator.

The activities carried out by workers in Plant 1 can be seen as follows:

1. PLC Operator
   Activities that included part of the work of the PLC operator can be seen as follows:
   a. Controlling Plant condition
   b. Filling logsheet and logbook
   c. Noting the results of laboratory analysis
2. Granulator Operator
   Activities that included part of the work of the operator granulator can be seen as follows:
   a. Checking the inlet and outlet chute granulator
   b. Checking burner
   c. Clean the work area
3. Process Screen Operator
   Activities that included part of the work of the operator screen process can be seen as follows:
   a. Process control screen
   b. Clean the work area
   c. Check chute outlet screen and polishing
4. Feeder and Recycle Operator
   Activities that included part of the work of the feeder and recycled the operator can be seen as follows:
   a. Checking the inlet feeder and recycle
   b. Clean the work area
   c. Controlling the smooth feeding of raw material
   d. Coordinating the feed material
5. Final Product and Cyclone Operator
   Activities that included part of the work of the final product and operator cyclone can be seen as follows:
   a. Check the outlet conveyor
   b. Physical checking product
   c. Feeding white clay
   d. Moving cooling bay tripper
   e. Controlling cyclone engine
   f. Controlling machine dryer
   g. Controlling cooler fan engine
   h. Clean the work area
6. Feeding Operator
   Activities that included part of the job of feeding the operator can be seen as follows:
   a. Monitoring the raw material feeding activity
   b. Tally feeding
   c. Filling logsheet and logbook feeding
   d. Work area

The entire activities outside the job description above, including the activities of non-productive.

4.2. Adequacy and Uniformity Test Data

Test the adequacy of the data is performed to determine whether the data collected is adequate or qualified specified accuracy. If the data collected is not sufficient, it is necessary to increase the amount of data. Test the adequacy of the data is done with the value of S is 5%; the calculation formula can be seen as follows:

$$N' = \frac{k^2\left(1 - \bar{p}\right)}{s^2 \bar{p}} \tag{1}$$

Where:
N '= the number of observations that need to be done
$\bar{p}$ = the percentage of productive
S = level of accuracy
k = the price index of the level of confidence is taken (1.96)
If N' ≤ N, the data is said to be sufficient.
The calculation of the adequacy of test data for PLC operator is as follows:

$$N' = \frac{1.96^2(1 - 0{,}8551)}{(0.05^2)(0{,}8551)} \approx 260.3896 = 261$$

The number of observations (N) is 276, and the results of the calculation (N ') is 261, it can be seen that N' ≤ N, the data for the PLC operator is said to be sufficient. The results of the adequacy of the data for other workers can be seen in Table 1.

Table 1. Test Data Sufficiency

Workers	N	N'	Information
PLC Operator	276	261	Sufficient
Granulator Operator	276	108	Sufficient
Process Screen Operator	276	108	Sufficient
Feeder and Recycle Operator	276	174	Sufficient
Final Product and Cyclone Operator	276	134	Sufficient
Feeding Operator	276	154	Sufficient

Uniformity test data is done in advance by defining UCL and LCL. UCL and LCL calculation can be done with the following formula:

$$UCL = \bar{P} + k\sqrt{\frac{\bar{p}\,(1 - \bar{p}\,)}{n}} \quad \text{.................................. (2)}$$

$$LCL = \bar{P} - k\sqrt{\frac{\bar{p}\,(1 - \bar{p}\,)}{n}} \quad \text{.................................. (3)}$$

$\bar{p}$ = The average percentage of productive time
n = Number of observations performed per cycle working time
k = Price index of the amount depends on the level of trust
k = 1.96 (95% confidence level)
UCL and LCL calculation PLC Operator are:

$$\bar{P} = \frac{0{,}8587 + 0{,}8478 + 0{,}8587}{3} = 0.8551$$

$$UCL = 0.8551 + 1.96\sqrt{\frac{0{,}8551(1 - 0{,}1449)}{276}} = 0.9560$$

$$LCL = 0.8551 - 1.96\sqrt{\frac{0,8551(1 - 0,1449)}{276}} = 0.7542$$

The value of the proportion of productive time during the observation is shown in the control map shown in figure 1.

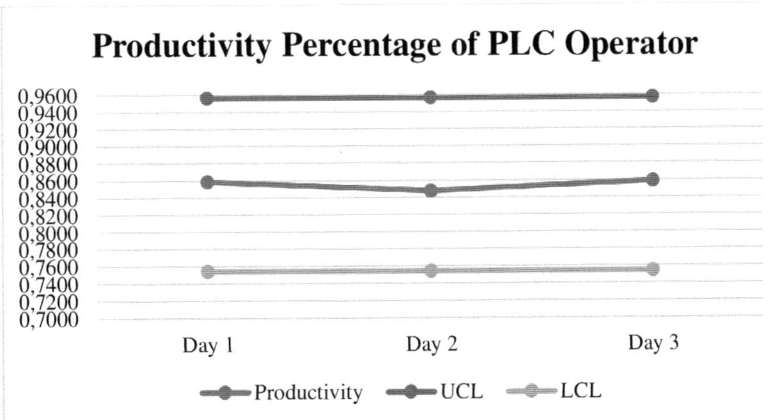

Figure 1. Operator PLC Data Uniformity

Based on the calculations above, the uniformity test data for other operators can be seen in table 2.

Table 2. Uniformity Test Data

Workers	% P Day 1	% P Day 2	% P Day 3	UCL	LCL	Information
PLC Operator	.8587	.8478	.8587	.9560	.7542	Uniform
Granulator Operator	.9348	.9348	.9348	1.0451	.8245	Uniform
Process Screen Operator	.9457	.9457	.9130	1.0451	.8245	Uniform
Feeder and Recycle Operator	.8587	.913	.9239	1.0045	.7925	Uniform
Final Product and Cyclone Operator	.9783	.8913	.8913	1.0289	.8117	Uniform
Feeding Operator	.9022	.9239	.9022	1.0167	.8021	Uniform

The results of calculations uniformity test data for 6 operators indicates that all data has been uniformly having been among the Upper Control Limit (UCL) and the Lower Control Limit (LCL).

4.3. Productivity, Rating Factor and Allowance

After testing the adequacy of the data and uniformity test data then calculating the percentage of productivity rating factor along with great value and allowance. The results of the calculation of the rate of productivity, rating factor, and the allowance of each worker can be seen in the following table 3.

Table 3. Percentage Productivity, Rating Factor Allowance PLC Operator

No.	Type of activity	Frequency		
		Day 1	Day 2	Day 3
1.	Control plant conditions	64	52	40
2.	Filling logsheet and logbook	9	18	26
3.	Noting the results of laboratory analysis	6	8	13
4.	Activities outside the job description	13	14	13
	Productivity (%)	**85.87**	**84.78**	**85.87**
	Rating Factor		**0.13**	

Allowance (%)	**14.2**

Table 4. Percentage Productivity, Rating Factor and Allowance Granulator Operator

No.	Type of activity	Frequency		
		Day 1	Day 2	Day 3
1.	Checking the inlet and outlet chute granulator	39	32	34
2.	Refuel	6	7	6
3.	Clean the work area	11	21	24
4.	Checking burner engine	30	26	22
5.	Activities outside jobdesc	6	6	6
	Productivity (%)	**93.48**	**93.48**	**93.48**
	Rating Factor		**0.04**	
	Allowance (%)		**43.4**	

Table 5. Percentage Productivity, Rating Factor, and Allowance Process Screen Operator

No.	Type of activity	Frequency		
		Day 1	Day 2	Day 3
1.	Check chute outlet screen and polishing	65	57	41
2.	Process control screen	16	16	24
3.	Clean the work area	6	14	19
4.	Activities outside the job description	5	5	8
	Productivity (%)	**94.57**	**94.57**	**91.30**
	Rating Factor		**0.07**	
	Allowance (%)		**38.5**	

Table 6. Percentage Productivity, Rating Factor, and Allowance Feeder and Recycle Operator

No.	Type of activity	Frequency		
		Day 1	Day 2	Day 3
1.	Controlling the smooth feeding of raw material	32	35	34
2.	Checking the inlet feeder and recycle	37	36	35
3.	Coordinating the feed material	2	6	11
4.	Work area	8	7	5
5.	Activities outside jobdesc	13	8	7
	Productivity (%)	**85.87**	**91.30**	**92.39**
	Rating Factor		**0.01**	
	Allowance (%)		**36.3**	

Table 7. Percentage Productivity, Rating Factor, and Allowance Final Product and Operator Cyclone

No.	Type of activity	Frequency		
		Day 1	Day 2	Day 3
1.	Check the outlet conveyor	22	11	46
2.	Controlling cooler fan engine	10	12	6
3.	Physical checking product	8	7	3
4.	Feeding white clay	15	11	4
5.	Engine control cyclone	8	14	6

6.	Controlling machine dryer	8	6	4
7.	Clean the work area	11	13	9
8.	Move cooling bay tripper	8	8	4
9.	Activities outside jobdesc	2	10	10
	Productivity (%)	**97.83**	**89.13**	**89.13**
	Rating Factor		**0.01**	
	Allowance (%)		**39.5**	

Table 8. Percentage of Productivity, Rating Factor and Operator Feeding Allowance

No.	Type of activity	Frequency		
		Day 1	**Day 2**	**Day 3**
1.	Controlling the raw material feeding activity	44	41	49
2.	Filling logsheet and logbook feeding	4	11	7
3.	Tally feeding	17	8	9
4.	Clean the work area	11	13	14
5.	Controlling feeding spillage	7	12	4
6.	Activities outside jobdesc	9	7	9
	Productivity (%)	**90.22**	**92.39**	**90.22**
	Rating Factor		**0.09**	
	Allowance (%)		**40.5**	

4.4. Workload Calculation with Work Load Analysis (WLA)

Workload, preferably close to 100% or under normal conditions. The workload of 100% means that for 8 hours workers were able to work continuously in normal conditions.

Operator Workload PLC

Workload Day 1

$$= (\% \text{ P x } (1 + \text{Rating Factor})) \text{ x } (1 + \text{Allowance})$$
$$= (0.8587 \text{ x } (1 + 0.13)) \text{ x } (1 + 0.1)$$
$$= 1.0673 \approx 106.73\%$$

Workload Day 2

$$= (\% \text{ P x } (1 + \text{Rating Factor})) \text{ x } (1 + \text{Allowance})$$
$$= (0.8478 \text{ x } (1 + 0.13)) \text{ x } (1 + 0.1)$$
$$= 1.0538 \approx 105.38\%$$

Workload Day 3

$$= (\% \text{ P x } (1 + \text{Rating Factor})) \text{ x } (1 + \text{Allowance})$$
$$= (0.8587 \text{ x } (1 + 0.13)) \text{ x } (1 + 0.1)$$
$$= 1.0673 \approx 106.73\%$$

On average Workload (% WL1)

$$= \frac{\text{Workload Day 1 + Workload Day 2 + Workload Day 3}}{\text{Number of Days}}$$
$$= \frac{106{,}73 + 105{,}38 + 106{,}73}{3} = 106.28\%$$

Table 9. Workload Each Operator

No.	Operator	Workload (%)			Average (%)
		Day 1	**Day 2**	**Day 3**	
1.	PLC Operator	106.73	105.38	106.73	106.28
2.	Granulator Operator	139.41	139.41	139.41	139.41
3.	Process Screen Operator	140.14	140.14	135.31	138.53
4.	Feeder and Recycle Operator	128.74	136.89	138.52	134.72
5.	Final Product and Cyclone Operator	150.11	136.77	136.77	141.22

6.	Feeding Operator	138.16	141.49	138.16	139.27

Based on the average workload per worker above, the overall productivity of workers' activities can be calculated as follows:
Total Operating Expenses

$= \%WL_1 + \%WL_2 + \%WL_3 + \%WL_4 + \%WL_5 + \%WL_6$
$= 106{,}28\% + 139{,}41\% + 138{,}53\% + 134{,}72\% + 141{,}22\% + 139{,}27\%$
$= 799{,}43\%$

On average Workload Each Workers (Actual)

$$= \frac{\text{Workload Total}}{\text{Number of workers}} = \frac{799{,}43}{6} = 133{,}24\%$$

4.5. Determination of Number of Workers Recommendation
Based on the above problems, the average workload per worker is 133.24% where the average workload of the workload exceeds the limit of 100% so that the necessary repairs workforce so that the workload of each worker can be reduced. Calculation of the recommended amount of labor can be seen as follows:
On average Workload Each Workers (Recommendations) = $\frac{\text{Workload Total}}{\text{Number of Workers}} = = 99.93\% \frac{799{,}43}{8}$
Based on the calculation above, it can be concluded that the amount of labor that recommended in Plant 1 is eight workers with an average workload per worker was 99.93%.

5. Conclusion

The conclusion of the discussion above is as follows.
1. The workload of PLC operator, granulator operator, process operator screen, feeder and recycle the operator, the final product and cyclone operator, and the operator feeding in a row is 106.28%, 139.41%, 138.53%, 134.72 %, 141.22%, and 139.27%. The average workload per worker is 133.24%.
2. Improvements were made to the number of workers at Plant 1 is the addition of a workforce of 2 people. With a total workforce of 8, then the average workload per worker in Plant 1 was 99.93%.
3. Every job done equalization jobdesc so with the addition of two workers, then jobdesc each operator has the same workload.

References

[1] Y Adityawarman. 2015 *Jurnal Manajemen dan Organisasi*, Vol VI, pp 35-36.
[2] Yanti Helianty. 2014 *Jurnal Online Institut Teknologi Nasional*, Vol. 1, pp 251.
[3] Tarwaka. 2004. *Ergonomics For Work Safety and Health and Productivity, Ergonomi untuk Keselamatan, Kesehatan Kerja dan Produktivitas,1st ed.*, (Surakarta: UNIBA Press)
[4] Ridwan Arif. "Analisa Beban kerja dan Jumlah Tenaga Kerja yang Optimal pada Bagian Produksi dengan Metode Workload Analysis (WLA) di. PT. Surabaya Perdana Rotopack". Edisi Ketiga. Jawa Timur: Universitas Pembangunan Nasional Veteran. 2012.
[5] Rahmat RF, Herly E T, Siregar B, Syahputra M F and Sitompul O S 2018 RFID presence monitoring system as an input to measure the workload of employee *International Conference on Computer Applications and Information Processing,* pp 1-6.
[6] G Dessler. 2006 *Human Resources Management. Pearson International Edition*
[7] P Dasgupta and Roy. 2013 *International Journal of Applied Research and Studies* (iJARS), 2-5.
[8] E Cahyasari. *Jurnal Ilmiah Administrasi Publik* (JIAP), Vol. 2, pp 59-64, 2016.

Acknowledgment

The author would like to thanks to the Universitas Sumatra Utara who has provided financial support to publish this paper. The authors express their gratitude to the leader of the compound fertilizer's industry who are willing to be the object of research. We appreciate the efforts of all those who have cooperated in conducting this study.

IOP Conf. Series: Materials Science and Engineering **648** (2019) 012016 doi:10.1088/1757-899X/648/1/012016

Reducing Waste in Spare Part Production Process with Lean Manufacturing Approach

Khawarita Siregar[1], Farida Ariani[2], Mangara M Tambunan[1]

[1]Faculty of Engineering, Universitas Sumatera Utara, Medan Indonesia
[2]Departement of Mechanical Engineering, Faculty of Engineering, Universitas Sumatera Utara, Medan, Indonesia

Email: khawaritasiregar@yahoo.co.id

Abstract. In the current era of globalization, competition in the industrial world is increasingly competitive, and then the manufacturing industry must be able to optimize its ability to compete. PT. ABC is a battery production manufacturing company that also produces engine parts for use on the company's production floor. The inability to meet the punch needs to be used on the battery production floor is one of the problems faced by the company. In the punch production process, some activities are not needed such as operators looking for irregularly located support devices and operators repeatedly measuring the results of object feeding in the manual turning process because there is no standard measurement, which causes the grace period to belong. This study aims to reduce waste in a series of production processes or not valuable adding activities and shortening the total waiting time using the Value Stream Mapping tool. To identify waste, you must first know which activities are of added value and not value added. One method proposed to fix this problem is Lean Manufacturing. The result of using the Lean Manufacturing method is the proposed improvement is to eliminate activities that are less efficient and can reduce production time by 2897 seconds and reduce production activities with 17 events so that the production process is more efficient than before. It can be concluded that the waste that occurs is over processing and movement caused by work methods and machinery.

1. Introduction

In the current era of globalization, competition in the industrial world is increasingly competitive. Therefore, the manufacturing industry must be able to compete. Competition in the industry both service industry and manufacturing industry is currently very fast at all, so it requires the company to improve the quality of products made. Improving the quality of the product will provide a guarantee to consumers that the company provides the best services to meet consumer demand at competitive prices in the market.

In the aircraft industry, generally 75% of the material used to make an aircraft component will be garbage, so we need a system efficiency of production on the other side. The efficiency of production systems that run very influential on the performance of the company. Thus, elimination or minimization of waste that occurs is one solution. The approach can be used to eliminate or minimize waste in a production system is lean manufacturing company. Lean manufacturing is a systematic approach to identify and find the cause of the waste and then minimizing or even eliminating it. This study aims to minimize waste in the production process RIB AT STN WFX000 part name in the area of medium prismatic machine 2. The results of the analysis of waste relationship matrix obtained two types of waste that is inventory and waiting, each of which has a weight 7,143 and 6,28 with detailed mapping tools are chosen process mapping activity with the score 198,286 and supply chain response matrix with a score of 139,429.

To increase productivity, companies must know what activities can increase added (value added), reduce various waste (waste), and shorten lead times.

Table 1. Comparison between the Number of Needs Punch on the Battery Production Floor and the Number of production Results Punch at the Workshop in 2016

Month	Production	Requests
1	25	29
2	28	30
3	55	56
4	120	123
5	15	17
6	30	31
7	15	15
8	45	48
9	40	43
10	41	42
11	15	16
12	42	43

From Table 1. It can be seen that the problem faced by companies is the lack of need punch on the production floor every month. Therefore, the company must make improvements by eliminating activities that are not needed in the production process. Thus, improvement by reducing and eliminating activities that do not have added value (not value-added activity) in the production process can improve the quality of the product. The approach that can be used to reduce and eliminate activities that do not have added value is Lean Manufacturing.

2. Research Method

The study was conducted at a battery production plant located in Medan City, North Sumatra. The research was conducted since February 2017

The stages of this research are as follows:

1. Data collection. The data used are two types, namely primary and secondary. The primary data is the real condition on the production floor and the secondary data, namely: The data collected are:
   a. Cycle time data for the making process *punch*.
   b. Actual work procedures.
2. Data Processing The steps for processing data are as follows:
   a. Establishment of the current state map.
   b. Analyze the current state map.
   c. Formation Future state map.

3. Results and Discussion

3.1. Test of Uniformity and Adequacy of Data

Test of data uniformity is used to determine whether there is data that is outside the control limit. To test the consistency of the data used the statistical method and the level of confidence and the level of accuracy desired by the gauge is a 95% confidence level and an accuracy rate of 5%.

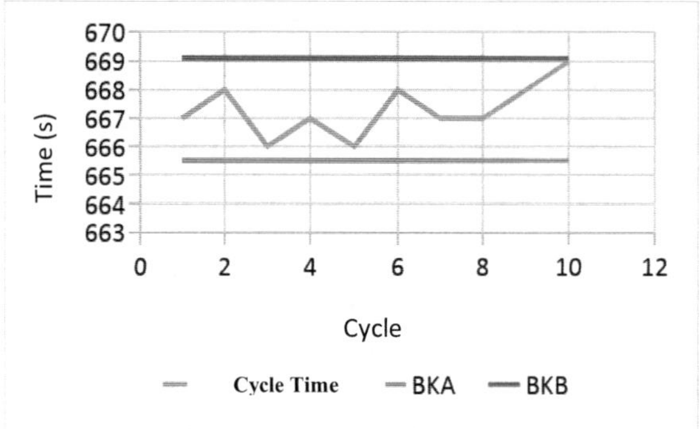

Figure 1. Time Control Map of Object Cutting Activities

Data adequacy tests were carried out to find out whether the data are taken already represented the population to be studied. This data adequacy test is carried out for each process. Calculation of test data adequacy for cutting objects can be seen in Table 2.

Table 2. Test of Sufficiency of Activity Data Cutting Objects

No.	Time (sec)	X ^ 2
1	667	444 889
2	2668	446 224
3	3666	443 556
4	4667	444 889
5	5666	443 556
6	6668	446 224
7	7667	444 889
8	8667	444 889
9	9668	446 224
10	669	447 561
Total	6673	4,452,901

$$N' = \left[\frac{40\sqrt{10(4452901) - (6673)^2}}{6673} \right]^2 = 2.4$$

Value N = 10> N '= 2.4 then it was concluded that the data when cutting objects was sufficient.

3.2. Calculation of Production Process Standard Time

Estimation of standard time is obtained by using cycle time data. Standard time is obtained by calculating the average time first. Summary of standard time calculation of production processes can be seen in Table 3.

Table 3. Summary of Production Processes Standard Time Calculation

Station	Standard Time	
	Seconds	**Minutes**
Cutting	928.54	15:48
lathing	338.06	5.63
CNC	1251.47	20.86
Galvanizing	587.89	35273.34

Test Strength	116.49	1.94
Finishing	2911.60	48.53

3.3. Making Current State Map

These elements are drawn in the SIPOC diagram which can be seen in Figure 2.

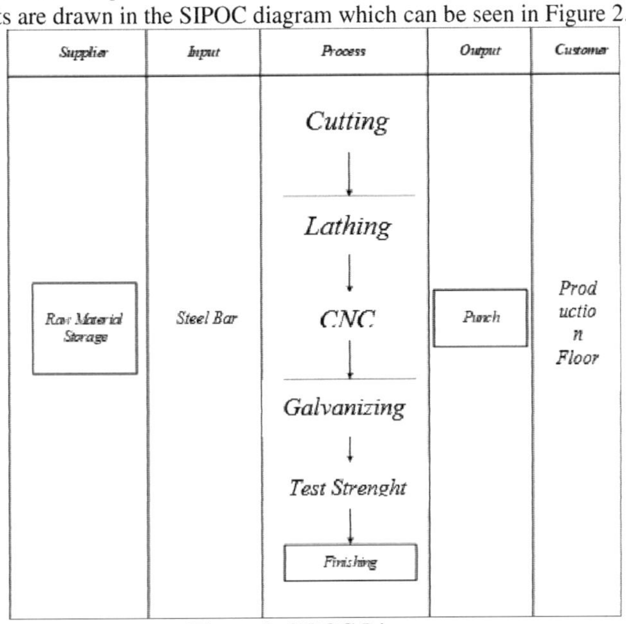

Figure 2. SIPOC Diagram

Table 4. Information for *Mapping*

Proses	Operator (Orang)	*Uptime (%)*	*C/O* (Menit)	*C/T* (Menit)	*Available time (Jam)*	*Scrap (%)*
Pemotongan	1	100	0	15.48	8	0
Pembubutan	1	100	0	5.63	8	1
CNC	1	100	0	20.86	8	4
Penyepuhan	1	100	0	587.89	8	0
Uji Kekuatan	1	100	0	1.94	8	0
Finishing	1	100	0	48.53	8	1

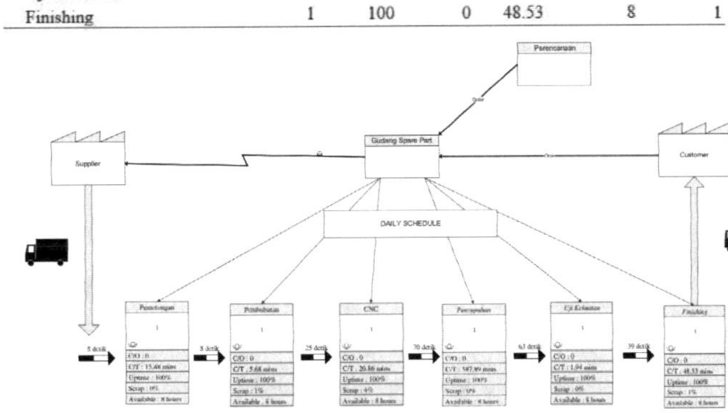

Figure 3. Current State Map

3.4. Selection tool

Value stream mapping with the largest total score according to VALSAT results will be used as the mapping to be able to identify waste in detail. This selection is based on the value stream mapping with the largest value that is most suitable for identifying waste in the value stream. Complete filling results of VALSAT can be seen in the following table.

Table 5. Recapitulation of Calculations VALSAT

Mapping Tool / Waste	*Process Activity Mapping*	*Supply Chain Response Matrix*	*Production Variety Funnel*	*Quality Filter Mapping*	*Demand Amplification Mapping*	*Decision Point Analysis*	*Physical Structure*
Overproduction	0^{0x1}	0^{0x3}	0^{0x3}	0^{0x1}	0^{0x3}	0^{0x3}	0^{0x3}
Waiting	27^{3x9}	27^{3x9}	3^{3x1}	0^{3x0}	9^{3x3}	9^{3x3}	0^{3x0}
Transport	9^{1x9}	0^{1x0}	0^{1x0}	0^{1x0}	0^{1x0}	0^{1x0}	1^{1x1}
Inappropriate Processing	9^{1x9}	0^{2x0}	3^{1x3}	1^{1x1}	0^{1x0}	1^{1x1}	0^{2x0}
Unnecessary Inventory	3^{1x3}	9^{1x9}	3^{1x3}	0^{1x0}	9^{1x9}	3^{1x3}	1^{1x1}
Unnecessary Motion	9^{1x9}	1^{1x1}	0^{3x0}	0^{3x0}	0^{3x0}	0^{3x0}	0^{3x0}
Defects	1^{1x1}	0^{1x0}	0^{1x0}	9^{1x9}	0^{1x0}	0^{1x0}	0^{1x0}
Total	**58**	**37**	**9**	**10**	**18**	**14**	**2**
Ranking	**1**	**2**	**6**	**5**	**3**	**4**	7

3. 5. Process Activity Mapping

In identifying activities non-value added for the company, observations are carried out directly into the company. With observations supported by interviews with workers and supervisors, it can be seen whether or not there is a waste in the company. Besides, it can be a used tool that can identify the presence of waste, namely Process Activity Mapping (PAM). The recapitulation of Process Activity Mapping (PAM) for punch can be seen in Table 6.

Table 6. Amount and Time of PAM Results for Making Punch

Activities	**Amount of**	**Time (seconds)**	**Percentage (%)**
Operation (O)	36	28025.9	95.49
Transportation (T)	14	555	1.89
Inspection (I)	6	77.9	0.27
Storage (S)	0	0	0
Delay (D)	2	689.4	2.35

3.6. Proposed Improvement Results

Proposed improvements made can reduce the *waste* that occurs in the making process punch and reduces the time and number of activities so that the production process runs more efficiently. Based on the comparison of the current state map and future state map, there are changes in production process time and production process activities which can be seen in Figure 4 and Table 7.

IOP Conf. Series: Materials Science and Engineering **648** (2019) 012016 doi:10.1088/1757-899X/648/1/012016

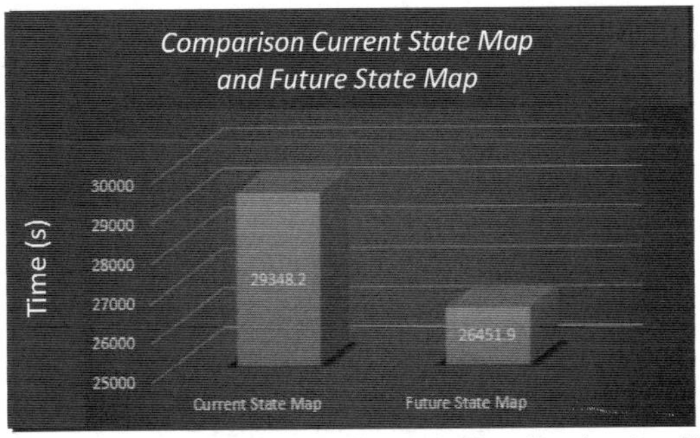

Figure 4. Comparison of *Production Process TimeProduction Process Time*

Table 7. Comparison of Before and After Repair

No.	Station	Time (Seconds) Before Improvement	Time (Seconds) After Improvement
1	Cutting	15:48	15:48
2	lathe	5.63	5.63
3	CNC	20.86	20.86
4	Gilding	587.89	570.47
5	Strength Test	1.94	1:18
6	*Finishing*	48.53	16.78

Figure 4. Based on a reduction in *production process time* amounted to 29348.2 seconds - 26 451, 2 seconds = 2897 seconds. This means that there is a time-saving in the production process of 2897 seconds. Reduction of time taken can help companies meet demand *punch* with *production process time* a shorter.

Table 8. Production Process Activities After Repairing

No.	Activity
1	The raw material is brought to cutting machine
2	The raw material is inserted into the cutting tool and cutting size adjustment
3	Tightens workpiece holder and measures to adjust blade location
4	Cut object
5	Turns off machine and releases objects from retaining
6	Brought objects into the lathe
7	Associated objects to chuck
8	smoothed the front end of the workpiece
9	Measured workpiece
10	Released objects from the chuck and reassembled
11	smoothed the rear end of the workpiece
12	Measured workpiece
13	Released from chuck
14	Brought to CNC
15	Measured workpiece
16	Conducted setup for the formation process
17	The inserted workpiece in the chuck

IOP Conf. Series: Materials Science and Engineering **648** (2019) 012016 doi:10.1088/1757-899X/648/1/012016

18	Conducted formation
19	Conducted setup for CNC finishing
20	Performed finishing on an object
21	Removed object from the chuck and taken to gilding stationBonded
22	object with wire
23	Inserting sand into the container
24	The object is put in the container
25	the container is inserted into the kitchen about
26	Turning on the engine and setting up temperature
27	Performing the plating process
28	The machine is turned off
29	The container is removed from the kitchen sepuh
30	The coolant
31	The object is transferred to the storage not removed
32	The wire is on the object
33	The object is smoothed
34	The object is brought to the CNC machine to test its strength (HRC test)
35	Conducted setup test machine HRC
36	Mounted objects on HRC test machine hook
37	Performed strength test
38	Objects released from the hook
39	Objects have taken to CNC machine
40	Performed *finishing* using the CNC machine
41	Punch brought to GSP

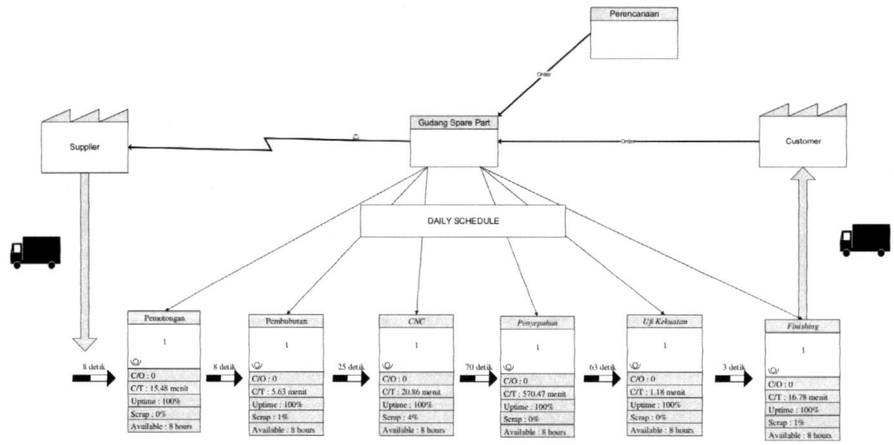

Figure 5. ***Future State Map***

4. Conclusion

Based on results, we can conclude that :

1. Waste that occurs is over processing and motion that is caused by work methods and machinery and also there is no value-creating time at CNC stations and gilding stations in the form of operators just waiting for the machine to operate.

2. The result of reducing waste in the process of making a punch is a reduction in time by 2897 seconds, from 29348.2 seconds to 26451.2 seconds and a reduction in production activities from 58 to 41 activities.
3. Proposed improvements for the manufacturing process, Punch namely eliminating the activity of taking tools from outside the work area, neatly arranging the tools that will be used so that they are easy to find and work on finishing not manually.

References

[1] Afferdhy arieffien. 2016. *Lean Manufacturing Approach to Minimize Waste Production Parts STN Wfx000 RIB at the Primatic Medium Machine 2 in PT X.* International Review of Management and Marketing.
[2] Capital, Melong. 2004. *Introduction to Lean Manufacturing*. Vietnamese.
[3] Imai, Masaaki. 1997. *Gemba Kaizen*. Publisher: PT. Binaman Pressindo Library.
[4] Liker, JK and Meier, D. 2006. *The Toyota Way Fieldbook*. US: Mc Grow-Hill.
[5] Rother, M. and Shook, J. 2003. *Learning To See, Value Stream Mapping To Create Young Value and Eliminate The Lean Enterprise Institute*.
[6] Sinulingga, Sukaria. 2011. *Research Methodology*. Medan: USU Press.
[7] Sutalaksana, Iftikar. 1979. *Technical Procedures for Working*. Bandung: ITB.
[8] Wignjosoebroto, Sritomo. 2000. *Ergonomics, Motion and Time Studies. Analysis Techniques for Work Productivity*. Surabaya: For Widya.
[9] Wilson, Lonie. 2010. *How To Implement Lean Manufacturing*. US: Mc Grow-Hill.

Optimization in Critical Work Center Using Theory of Constraints

R M Sari[1], K Syahputri[1], I Rizkya[1], N Siboro[1]

[1]Department of Industrial Engineering, Faculty of Engineering, Universitas Sumatera Utara, Medan, Indonesia

Email : rahmi.m.sari@usu.ac.id, khalida@usu.ac.id

Abstract. Many companies experience problems or problems in the production process. One of the problems disturb the production process is the bottleneck. The pellet industry has problems in pellets production. The capacity imbalance available between the drying station and the milling station causes the pellet making process to accumulate resulting in a bottleneck. Therefore, an analysis is carried can be done to overcome these problems by overtime. Overtime is carried out for 1 hour on each shift and 2 hours on certain shifts in July 2019 period and September 2019 and the number of working hours becomes 18 working hours and 19 working hours from the previous one which is only 16 working hours at the milling work station because it is seen from a negative variance value and the percentage of load is below 100%. This shows that overtime can be effective to eliminate constraints.

1. Introduction

The development of the industrial world at this time requires companies to produce perfectly. This is because in the increasing of industrial world competition, productivity is one of the factors determines a company to be successful [1]. However, in reality many companies experience problems in the production process such as in the pellet industry.

Pellet industry produces products through five work stations including drying, milling, mixing, pelleting and packing work stations. The five work stations have different production capacities. Drying work stations have a greater production capacity compared to milling work stations which is cause bottleneck. The bottleneck occurs will result in delays in the production process and have an impact on the company's performance. Therefore, this problem must be resolved and work stations can be optimized and bottleneck does not happen.

Bottleneck is defined as a problem that occurs at work stations that limits the efficiency of the entire process [2]. This can lead to situations where the work station before the bottleneck can complete processing so that it cannot continue the material, because the works that follow it are still involved in the previous processing [3]. This problem must be solved in order to increase work productivity and production profits. Bottlenecks occur if the capacity available in an industry is insufficient to meet product demand from consumers [4]. Bottleneck problems overcome by various methods, one method is theory of constraint.

Theory of Constraints (TOC) was first known as the Theory of production introduced by Goldratt and Cox [5,6]. TOC is basically eliminating the restrictions of the system (or bottlenecks) that prevents production flow from completing the process according to demand [7] TOC is a method that is concerned with limiting capacity in manufacturing processes by its capacity-constraints resources. TOC can be used in both manufacturing and service companies and in several managerial areas, such as: project management, quality management, information systems. This method is especially widespread in production planning [8]. There is several strengths of TOC method that is in contrast to other improvement approaches that tends to optimize performance in each area of the system (such as TQM, Six Sigma and Lean) a that is on the basis of systemic thinking, focusing efforts on increasing the important components of the system [9].

Many research has been done in overcoming bottleneck problems. Imam and Andrie (2013) conducted a study by analyzing capacity constrained resources using the theory of constraint approach to optimize production flow. The use of Theory of Constraint in this research is used to remove constraints by maximizing the area of the work station constraints, optimizing resources, increasing production performance and making the optimal number of product combinations [10]. Another study was also conducted by Ercument et al. (2015) in optimizing production in the furniture industry [11]. The purpose of this study is to provide a solution to the contrains that occur in the work station to be able to complete their targets so that by increasing the profitability of the company. The theory of constraints has been carried out on furniture companies operating in the Mediterranean Region. The result is that, there are capacity constraints that occur in the production process in the company and profitability will increase by 42% after the elimination of these constraints. However, this research is closely related to the optimization of resources in increasing company productivity. This study aims to optimize resources in overcoming bottlenecks that disturbs industrial productivity.

2. Methodology

The research was conducted in one of the pellet manufacturing industries located in Medan City. The research was conducted by observing the real conditions in the industry. From the results of direct observation, the formulation of the problem is determined according to the conditions and the research objectives can be applied. The research objectives determined are as a solution to the problems that happen. Furthermore, data collection is needed as input in solving bottleneck problems. Data is obtained through observations on the pellet production process and the results of interviews with workers and employees in the pellet industry. Based on these data, an analysis of the problems that occur is done by optimizing resources to eliminate bottlenecks. The analysis is carried out with several stages, namely forecasting product demand and calculating product capacity with rough cut capacity planning. Forecasting is done with the aim to estimate the amount in the next period. Calculation of forecasting is done by the time series method and consider the estimation of errors. While the calculation of capacity with rough cut capacity planning is done to see which work stations have bottlenecks. After it was discovered that the work station had a bottleneck, then a problem was solved by using theory of constraint. Overall, the philosophy of TOC management is directed at helping organizations sustainably achieve their goals

The underlying three measures are: throughput, operating expense, and inventory [12]. TOC is carried out in several stages starting with identifying company constraints. In this case the constraints in question are in the form of identifying a bottleneck work station. Next is the exploitation of binding constraints. At this stage alternatives that can solve problems so that they can optimize work stations that experience bottlenecks. The final step is the subordination of solutions to overcome existing constraints [1]. With these stages, a solution will be obtained to overcome the bottleneck problem that is currently an obstacle in the industry.

3. Result and Discussion

3.1. Forecasting of Product Demand

Forecasting the number demand of pellet products for the next one year is based on demand data from January 2018 to December 2018 with the time series method. Forecasting results are selected

IOP Conf. Series: Materials Science and Engineering **648** (2019) 012017 doi:10.1088/1757-899X/648/1/012017

based on the smallest error estimates. The results for the number of pellet demands can be seen in Figure 1.

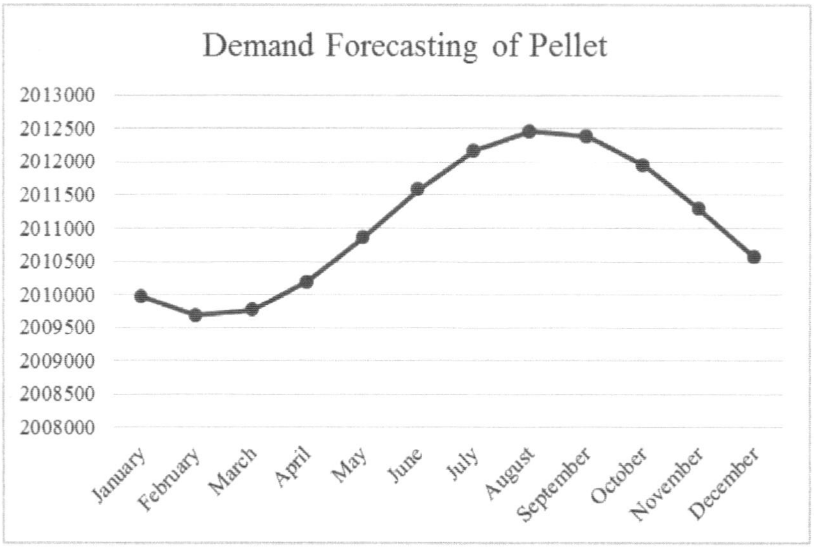

Figure 1. Result of Pellet Products Forecasting

Based on the figure 1, it can be seen that product demand are fluctuating for each order period. For this reason, it is necessary to calculate the capacity needed to fulfill the demand and the bottleneck can be avoided.

3.2. Determination of Rough-Cut Capacity Planning

Rough Cut Capacity Planning is used to calculate the capacity requirements roughly and compare them with available capacity [13]. Calculations for the capacity needed are carried out in determining the capacity needed by each work station that is able to complete the product according to the forecasting results Calculation of rough-cut capacity planning is carried out based on the required capacity (CR) and available capacity (CA) on milling work station. Results the calculation of work station experiences bottleneck can be seen in Table 1.

Table 1. Recapitulation of Bottleneck and Non Bottleneck Work Stations

Work Station	Category
Drying	Non Bottleneck
Milling	Bottleneck
Mixing	Non Bottleneck
Pelleting	Non Bottleneck
Packing	Non Bottleneck

Based on the table 1 above, it can be seen that at the milling station there is a bottleneck and the capacity improvement is needed.

3.3. Recommendation Using Theory of Constraint

Recommendation is done using theory of constraint. Improvement starts by problem identification and done by bottleneck work station identification. Based on the result of identification obtained that bottleneck happen at milling station as in Table 2.

Table 2. Recapitulation of RCCP Calculation on Bottleneck Determination

Work Station	CR (second)	CA (second)	Variance	Workload Percentage (%)	Category
Mixing	2728872	3720039	-991167	0.73	Bottleneck
	2728872	3720039	-991167	0.73	Bottleneck
	2728872	3863117	-1134245	0.71	Bottleneck
	2730231	3720039	-989808	0.73	Bottleneck
	2730231	3863117	-1132886	0.71	Bottleneck
	2731590	3720039	-988449	0.73	Bottleneck
	2732949	3576960	-844011	0.76	Bottleneck
	2732949	3720039	-987090	0.73	Bottleneck
	2732949	3576960	-844011	0.76	Bottleneck
	2731590	3720039	-988449	0.73	Bottleneck
	2731590	3720039	-988449	0.73	Bottleneck
	2730231	3863117	-1132886	0.71	Bottleneck

Based on the table above, it can be seen that the milling station has a bottleneck due to insufficient capacity available to produce in accordance with the required capacity and alternatives are needed to overcome these problems. Alternative solution to optimize the bottleneck work station is overtime. *Overtime* is done for 1 hour on each shift and 2 hours on certain shift in July 2019 period and September 2019 the number of working hours are 18 hours and 19 hours of working hours from the previous working hours only about 16 hours. Result by using overtime shows that there is no bottleneck on bottleneck work station because it seen from negative variance value and load percentage below to 100%. This condition shows that using overtime can be an alternative to eliminate constraint.

4. Conclusion

Manufacturing industry of pellet experienced a bottleneck at the mixing station due to the capacity imbalance available at the industry. Problem solving is done by using theory *of constraints* principle with solution alternative of *overtime* for 1 hour (9 hours/shift) and 2 hours on certain shift. This solution results *non-bottleneck* work station in all of period and shows that overtime can be an alternative to eliminate constraint.

Acknowledgement

The author would like to thank the Universitas Sumatera Utara for supporting finance in publishing this article and the pellet company that is willing to be the object of this research. We appreciate the efforts of all participants in conducting this research.

References

[1] Anita M and Dodo A 2017 *Seminar dan Konferensi Nasional IDEC* (Surakarta, 8-9 Mei)
[2] Betterton C E and Silver S J 2012 *International Journal of Production Research* **50** (15), pp 4158-4174
[3] Mateusz K 2016 *Economic and Management* **8**, pp 103-122
[4] Himanshu K and Ashish S 2014 *International Journal of Mechanical Engineering and Robotics*

Research
[5] Goldratt E M and Cox J 2014 *The Goal: A Process of Ongoing Improvement* (New York: North River Press)
[6] Radoslaw W, Bozena S Z, and Michal Z 2017 *3rd International Conference on Social, Education and Management Engineering*
[7] Gupta M and Snyder D 2009 *International Journal of Production Research* **47** (13), pp. 3705–3739
[8] Alsmadi M, Almani A, and Khan Z 2014 *International Journal of Quality & Reliability Management* **8**, pp. 906-920
[9] Victor G A E, Pedro G V, and Maria M G Z 2016 *European Research on Management and Business Economics* **22**, pp. 139–146
[10] Imam S and Andrie W A 2013 *Analisis Capacity Constrained Resources Guna Mengoptimalkan Aliran Produksi Dengan Pendekatan Theory Of Constraints* (Yogyakarta: IST AKPRIND)
[11] Ercument O, Ata K, and Jekaterina K 2015 *Intellectual Economics* **9**, pp. 138-149
[12] Trojanowska J and Pajak E 2010 *7th International DAAAM Baltic Conference* (Estonia: Tallinn, 22-24 April)
[13] Sukaria S 2009 *Perencanaan dan Pengendalian Produksi* (Yogyakarta : Graha Ilmu)

Design of A Convolutional Neural Network System to Increase Diagnostic Efficiency of Alzheimer's Disease

Faransi Al-azdi[1], R Passarella[1], A Susanto[1], C Caroline[1], RA Dwi Puspa[2], T W Yudha[2]
[1]Department of Computer Engineering, Faculty of Computer Science, Universitas Sriwijaya Palembang, Indonesia
[2]Faculty of Medicine, Universitas Sriwijaya Palembang, Indonesia

*Email: passarella.rossi@gmail.com

Abstract. The most common degenerative neural disease, Alzheimer's disease (AD), is insidious and almost always requires imaging modalities to be diagnosed early. MRI is the most common one used, but requires timely interpretation. Here we develop a convolutional neural network (CNN)-based system that determines whether a brain MR image has AD or normal. First, feature extraction is performed to separate various parts of the brain. Then, the data is processed to differentiate normal brain from AD brain, solely using MR image. Finally, the neural network is supplemented using data from the patient's history and physical examination. In this first phase, we were able to extract features from the brain MR image, initially by masking the image and separating the white matter, grey matter, and cerebrospinal fluid called the grey level co-occurrence method (GLCM). This method is able to using a convolutional neural network.

1. Introduction

The nervous system's the most common degenerative disease, Alzheimer's disease, is a disease which is caused by damage to the brain cells associated with cognitive processes of reasoning, memorising, and actualising personality. Dementia or senility occurring in the elderly is assumed as a natural course of life by the majority of Indonesian because of its relationship to the aging process. In fact, there is a type of abnormal dementia namely Alzheimer's disease. An estimated 46 million people are Alzheimer's disease patients in worldwide, and one million of them live in Indonesia. This number is expected to increase dramatically to double in 2030[1].

Neuroscientists are fascinated in learning Alzheimer's disease (AD), not only because it is the most common of the degenerative central nervous disease, but also because of its insidious clinical progression, where the first symptoms are purely cognitive. This disorder affects the limbic system and the association cortices. Thus, the patients have impaired novel memory generation, first of trivial, and then important details of life. However, interestingly, motor and sensory functions of AD patients are often intact and show no pathological signs. The brain cortex of the patient shrivels up and shrinks severely, particularly in the hippocampus area, which is involved in thinking, reasoning and creating new memories. Brain ventricles, which produce cerebrospinal fluid, also become larger in an AD patient.

A well-timed diagnosis of this disease is crucial and requires good clinical assessment based on the patient's medical history. Several neuropsychological tests can be used to help to establish the diagnosis of AD, such as mini-mental state examination (MMSE), neuropsychiatric inventory questionnaire, clinical dementia rating and other pathological evaluations. National Institute of Aging Alzheimer's Association developed the first clinical criteria for AD diagnosis. Imaging tests, such as MRI, are required to accurately diagnose AD, but it needs an expert (neurologist or radiologist) to interpret the features, which often consumes a lot of time and money.

In the present, computer-aided diagnostic systems have been and are being developed by several researchers for accurate detection of AD. Some of the earliest expert systems are rule-based, and had been in development since the 1970s. In the 1990s, supervised models began replacing these rule-based systems, being trained with feature vectors taken from medical sources like computer tomography or MRI.. In this paper, we attempted to create a MRI-based expertise system that would allow a neurologist

IOP Conf. Series: Materials Science and Engineering **648** (2019) 012018 doi:10.1088/1757-899X/648/1/012018

or radiologist to interpret results quickly, that we hope will be able to increase the efficiency of AD diagnosis and therapy, whose number of patients are climbing at an ever-rising rate.

2. Methodology

2.1. Data

This project uses data from brain Magnetic Resonance Imaging (MRI). MRI is often the near gold-standard method when soft tissue delineation is necessary [2] (the gold standard for AD is brain autopsy, which would only be possible post-mortem, after the patient's death).

The Magnetic Resonance Imaging data for the model system has been earned from the Open Access Series of Imaging Studies (OASIS) dataset. OASIS is a freely distributed collection of neural imaging datasets made for the scientific development. The distributors hoped that the free dissemination would propel future discoveries in neuroscience. Currently, there are three OASIS datasets, with the first two, OASIS-Cross sectional and OASIS-Longitudinal have been used for data analyses, creation of anatomical reference, and development of algorithms to segmentate parts of the brain [3]. Its dataset also contains additional crucial information including age, sex, main complaint, systolic blood pressure, diastolic blood pressure, pulse, and axial head length and head width.

In the rst phase of system design, only MRI scan images is used, using the OASIS database as sources. Later, in advanced phases, we will be taking into account history-taking (anamnesis) and the patient's physical examination results as parameters of the dataset to be processed and programmed into the diagnostic system. The parameters can be seen in Table1. Then, from these, data testing is carried out and training data using Convolutional Neural Network.Table1 is the parameter that will be used in the system diagnose. In row of Code, A refer to result of history-taking (anamnesis), P refer to result of physical examination, and M refer to MRI result. The example of image in OASIS dataset shows in Figure1.

2.2. Preprocessing

The dataset's images have been processed by skull stripping and normalisation, so that the parts of the brain keep aligned properly at the relatively same coordinates, no matter how the image is transformed [4].

Brain image features consist of: White Matter (WM), Grey Matter (GM) and Cerebrospinal Fluid (CSF). The data are prepared (masked) to strengthen contrast and make a clear distinction between the WM, GM and CSF [5]. Image contrast is the difference in brightness between objects or region [6]. An image lacks contrast when it does not have strong difference between black and white. In a image, brightness refers to the overall its lightness or darkness. To change the contrast in an image, contras-stretching is performed by the Adjust Contrast tool. The better result of contrast enhancement could provide better segmentation accuracy at the anatomical structure of the brain, and digital image processing [7].The proposed design is showing in Figure 2.

Table 1. Parameters

No	Code	Parameters	Example
1	A	Name	Jhon
2	A	Sex	Male/M
3	A	Age	73
4	A	Main Complaint	Dementia
5	A	History of Headaches	Yes
6	A	History of Di culty Remembering the way home	No
7	A	History of Head Trauma	No
8	A	Family Medical History	Yes
9	A	Glasgow coma scale	15
10	P	Systolic Blood Pressure	120
11	P	Diastolic Blood Pressure	80
12	P	Pulse Rate	80
13	P	Respiratory Rate	16
14	P	Cranial Nervous Examination	Normal
15	P	Meningeal Re exes	Abnormal
16	P	Motor/Muscle Tones	Normal

17	P	Re exes	Normal
18	P	Sensory Function	Normal
19	P	Autonomous Nervous Function	Normal
20	P	Coordination	Normal
21	P	MMSE	18
22	M	Head length	20
23	M	Head Width	16
24	M	Cerebrum Density	Kg/m^3
25	M	Cerebellum Density	Kg/m^3
26	M	Ratio of Black Area/White Area	0.1

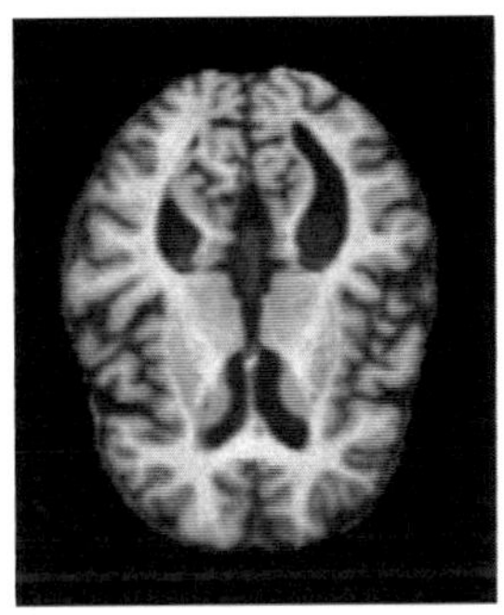

Figure 1. Sample Image in the OASIS Dataset

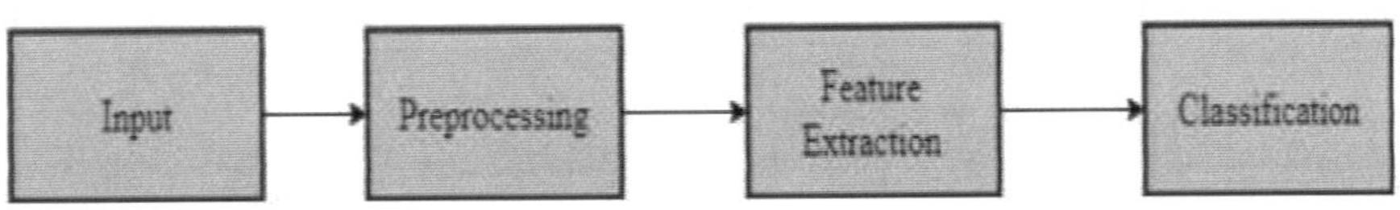

Figure 2. Block Diagram System

2.3. Features Extraction

After thresholding, Grey Matter, White Matter, and CSF were extracted. This paper uses the following feature extraction by using Grey Level Co-Occurrence (GLCM). GLCM is is a matrix (number table) that represents the grey levels of each pixels in the image. The number of grey levels, G, is equivalent to the rows and columns. P is a matrix element (i,j) and represents the relative frequency in which two adjacent or non-adjacent pixels separated by the distance (x,y) , occur wiithin a specific neighbourhood. One of them has the intesnity 'i' and the other 'j' [8]. The equation that used in this phase by using eq(1), eq(2), eq(3) and eq(4).

$$Contrast = \sum_{i,j=0}^{N-1} P_{ij}(i-j)^2 \tag{1}$$

$$Correlation = \sum_{i,j=0}^{N-1} P_{ij}\frac{(i-\mu)(j-\mu)}{\sigma} \tag{2}$$

$$Homogeneity = \sum_{i,j=0}^{N-1} P_{ij}\frac{(P_{ij}}{1+(i-j)^2} \tag{3}$$

$$Entropy = \sum_{i,j=0}^{N-1} -ln(P_{ij})P_{ij} \tag{4}$$

2.4. Convolutional Neural Network

CNN is an algorithm which can receive input image, assign importance to various aspect/object in the image and be able to di erentiate one from the other(Figure3) .

A CNN is a type of feed-forward neural network (FNN) which has neurons that is inspired by an animal's visual cortex for all computer vision tasks [9]. The goal of a CNN is to learn-order features in the data via convolution. CNN's topology utilises spatial relationships to reduce the number of parameters which have to be learned [10]. In relation to the convolutional network, is to extract features from the input image. the major objective of convolution. CNN is inspired by the way human vision works, therefore layers of a convolutional network have neurons dispersed in three dimensions: height, width, and depth [11].

2.4.1. Transfer Learning

Transfer learning is the action of improving the learning in a novel task by tweaking and re-adjusting weights between neurons to classify an object of the same type but different category. Transfer methods is given huge ammount of resources to train CNN models or even larger datasets which the model are trained. Basically, it is the process of copying knowledge from an already trained network to a new network to solve similar problems [9]. In transfer learning, there are two forms of transfer learning which are inductive learning and inductive transfer that narrowed by using a model on not related task. For using transfer learning, we can use on two approaches such as develop model and pre-trained model. Furthermore, transfer learning has several models for image data especially MRI scan is VGG 16 model, ResNet, Inception from Google, and AlexNet.

2.4.2. AlexNet

The first major breakthrough in the architecture of CNN came in seven years ago. This award-winning CNN architecture is called AlexNet. At first run, a ReLU activation function and a dropout of 0.5 were used in the network to ght to over t [9]. AlexNet is an enormous network structure with 60 million parameters and 650.000 neurons.

The rst convolutional layers filter the 224 x 224 x 3 input images with 96 kernels of size 11 x 11 x 3, with a stride of 4 pixels (this is the distance between the receptive field centers of neighboring neurons in a kernel map).The second convolutional layer takes as input the

IOP Conf. Series: Materials Science and Engineering **648** (2019) 012018 doi:10.1088/1757-899X/648/1/012018

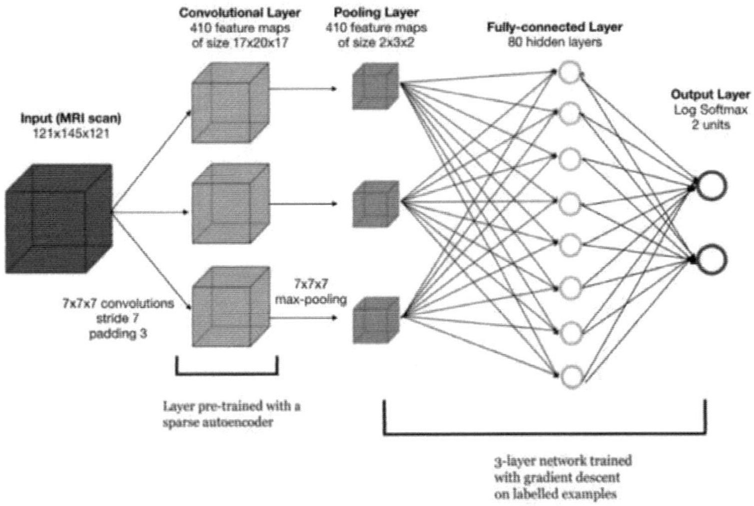

Figure 3. The Architecture of the naural network used for 2-way classi cation

(response-normalised and pooled) output of the first convolutional layer. The third, fourth, and fifth convolutional layers are connected to one another without any interventional pooling or normalisation layers. The third convolutional layer has 384 kernels which is connected to the (normalised, pooled) outputs of the second convolutional layer. The fourth convolutional layer has 384 kernels, and the fifth convolutional layer has 256 kernels. The fully-connected layers have 4096 neurons each.

3. Discusion

In this study method, we used brain images from OASIS dataset to analyze the method performance. The random datasets were marked for binary classi cation and the percentage data 75% for data training and 25% for data testing purpose. The dataset were preprocessing before through training and testing. The architecture of neural network are using Alexnet architecture with ve layer of convolution. Compare to another journal result, the study method mostly using ADNI database and LeNet or GoogleNet architecture. According to [10] used architecture of adopted LeNet and the average result for testing accuracy is 97.4%. In same study they also used archinture of adopted GoogleNet and the highest result for testing accuracy is 98.84%.Other reseach by [12] used transfer learning and BellCNN got the result for nal testing accuary is 85.7%.

4. Conclusion

In this study we present a method of feature extraction from brain MR image input. The method is able to separate the white matter, grey matter, and the cerebrosipnal uid from each other.

Acknowledgement

This research was supported by Ministry of Research, Technology and Higher Education under scheme Program Kreativitas Mahasiswa Karsa Cipta. We would like to say thank to Rossi Passarella for assistance as head of Computer Engineering department and corresponding author for comment that greatly improved & his help in preparation of this manuscript.

References

[1] "Menkes: Lansia Yang Sehat, Lasia yang Jauh dari Demensia.," Kementrian Kesehatan RI, 2016. [Online]. Available: http://www.depkes.go.id/article/print/16031000003/menkes-lansia-yangsehat-lansia-yang-jauh-dari-demensia.html . [Accessed 31 October 2018].

[2] El-Dahshan, E.S.A., Hosny, T. and Salem, A.B.M., 2010. Hybrid intelligent techniques for MRI brain images classi cation. Digital Signal Processing, 20(2), pp.433-441.

[3] Marcus, D.S., Wang, T.H., Parker, J., Csernansky, J.G., Morris, J.C. and Buckner, R.L., 2007. Open Access Series of Imaging Studies (OASIS): cross-sectional MRI data in young, middle aged, nondemented, and demented older adults. Journal of cognitive neuroscience, 19(9), pp.1498-1507.

[4] Payan, A. and Montana, G., 2015. Predicting Alzheimer's disease: a neuroimaging study with 3D convolutional neural networks. arXiv preprint arXiv:1502.02506.

[5] Davatzikos, C., Fan, Y., Wu, X., Shen, D. and Resnick, S.M., 2008. Detection of prodromal Alzheimer's disease via pattern classi cation of magnetic resonance imaging. Neurobiology of aging, 29(4), pp.514-523.

[6] P. Steven W. Smith.2011.Chapter 23 :Image Formation & Display. (Online). Available: https://www.dspguide.com/ch23/5.htm. (Accessed 16 April 2019).

[7] Chen, C.M., Chen, C.C., Wu, M.C., Horng, G., Wu, H.C., Hsueh, S.H. and Ho, H.Y., 2015. Automatic contrast enhancement of brain MR images using hierarchical correlation histogram analysis. Journal of medical and biological engineering, 35(6), pp.724-734.

[8] Mohanaiah, P., Sathyanarayana, P. and GuruKumar, L., 2013. Image texture feature extraction using GLCM approach. International journal of scienti c and research publications, **3**(5), p.1.

[9] Sewak, M., Karim, M.R. and Pujari, P., 2018. Practical Convolutional Neural Networks: Implement advanced deep learning models using Python. Packt Publishing Ltd.

[10] Sarraf, S. and To ghi, G., 2016. DeepAD: Alzheimers disease classi cation via deep convolutional neural networks using MRI and fMRI. BioRxiv, p.070441.

[11] Buduma, N. and Locascio, N., 2017. Fundamentals of deep learning: Designing next-generation machine intelligence algorithms. " O'Reilly Media, Inc.".

[12] Awate, G., Bangare, S., Pradeepini, G. and Patil, S., 2018. Detection of Alzheimers Disease from MRI using Convolutional Neural Network with Tensor ow. arXiv preprint arXiv:1806.10170.

IOP Conf. Series: Materials Science and Engineering **648** (2019) 012019 doi:10.1088/1757-899X/648/1/012019

Workload of Workforce in Fertilizing Industry: An Analysis

Rahmi M. Sari*[1], Ukurta Tarigan[1], Indah Rizkya[1], Elvira[1]

Department of Industrial Engineering, Faculty of Engineering, Universitas Sumatera Utara, Medan, Indonesia

*Email : rahmi.m.sari@usu.ac.id

Abstract. *Analysis of the number of workers in a company is very important. This is very necessary to determine the optimal number of workers. This research has been done in a company engaged for processing single and compound fertilizers. The company trades fertilizer as its first business which is a subsidiary one of the largest agribusiness company in Indonesia. In this study an analysis of the needs of the number of workers was carried out based on the job description of each worker at packaging station. Thus, the workload is calculated for each worker at the packaging station using the Work Load Analysis (WLA) method to calculate the percentage of productivity using the Work Sampling method. Next, the determination of the Rating Factor value using the Westinghouse method and determining the value of Allowance using the Industrial Labour Organization (ILO) table. The workload calculation results obtained that the workload received by 5 operators is high because the value is above 100%. The proposed improvement recommendations given related to the high workload condition is by adding 2 more workers so that the number of workers recommended at packaging station is 7 workers.*

1. Introduction

Humans are the main assets in the organization, so that human resources (HR) must be managed and utilized in a balanced and humane manner. HR does not mean just referring to people but the ability of the employee. Talent, competence, ideas, energy from employees, productivity, and performance of the organization are collectively identified as HR of a company [1]. Human resource management is the science and art that regulates the relationship and role of the workforce, so that it can effectively and efficiently help realize the goal [2]. Every organization will always strive to improve the work productivity of its employees, with the hope that what will be achieved by the company will be achieved [3]. The role of human resources (HR) in a company is very important because HR conducts all activities in order to meet the objectives of the company. With the effectiveness, efficiency and productivity, company can find out how to optimize the resources used and can find out the achievement of targets that have been carried out by the company. With human resource planning the company can determine the number of employees needed to run the company's system efficiently [4]. Work productivity is a measure that shows the consideration between input and output issued by the company and the role of labor that has a union of time. The productivity measurement is carried out by the company with the aim of knowing whether it is successful or not in improving the quality of work productivity of employees. If there is an increase in productivity, it means that the company is able to improve the quality of the company's operational activities [6]. Employee productivity can also be influenced by workload [7]. The high and low workload depends on the level of complexity of work procedures, the demands of work, the tempo of work, and job responsibilities that are not the same. This causes the speed of completing tasks and the productivity of each employee is not the same according to the workload of each [8]. The workload has a significant influence on employee work productivity. Training is a planned effort from the company to improve the knowledge, skills and abilities of employees [9]. Training or training is a systematic effort of the company to increase all knowledge, skills and attitudes of employees through the learning process so that they are optimal in carrying out their functions and job duties [10]. Training or training can affect employee work productivity. Training can prepare employees to have the knowledge, expertise and ability to be able to adapt and anticipate changes to the business environment [11]. There are several factors that influence employee performance [12] intrinsic factor and extrinsic factor.

Workload is a set or number of activities that must be completed by an organizational unit or office holder within a certain period of time. Workload variables use indicators namely work time, number of jobs, internal body factors, and external factors of the body [13]. Job analysis is always needed when the organization makes work plans for employees.

Workload analysis is a process to determine the total time spent completing a particular job. In other words, workload analysis aims to ensure the number of employees needed is in accordance with several workloads and specific responsibilities given to employees. This too, is an activity carried out to identify the number and qualifications of employees needed to realize organizational goals. In addition, workload measurements are also considered as management techniques to obtain information about certain positions or jobs. This is solved through a systematic research and evaluation process. Workload analysis calculates the time and ability of employees to complete tasks. Therefore, workload analysis will be important in studying the workload of each position and the lowest work unit. Finally, the exact amount of work volume imposed on a unit can be known, and this will later be useful in interpreting strategic policies in the future [14].

The method used in this study is the Workload Analysis (WLA) method, because this method is suitable for analysing the workload of workers. WLA is one way that can be used to calculate the amount of workload caused by activities carried out. This study will analyse several causes of the magnitude of the workload and determine remedial solutions to reduce the high workload. In addition, the workload that has been received by workers can also be used to determine the number of workers that the company needs to have [15].

2. Methodology

This study examines the workload conditions of workers using the workload analysis method to calculate workload and number of workers. The steps taken in carrying out this research are as follows. First, field studies are conducted to obtain information about the general description and actual conditions of the company. Second, identification of the problem based on existing problems. To identify the problem, cause-and-effect diagram has been used. This diagram shows the root cause of the workload problems. Third, specify the research objectives to determine the research goals. In this research, the objective is to measure the exact workload for each workforce in the company. Fourth, collecting the data from the organization including the number of workers, job description and so forth. After collecting the data, then do the data processing. In this step, using the WLA method to determine the number of workers. The analysis carried out in this study is related to the number of workers which will compare the number of workers currently with the number of workers based on their workload.

3. Result

Cause-and-effect diagram due to mismatch of workload on Bagging station as follows:

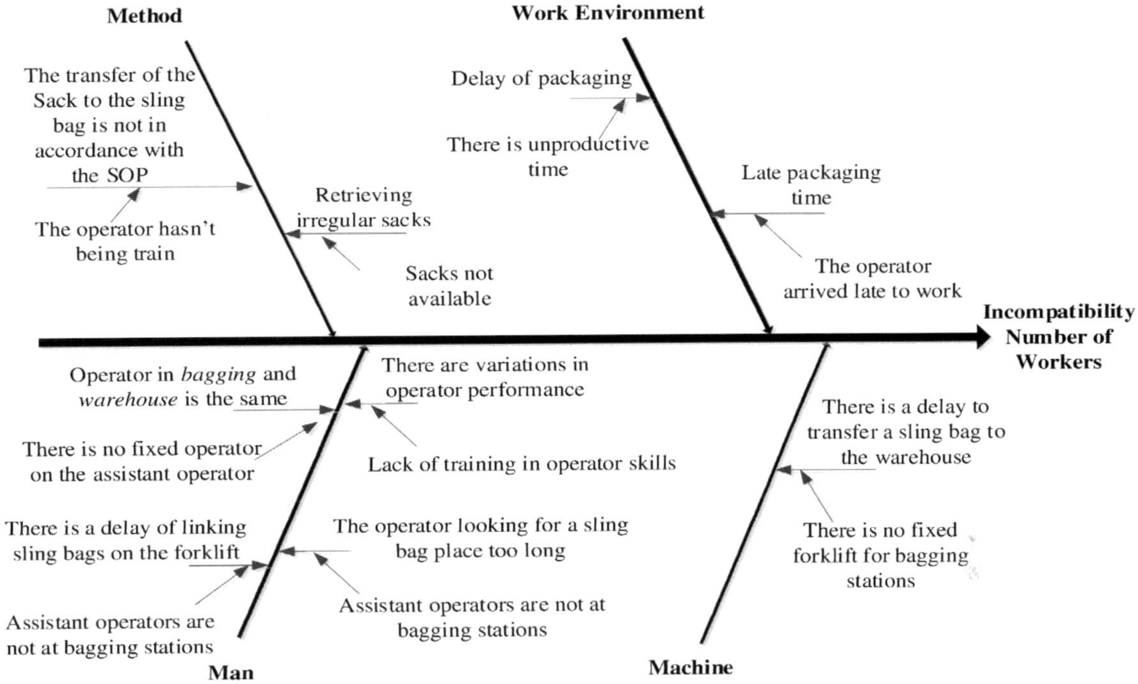

Figure 1. Cause and Effect Diagram

3.1 Workload Measurement With Work Load Analysis (WLA)

A good workload, preferably close to 100% or under normal conditions. The 100% workload means that for 8 hours of work the worker is able to work continuously in normal conditions. The following is the calculation of workload for each filling and recapitulation of workload for each workforce can be seen in Table 1.

Filling Operator Workload

Workload Day 1

$= (\%P \times (1 + \textit{Rating Factor})) \times (1 + \textit{Allowance})$

$= (0{,}75 \times (1 + 0{,}22)) \times (1 + 0{,}20)$

$= 1{,}0980 \approx 109{,}80\%$

Workload Day 2

$= (\%P \times (1 + \textit{Rating Factor})) \times (1 + \textit{Allowance})$

$= (0{,}8804 \times (1 + 0{,}22)) \times (1 + 0{,}20)$

$= 1{,}2310 \approx 123{,}10\%$

Workload Day 3

$= (\%P \times (1 + \textit{Rating Factor})) \times (1 + \textit{Allowance})$

$= (0{,}7826 \times (1 + 0{,}22) \times (1 + 0{,}20)$

$= 1{,}1457 \approx 114{,}57\%$

Table 1. Workload Each Operator

No	Operator	Workload (%)		
		Day 1	Day 2	Day 3
1.	Filing Operator	109,80	123,10	114,57
2.	Binding Operator	102,59	123,10	126,26
3.	Sewing Operator	107,77	187,81	124,09
4.	Transfer Operator	178,86	187,81	176,63

5.	Operator Assistant	89,25	133,06	126,57

The average workload calculation for filling operators is as follows

$$= \text{Workload Average } (\%WL_1)$$

$$= \frac{\text{Workload Day 1} + \text{Workload Day 2} + \text{Workload Day 3}}{\text{Amount of Days}}$$

$$= \frac{109{,}80 + 123{,}10 + 114{,}57}{3} = 115{,}82\%$$

Table 2 shows the average workload of operators in each activity.

Table 2. Average workload

No	Operator	Workload Average (%)
1.	Filing Operator	115,82
2.	Binding Operator	117,32
3.	Sewing Operator	139,89
4.	Transfer Operator	180,99
5.	Operator Assistant	116,29

Based on the average workload of each worker above, the overall productivity of operator activities can be calculated as follows:

Total Workload

$$= \%WL_1 + \%WL_2 + \%WL_3 + \%WL_4 + \%WL_5$$

$$= 115{,}82\% + 117{,}32\% + 139{,}89\% + 180{,}99\% + 116{,}29\%$$

$$= 670{,}31\%$$

Average Workload of Each Worker (Actual)

$$= \frac{\text{Amount of Workload}}{\text{Number of Workes}} = \frac{670{,}31}{5} = 134{,}06\%$$

3.2 Data Analysis and Discussion

The number of workers is one of the problems in Bagging due to the excessive amount of workforce will make the productivity of each worker decrease while the number of workers who are less will make the productivity of each worker After calculating the workload for each operator, the workload for filling operators is obtained. binding operators, sewing operators, transfer operators and auxiliary operators are 115.82%, 117.32%, 139.89, 180.99% and 116.29% respectively. The average workload of each worker is 134.06%, where the average workload exceeds the workload limit of 100% so that the number of workers is needed so that the workload of each worker can be reduced. The calculation of the number of workers recommended can be seen as follows:

Average Workload for Each Worker (Recommended)

$$= \frac{\text{Amount of Workload}}{\text{Number of Workes}} = \frac{670{,}31}{7} = 95{,}76\%$$

Based on the results of the above calculations, it can be concluded that the number of workers recommended at Bagging station is 7 workers with an average workload of each worker is 95.76%.

4 Conclusion

Based on the calculation of workload on bagging stations for charging operators, binding operators, sewing operators, transfer operators and auxiliary operators respectively 115.82%, 117.32%, 139.89%, 180.99% and 116.29% . The average workload of each operator is 134.06% (Exceeding the workload limit) and improvements are made to the work system at bagging stations by increasing the number of workers to 7 people so that the average workload of each operator can be minimized to 95.76% (Does not exceed the workload limit).

References

[1] Yudha A 2015 Pengaruh Beban Kerja Terhadap Kinerja Karyawan PT. Bank Rakyat Indonesia (Pesero) Tbk Cabang Krekot (The relation between workload anda employee performance in PT. Bank Rakyat Indonesia) *Jurnal Manajemen dan Organisasi*, Vol VI, pp 35-36. 2015.

[2] Malayu S P H. 2003. *Manajemen Sumber Daya Manusia*. Cet ke-5. Jakarta:Bumi Aksara.

[3] Mathis R L and Jackson J H. 2000. *Manajemen Sumber Daya Manusia Buku I.* Terjemahan oleh Jimmy Sadeli dan Bayu Prawira Hie. 2001. Jakarta : Salemba Empat.

[4] Yanti H 2014 Analisis Kebutuhan Jumlah Pegawai berdasarkan Analisis Beban Kerja (The analysis man power requirement based on existing workload) *Jurnal Online Institut Teknologi Nasional*, **1**, pp 251. 2014.

[5] Erlita C 2016 The Management of Local Goverment Apparatus Resource Based on Job and Workload Analysis*" Jurnal Ilmiah Administrasi Publik (JIAP),* **2**, pp 59-64.

[6] Sunyoto D 2012. *Manajemen Sumber Daya Manusia*. Yogyakarta: CAPS (Center for Academic Publishing Service).

[7] Rahmat RF, Herly E T, Siregar B, Syahputra M F and Sitompul O S 2018 RFID presence monitoring system as an input to measure the workload of employee *International Conference on Computer Applications and Information Processing,* pp 1-6.

[8] Munandar S A 2011. *Stress dan Keselamatan Kerja, Psikologi Industri dan Organisasi.* Jakarta: Universitas Indonesia

[9] Afia and Meylia. 2014.Kontribusi Beban Kerja, Disiplin Kerja, Hubungan Dengan Teman Sekerja Terhadap Produktivitas Kerja di PT. Viccon Modern Industry (The contribution of workload, work discipline, relation with other employee to work productivity in PT. Viccon Modern Industry) *Jurnal Ilmiah Fakultas Ekonomi,* Unesa Kampus Ketintang, Surabaya.

[10] Purwanto and Wahyuddin 2007. Pengaruh Faktor-faktor Kepuasan Kerja Terhadap Kinerja Karyawan Pusat Pendidikan Komputer Akuntansi Imka di Surakarta. *Jurnal Akuntansi dan Manajemen.*

[11] Fuanida, Afini. "*Pengaruh Pelatihan, Disiplin Kerja, dan Motivasi Terhadap Produktivitas Kerja Karyawan CV. Sapu Dunia Semarang"*. Jurnal Ilmu Administrasi Bisnis Jurusan Ilmu Administrasi Bisnis Semarang.

[12] Mangkuprawira S, Hubeis AV. 2007. *Manajemen Mutu Sumber Daya Manusia*. Jakarta (ID): Ghalia Indonesia.

[13] Ridwan A 2012 Analisa Beban kerja dan Jumlah Tenaga Kerja yang Optimal pada Bagian Produksi dengan Metode Workload Analysis (WLA) di. PT. Surabaya Perdana Rotopack *Universitas Pembangunan Nasional Veteran*.

[14] Raissa P N W 2012 Analisis Beban Kerja dengan Metode Workload Analysis Sebagai Pertimbangan Pemberian Insentif Pekerja (Studi Kasus di Bidang PPIP PT. Barata Indonesia (Persero) Gressik *Jurnal Rekayasa dan Manajemen Sistem Industri*, **2**

[15] Raras Mayang Arsi. 2012 "Analisis Beban Kerja untuk Menentukan Jumlah Optimal Karyawan dan Pemetaan Kompetensi Karyawan Berdasar Pada Job Description". *Jurnal Teknik ITS*, **1**

Acknowledgement

The author would like to thanks to the Universitas Sumatera Utara who has provided financial support to publish this paper. The authors express their gratitude to the leader of the fertilizer industry who are willing to be the object of research. We appreciate the efforts of all those who have cooperated in conducting this study.

IOP Conf. Series: Materials Science and Engineering **648** (2019) 012020 doi:10.1088/1757-899X/648/1/012020

Study of SMS gateway service application at the one stop agency of Medan city investment and service for licensing customer satisfaction

M A Nasution[1*] and S Y Regif[2]

[1]Department of Business Administration, Universitas Sumatera Utara, Medan, Indonesia
[2]Department of Pancasila and Citizenship Education, STKIP Budidaya, Binjai, Indonesia

*Email: m.arifin.nasution@usu.ac.id

Abstract. This research is about the Index of Public Satisfaction on the management of public services that are managed specifically for its users through the application of a service information system called the SMS Gateway. This application system provides information on licensing services in the Medan City of One Stop Agency Service and Investment Service in fulfilling requests from licensing service users without having to come to the DPMPTSP Medan office to get information on the progress of permits that have been entered in the licensing application. This research has an important objective, namely to provide a measurable analysis of the ability of state civil apparatus to manage access to be information technology progress for the benefit of the community, especially users of SMS Gateway, which is based on the assessment of community satisfaction and demands for the development of the industrial era 4.0. Based on information received from the DPMPTSP in Medan City, which was every day at the customer hall serves no less than 40 people who need licensing services within the workday span of 3 months, Monday - Friday totalling 40 x 60 days = 2400 populations with the number of samples 331 people.

1. Introduction

Entwistle states that effective public administration services will be able to save budget so that the state budget can be allocated to needs that are directly related to people's welfare[1]. In addition, the effectiveness of administrative services will accelerate the fulfillment of administrative needs to accelerate the economic growth of the community.

The Medan One-Stop Integrated Investment and Service Office (DPMPTSP) is a licensing and non-licensing service in the form of a joint office that provides several types of public services to the community, The functions of the regional device for investment affairs as stipulated in Domestic Government Regulations 100 of 2016 are investment planning, development of investment climate, investment promotion, investment licensing, investment control and implementation as well as processing of investment data and information.

Regulations of the Mayor of Medan Number 69 of 2017 are concerning the Duties and Functions of the Planting Service. Medan City's One Stop Integrated Capital and Services, it is stated that the Investment Office and PTSP are implementing elements of government affairs in the field of investment and integrated one-stop services led by the Head of the Office who is under and responsible to the Mayor through the Regional Secretary. The Office has the duty and obligation to assist the Mayor in implementing government affairs in the field of investment and one-stop integrated services.

To support performance in carrying out is its main tasks and functions, namely to improve public services to the applicants and related parties who need data and information from the DPMPTSP of the Medan City Government. One of the planned improvements in the service is the development of the mobile website DPMPTSP, namely the Preparation of Media Licensing Process Information in the form of SMS gateways at the Medan City Government's One Stop Investment Service, which are part of further services and are built to further improve service to applicants get the data and information they need. It is projected that this service is expected to improve service to the community as existing applicants and improve the system of government work that can run more effectively and efficiently in accordance with the principles of good governance and e-government. With the presentation of the communication media application in the form of an SMS gateway through the cellular telephone, it will produce a different model of interaction between the local government and the community intended to shorten the distance between the government and the community. It will naturally increase transparency and accountability.

Through the SMS gateway application system, the public as users of licensing services after filing in Medan City DPMPTSP, the community only needs to use short message service (SMS) via their cellular phone. Similarly is to monitor and to find out the requirements for processing permits, simply by SMS. Eventhough SMS is less conventional than Android application that also developed in Medan City for Tax measurement[2], we can say it is still useful enough. Through the application, if there is any notification submitted to the community, DPMPTSP simply uses SMS. Based on the growth of household access to cellphones issued by Balitbang Communication and Information Technology in 2016, 84.4% of 55.3 million households in both urban and rural areas have and use wireless mobile phones in other words that the use of bureaucratic services through technology access information is considered to be applicable because most people already understand its use. The results of this study will become one of the benchmarks for achieving and improving access to product services that will be carried out by Medan City DPMPTSP in improving the quality of services for people who use licensing services and at the same time adjust to the grand design of ASN 2020-2024 development.

2. Research Method

The research design used in this study was a combination of qualitative and quantitative, which was trying to provide an accurate description or explanation of the problems under study, interpreting and explaining data that exists systematically based on facts that appear or as they were. The combination method or also commonly called combined research method was a research approach that combines or connects between quantitative and qualitative research methods[3].

Combined research methods was useful if quantitative methods or qualitative methods alone were not accurate enough to be used to understand research problems, or by using quantitative and qualitative methods in combination will be able to get the best understanding compared to one method.

3. Data Acquisition

Retrieving data with observation instruments, interviews, field notes and document usage. The source of primary data was data obtained directly with the informant interview technique or direct source. Primary sources were data sources that directly provide data to data collectors[4]. The primary data source in this study was the respondents / service users (customers) who used integrated licensing services.

Population was a generalization area consisting of: objects / subjects that have certain qualities and characteristics[4]. The population in this study was the people who had obtained licensing services at the Medan Investment and Integrated Licensing Service Office. Because in this study using a questionnaire was as a data collection tool, this study also used research respondents. Based on information received from the Medan City DPMPTSP, which was every day at the customer hall serves no less than 40 people who need licensing services within the workday span of 3 months, Monday - Friday totaling 40 x 60 days = 2400 populations.

Table1. Table of Krejcie and Morgan

Population (N)	Sample (n)	Population (N)	Sample (n)	Population (N)	Sample (n)
55	48	320	175	2200	327
60	52	340	181	2400	331
65	56	360	186	2600	335

The sampling technique can be adjusted to the type of service, the purpose of the survey and the data to be obtained. Respondents were randomly selected according to the coverage area of each service unit. Whereas for sample and population magnitudes use a sample table from Krejcie and Morgan or calculated using the formula:

$$S = \{ \lambda^2 . N. P. Q\}/ \{d^2 (N-1) + \lambda^2 . P. Q \} \quad (1)$$

Where :

S = number of samples

λ2 = lambda (multiplier) with

dk = 1, the error rate can be 1%, 5%, 10%

N = population

P (normal population spread) = Q = 0.5 d = 0.0

The analysis technique was used in making the frequency distribution table set the classification of answers as much as 4 classes, with the consideration that the determination of the classification of answers was adjusted to the number of categories / classes used in making questionnaires based on a Likert scale, namely:

Table 2. Table of Likert scale

Value of Perception	Value Of Interval (NI)	Value of Conversion Interval(NIK)	Service Quality (X)	Performance of The Service Unit (Y)
1	1,00 – 2,5996	25,00 – 64,99	D	Not Good
2	2,60 – 3,064	65,00 – 76,60	C	Rather Good
3	3,0644 – 3,532	76,61 – 88,30	B	Good
4	3,5324 – 4,00	88,31 – 100,00	A	Very Good

With the kind of indicators through the Elements of the Community Satisfaction Survey in this regulation include:

1. Requirements
2. Systems, Mechanisms, and Procedures
3. Time of Settlement
4. Fees / Rates *)
5. Product Type of Service Specifications
6. Implementing Competencies **)
7. Implementing Behavior **)
8. Handling Complaints, Suggestions and Feedback
9. Facilities and infrastructure

4. Results and Discussion

Considering the basis of the evaluation of elements of Community Satisfaction as stipulated in ministerial regulations, a number of measures that were benchmarks and described in accordance with Minister of PAN-RB Regulation No. 14 of 2017 include[5]:

4.1. Requirements

Requirements were conditions that must be met in the management of a type of service, both technical and administrative requirements. From the results of the analysis obtained an average score of 3.20 is in the interval score of 3.0644 – 3.532 in the category "GOOD".

Table 3. Table of Requirements

No.	Answers	Score	Frequency		
			F	f.x	%
1.	Very Good	4	134	536	40
2.	Good	3	141	423	42
3.	Rather Good	2	43	86	13
4.	Not Good	1	13	13	5
	Total		331	1058	100

4.2. Systems, Mechanisms and Procedures

Procedures were procedures for standardized services for providers and recipients of services, including complaints. From the results of the analysis obtained an average score of 3.18 is in the interval score of 3.0644 – 3.532 in the category "GOOD".

Table 4. Systems, Mechanisms and Procedures

No.	Answer	Score	Frequency		
			F	f.x	%
1.	Very Good	4	132	528	40
2.	Good	3	136	408	41
3.	Rather Good	2	52	104	16
4.	Not Good	1	11	11	3
	Total		331	1051	100

4.3. Service Time

Service time was the period of time needed to complete the entire service process of each type of service. From the results of the analysis obtained an average score of 3.15 in the interval score of 3.0644 – 3.532 in the category "FAST".

Table 5. Service Time

NO	Answers	Score	Frequency		
			F	f.x	%
1.	Very Good	4	140	560	36
2.	Good	3	130	390	33
3.	Rather Good	2	33	66	16
4.	Not Good	1	28	28	15
	Total		331	1044	100

4.4. Fees / Rates

Fees / Tariffs are fees charged to people receiving SMS GATEWAY services in Medan City Investment and One Stop Services in managing and / or obtaining services from operators, the amount of which was based on an agreement between the organizer and the community with the delivery rates announced via GATEWAY SMS. From the results of the analysis obtained an average score of 3.27 in the interval score of 3.0644 – 3.532 in the category "GOOD".

Table 6. Fees / Rates

NO	Answers	Score	Frequency		
			F	f.x	%

1.	Very Good	4	142	568	41
2.	Good	3	147	441	44
3.	Rather Good	2	31	62	10
4.	Not Good	1	11	11	5
	Total		331	1082	100

4.5. Product Type of Service Specifications

The SMS Gateway service product was the result of each type of service specification. From the results of the analysis, the average score of 3.17 was in the interval score of 3.0644 – 3.532 in the category "GOOD".

Table 7. Product Type of Service Specifications

No.	Answers	Score	Frequency		
			F	f.x	%
1.	Very Good	4	135	540	41
2.	Good	3	141	423	43
3.	Rather Good	2	33	66	9
4.	Not Good	1	22	22	7
	Total		331	1051	100

4.6. Implementing Competencies

Implementing competence was the ability that must be possessed by the executor including knowledge, expertise, skills, and experience. From the results of the analysis was obtained an average score of 3.27 in the interval score of 3.0644 – 3.532 in the category "GOOD".

Table 8. Implementing Competencies

NO	Answers	Score	Frequency		
			F	f. x	%
1.	Very Good	4	152	608	46
2.	Good	3	136	408	41
3.	Rather Good	2	22	44	13
4.	Not Good	1	21	21	10
	Total		331	1081	100

4.7. Implementing Behavior

Implementing behavior was the attitude of officers in providing services. From the results of the analysis obtained an average score of 3.18 was in the interval score of 3.0644 – 3.532 in the category "GOOD".

Table 9. Implementing Behavior

No.	Answers	Score	Frequency		
			F	f. x	%
1.	Very Good	4	151	604	45
2.	Good	3	114	342	34
3.	Rather Good	2	43	86	13
4.	Not Good	1	23	23	8
	Total		331	1055	100

4.8. Handling Complaints, Suggestions and Feedback

Complaint handling, suggestions and input were procedures for implementing complaints handling and follow-up. From the results of the analysis, the average score was 3.22 in the interval score of 3.0644 – 3.532 in the category "GOOD".

Table 10. Handling Complaints, Suggestions and Feedback

NO	Answers	Score	Frequency		
			F	f.x	%
1.	Very Good	4	146	584	44
2.	Good	3	119	357	36
3.	Rather Good	2	61	122	18
4.	Not Good	1	5	5	2
	Total		331	1068	100

4.9. Facilities and infrastructure

Facilities and infrastructure were media that were used or become places and support for implementing services. From the results of the analysis obtained the value of the score is equal to 3.1 in the interval score of 3.0644 – 3.532 in the category "GOOD".

Table 11. Facilities and infrastructure

No.	Answers	Score	Frequency		
			F	f. x	%
1.	Very Good	4	102	408	31
2.	Good	3	165	495	50
3.	Rather Good	2	61	122	18
4.	Not Good	1	3	3	1
	Total		331	1028	100

Table 12. Table Of IKM Results

No.	Indicators	The Average of Score	Category
1.	Requirements	3.2	Good
2.	Systems, Mechanisms and Procedures	3.18	Good
3.	Time Services	3.15	Good
4.	Fees / Rates	3.27	Good
5.	Product Type of Service Specifications	3.17	Good
6.	Implementing Competencies	3.27	Good
7.	Implementing Behavior	3.18	Good
8.	Handling Complaints, Suggestions and Feedback	3.22	Good
9.	Facilities and Infrastructures	3.10	Good
	Average	3.19	Good

5. Conclusion

From the discussion of questionnaires to primary data totaling 331 questionnaires to see the results of the community satisfaction survey using the SMS GATEWAY service in the Investment and One-Stop Services and the results of data analysis conducted by the table it was known that the Community Satisfaction Index (SMS) GATEWAY at Investment and Integrated One-Stop Service about 77.85% were in the "GOOD" category (in the interval 76.61 to 88.30). The IKM results mentioned above consist of nine size indicators.

Service Quality at the use of the GATEWAY SMS Investment Office and the One-Stop Integrated Services in the City of Medan by the applicant shows a weight of 3.19 with the category of GOOD with Type B.

The Impact, the service of the SMS GATEWAY application provides an orientation to change the relationship of licensing services better to the community / customer so that in the future the ease and magnitude of investment in the city of Medan can provide meaningful development sensitive to the economic development of industrial revolution 4.0. It was responsive to technology and information.

6. Suggestion

Based on the results of the study, the suggestions that can be given are as follows:

1. The involvement of all relevant elements in the Regional Government and the Community will help the development of the SMS GATEWAY program that has targeted outlook and objectives in the process of adaptation of bureaucratic and community capabilities in the work culture of industrial revolution 4.0
2. Supervision and development of a work culture that uses technological advances at least cooperates with academics in the performance of government apparatus in every service that is based on service orientation as a consequence of responsibility.
3. Understanding the NPS (New Public Service) at each of each DPMPTSP Medan City employee with the Target:
    a. Encouraging community participation as users of SMS Gateway services in assessing the performance of service providers.
    b. Encouraging SMS Gateway service providers to improve the quality of public services.
    c. Encouraging SMS Gateway service providers to be more innovative in providing public services.
4. Maintaining the spirit of cooperation and compliance of the working group needs to be improved, so that the work discipline is maintained consistently creating effective and efficient services.

References

[1] Entwistle T and Martin S 2005 From competition to collaboration in public service delivery: A new agenda for research *Public Adm.*

[2] Rahmat R F, Purnamawati S, Saito S, Ichwan M F, Lubis T M 2019 Android-based automatic detection and measurement system of highway billboard for tax calculation in Indonesia *International Journal of Electrical Enginnering and Computer Science* **14** (2), pp. 877-886.

[3] Creswell J W 2014 *Research Design- Qualitative, Quantitative, and Mixed Approaches.*

[4] Sugiyono M 2014 Educational Research Methods Quantitative, Qualitative Approach and R&D *Bandung Alf.*

[5] KEMENPANRB 2017 Peraturan Menteri PEndayagunaan Aparatue Negara Dan Reformasi Birokrasi Republik Indonesia Nomor 14 Tahun 2017 *Pedoman Penyusunan Survei Kepuasan Masy. Unit Penyelenggara Pelayanan Publik*

IOP Conf. Series: Materials Science and Engineering **648** (2019) 012021 doi:10.1088/1757-899X/648/1/012021

Analysis of The Community Satisfaction Index on Medan Rumah Kita Public Service Application

M A Nasution[1*] and S Y Regif[2]

[1]Department of Business Administration, Universitas Sumatera Utara, Medan, Indonesia
[2]Department of Pancasila and Citizenship Education, STKIP Budidaya, Binjai, Indonesia

*Email: m.arifin.nasution@usu.ac.id

Abstract. This study is about the Public Satisfaction index of the management of public service applications managed by the Office of Communication and Information of the City of Medan using a Play Store-based system on smart cell phones or android. This application system provides information, facilities and service categories for the people of Medan City in submitting complaints to problems of public facilities. The management of this application is called Medan Rumah Kita or MRK. The main objective is to fulfil the NPS (New Public Service) as part of the implementation managed by the Medan City government with the main goal being community satisfaction in the city of Medan. The MRK (Medan Rumah Kita) application, which began in 2016, is expected to increase the contribution of the role of the community in improving the system of government work that is more effective and efficient in accordance with the principles of good governance and e-government. This study has an important objective, namely to provide a measurable analysis of the ability of state civil apparatus in managing the Medan Rumah Kita application for the public interest so that it becomes one of the stages and achievements that become a reference for improving access to public services in Medan City. The population in this study were people who had downloaded and used Medan Rumah Kita (MRK) application. This study is using a questionnaire as a data collection tool, this study also uses research respondents which is gathered from the Play Store application that consist of 1500 people who have downloaded MRK application, with the total sample is 306 people. The result shows that 75.95% of the respondents are in the "GOOD LESS" category (at intervals of 2.60 - 3.064 or 65.00 - 76.60).

1. Introduction

The Medan City Government has launched the MRK (Medan Rumah Kita), an application that allows citizens to report their complaints to government officials by cooperating with the private sector and operationalized by the Medan City Communication and Information Service which began in 2016. This application can be downloaded on the android play store. Medan Rumah Kita application or abbreviated as MRK is a social media application to report city problems to local governments, the private sector or share information among citizens in the surrounding area for the creation of smart cities. Reports of citizens entering through the MRK application will be forwarded to the relevant parties and their progress can still be monitored, by the Medan City Community who feel concerned about the surrounding environment by reporting the problem through this digital platform, which will be forwarded to the authorities. The commitment of the Medan City Government to create Medan even better through MRK, has shown that the Pemko Medan has followed the trend of e-government. E-government is used as a form of local government innovation in improving public services[1]. This is certainly in line with the demands of industry 4.0. sensitive to the advancement of technology and information[2] especially in City based industry 4.0 while there are several technology advancement that has been research and try to be implies in Medan city, such as Android Tax Calculation [3] and Medan Virtual Asset Augmented Reality [4].

Based on the initial observations that have been made on the MRK application, there are several problems encountered in the service department that question the accuracy, activeness of the application and follow-up of the authorized apparatus. This will also contradict the grand development design of ASN 2020 - 2024 which contains the direction of government policy on all aspects of ASN management as mandated in Law Number 5 of 2014 concerning ASN with one of its conditions being professional / competent in its field specifically mentioned in article 1 paragraph 13 of Government Regulation No.11 of 2017 namely Technical Competence is knowledge, skills, and attitudes / behaviors that can be observed, measured, and developed that are specifically related to the technical field of position.

There are still weaknesses which in general from the follow up of the implementation of the MRK application which shows that the government apparatus service has not been effective. Moreover, the emergence of uninhabitable city index reports in early 2018 ago with the results of a survey showing cities that were considered unfit for habitation by their own citizens. These cities were Pontianak (62.0 percent), Depok (61.8 percent), Mataram (61.6 percent), Tangerang (61.1 percent), Banda Aceh (60.9 percent), Pekanbaru (57, 8 percent), Samarinda (56.9 percent), Bandar Lampung (56.4 percent), Medan (56.2 percent), and Makassar (55.7 percent). Pontianak and Medan consistently rank lower. The five lowest aspects that most people feel are lacking are the availability of transportation, safety, dirty water and drainage management, pedestrian facilities, and information on development and community participation. The Association of Planning Experts (IAP) through the IDN Times revealed this in the title of the news, These are 7 of the most livable and non-habitable cities in Indonesia.

Based on the above, a study on IKM (Community Satisfaction Index) was carried out in using the MRK (Medan Rumah Kita) application, the results of which could be a reference for the Medan City Communication and Information Office in particular and the Medan City Government in general to foster, maintain and manage government in the area.

2. Research Methods

This research was carried out in the city of Medan with the scope of product applications, access and service facilities of the Medan City Communication and Information Service namely the MRK (Medan Rumah Kita) application available on the Play Store application. Access that is managed by the agency has a role as the main supporting facility between users of information services up to date in response to community complaints in accordance with poor conditions in various environments in the jurisdiction of Medan City.

The research design used in this study is a combination of qualitative and quantitative, which is trying to provide an accurate description or explanation of the problems under study, interpreting and explaining data that exists systematically based on facts that appear or as they are. Combination method or also commonly called combined research method is a research approach that combines or connects between quantitative and qualitative research methods[5].

Combined research methods will be useful if quantitative methods or qualitative methods alone are not accurate enough to be used to understand research problems, or by using quantitative and qualitative methods in combination will be able to get the best understanding compared to one method.

3. Information and Data Collection

Retrieving data with observation instruments, interviews, field notes and document usage. The source of primary data is data obtained directly with the Informationt interview technique or direct source. Primary sources are data sources that directly provide data to data collectors[4]. The primary data source in this study is the respondents / service users (customers) who using the MRK (Medan Rumah Kita) application.

Population is a generalization area consisting of: objects / subjects that have certain qualities and characteristics[4]. The population in this study were people who had used the application of the MRK

(Medan Rumah Kita) service at the Medan City Communication and Information Service. Because in this study using a questionnaire as a data collection tool, this study also uses research respondents. Based on data and information taken from the Medan Rumah Kita (MRK) play store application, there are 1500 populations that download the application. in other words the number of samples is 306 people selected and spread on 21 Sub-District area of Medan city with 14 – 15 samples of each Sub-District has been chosen to download the application and use it in the span of the trial period and then the response is taken without being influenced by the researcher in making a decision of they answer, adjusted for the following table:

Table 1. The Tabel of Krejcie and Morgan

Population (N)	Sampel (n)	Population (N)	Sampel (n)	population (N)	Sampel (n)
20	19	240	148	1400	302
25	24	250	152	1500	306
30	28	260	155	1600	310

With sampling techniques can be adjusted to the type of service, the purpose of the survey and the data you want to obtain. Respondents were randomly selected according to the coverage area of each service unit. Whereas for sample and population magnitudes use a sample table from Krejcie and Morgan or calculated using the formula:

$$S = \{ \lambda^2 . N. P. Q\} / \{d^2 (N-1) + \lambda^2 . P. Q \}$$

Where are:

S = number of samples
λ2 = lambda (multiplication factor) with
dk = 1, level of error can be 1 %, 5 %, 10 %
N = population
P (population spreads normally) = Q = 0,5
d= 0,0

Sampling is based on consideration of the limitations of observing all samples, for time and cost efficiency to generate generalizations on the population and reduce research errors in sampling. In order for the samples taken in this study to represent the population, it can be determined the number of samples calculated using the Morgan and Krejcie formulas which are regulated in Minister of Administrative Reform and Bureaucratic Reform of the Republic of Indonesia Number 14 of 2017 concerning Guidelines for Preparing the Community Satisfaction Survey Public Service Organizing Unit[6].

The analysis technique used is in making the frequency distribution table set the classification of answers as much as 4 classes, with the consideration that the determination of the classification of answers is adjusted to the number of categories / classes used in making questionnaires based on a Likert scale, namely:

Table 2. Likert Scale

PERCEPTION VALUE	VALUE INTERVAL (NI)	CONVERSION INTERVAL VALUE (NIK)	QUALITY OF SERVICE (x)	PERFORMANCE OF THE SERVICE UNIT (y)
1	1,00 – 2,5996	25,00 – 64,99	D	Not Good
2	2,60 – 3,064	65,00 – 76,60	C	Less Well
3	3,0644 – 3,532	76,61 – 88,30	B	Good
4	3,5324 – 4,00	88,31 – 100,00	A	Very Well

With size indicators through the Elements of the Community Satisfaction Survey in this regulation include:

1. Requirements

2. Systems, Mechanisms and Procedures
3. Completion Time
4. Fees / Rates *)
5. Product Specifications Type of Service
6. Implementing Competence **)
7. Implementing Behavior **)
8. Handling Complaints, Suggestions and Feedback
9. Facilities and infrastructure

4. Results and Discussion

Considering the basis of the evaluation of elements of Community Satisfaction as stipulated in ministerial regulations, a number of measures that are benchmarks and described in accordance with Minister of PAN-RB Regulation No. 14 of 2017 include:

4.1. Requirements

Requirements are conditions that must be met in the management of a type of service, both technical and administrative requirements. From the results of the analysis obtained an average score of 3.20 is in the interval score of 3,0644 - 3,532 in the category "GOOD".

Table 3. Requirements

No	The Answer	Score	Frequency		
			F	f.x	%
1.	Very well	4	92	368	30
2.	good	3	185	555	60
3.	Less well	2	27	54	9
4.	Not good	1	2	2	1
	Total		306	980	100

4.2. Systems, Mechanisms and Procedures

Procedure is a standardized service procedure for providers and recipients of services, including notification and delivery of events. From the results of the analysis obtained an average score of 2.46 is in the interval score 1.00 - 2.5996 category "not good".

Table 4. Systems, Mechanisms and Procedures

NO	The Answer	Score	Frequency		
			F	f.x	%
1.	Very well	4	30	120	9
2.	Good	3	105	315	35
3.	Less well	2	147	294	48
4.	Not good	1	24	24	8
	Total		306	743	100

4.3. Service Time

Service time is the period of time needed to complete the entire service process of each type of service. From the results of the analysis obtained an average score of 2.94 in the interval score of 2.60 - 3.064 in the category "GOOD LESS."

Table 5. Service Time

NO	The Answer	Sore	Frequency		
			F	f.x	%
1.	Very well	4	115	460	36
2.	good	3	100	300	33

3.	Less well	2	50	100	16
4.	Not good	1	41	41	15
	Total		306	964	100

4.4. Fees / Rates

Fees / rates are fees charged to the people who use the MRK (Medan Rumah Kita) application by the Medan City Communication and Information Service in downloading and using further fees are not charged anything. From the results of the analysis obtained an average score of 3.20 is in the interval score of 3,0644 - 3,532 in the category "GOOD".

Table 6. Fees / Rates

NO	The Answer	Score	Frequency		
			F	f.x	%
1.	Very well	4	92	368	30
2.	Good	3	185	555	60
3.	Less well	2	27	54	9
4.	Not good	1	2	2	1
	Total		306	980	100

4.5. Product Specifications Type of Service

Product specifications of the type of service are the categories of services presented and in accordance with the provisions set by the Medan Rumah Kita service by the Medan City Communication and Information Agency. MRK service products are the result of each type of service specification. From the results of the analysis it was obtained that the average score of 3.15 was in the interval score of 3,0644 - 3,532 in the category "GOOD".

Table 7. Product Specifications Type of Service

NO	The Answer	Score	Frequency		
			F	f.x	%
1.	Very well	4	74	296	23
2.	Good	3	196	588	63
3.	Less well	2	39	78	13
4.	Not good	1	3	3	1
	Total		306	965	100

4.6. Implementing Competencies

Implementing competence is the ability that must be possessed by the executor including knowledge, expertise, skills, and experience. From the results of the analysis obtained an average score of 2.96 is in the interval score of 2.60 - 3.064 in the category "GOOD LESS".

Table 8. Implementing Competencies

NO	The Answer	Score	Frequency		
			F	f.x	%
1.	Very well	4	98	392	32
2.	Good	3	123	369	40
3.	Less well	2	62	124	20
4.	Not good	1	23	23	8
	Total		306	908	100

4.7. Implementing Behavior

Implementing behavior is the attitude of officers in providing services. From the results of the analysis obtained an average score of 3.14 is in the interval score of 3,0644 - 3,532 in the category "GOOD".

Table 9. Implementing Behavior

NO	The Answer	Score	Frequency		
			F	f.x	%
1.	Very well	4	141	564	45
2.	Good	3	104	312	34
3.	Less well	2	25	50	8
4.	Not good	1	35	35	13
	Total		306	961	100

4.8. Handling Complaints, Suggestions and Feedback

Complaint handling, suggestions and input are procedures for implementing complaints handling and follow-up. From the results of the analysis, the average score of 2.46 is in the interval score of 1.00 - 2.5996 in the category "Not Good".

Table 10. Handling Complaints, Suggestions and Feedback

NO	The Answer	Score	Frequency		
			F	f.x	%
1.	Very well	4	40	160	13
2.	Good	3	60	240	20
3.	Less well	2	147	292	48
4.	Not good	1	59	59	19
	Total		306	751	100

4.9. Facilities and infrastructure

Facilities and infrastructure are media that are used or become places and support for implementing services. From the results of the analysis obtained an average score of 3.13 in the interval score of 3,0644 - 3,532 in the category "GOOD".

Table 11. Facilities and infrastructure

NO	The Answer	Score	Frequency		
			F	f.x	%
1.	Very well	4	102	408	31
2.	Good	3	142	426	50
3.	Less well	2	60	120	18
4.	Not good	1	3	3	1
	Total		306	957	100

5. Conclusion

From the discussion of the questionnaire which became the primary data totaling 306 questionnaires to see the results of the community satisfaction survey using the MEDAN RUMAH KITA Application (MRK) in the Medan City Communication and Information Service and the results of data analysis that the table had done it was known that the Community Satisfaction Index (IKM) on MRK applications in the Medan City Communication and Information Agency, 75.95% is in the "GOOD LESS" category (at intervals of 2.60 - 3.064 or 65.00 - 76.60). The IKM results mentioned above consist of nine size indicators.

Table 12. Table Of IKM Result

No.	Element	Average Score	Category

1.	Requirements	3,20	Good
2.	Systems, Mechanisms and Procedures	2,46	Not good
3.	Service Time	2,94	Less well
4.	fees/rates	3,20	Good
5.	Product Type Specifications Service	3,15	Good
6.	Implementing Competencies	2,96	Less well
7.	Implementing Behavior	3,14	Good
8.	Handling Complaints, Suggestions and Feedback	2,46	Not good
9.	Facilities and infrastructure	3,14	Good
	Average	2,96	LESS WELL

Complaints System Service Quality Through the Medan Rumah Kita (MRK) Application at the Medan City Communication and Information Service by users of the 2.96 MRK application service with the GOOD LESS category with Type C. It is said with Type C based on the accumulated choice of MRK application users (Medan Rumah We) feel unsatisfied with the performance and response of the SKPD related to public complaints regarding the service categories available in the MRK application, so that they give an unfavorable appreciation. some other things still need to be improved in the delivery of services through the Medan Rumah Kita Application (MRK), especially the responsiveness and competency of the relevant apparatus, because the applications that have been rolled out are the responsibility of managers and implementers in the field.
The Service Quality of Medan Rumah Kita Application (MRK) can be concluded as follows:

1. The quality of the Medan Rumah Kita Application (MRK) there are several things that need to be considered by the MRK service program implementers after complaints by application users that should be responded to quickly because complaints are made through technology applications that are not limited by time so that the apparatus can move according to the situation and conditions that have been conveyed.
2. Talking about the Accuracy of Service Goals through the application (Medan Rumah Kita) MRK, some indicators must be continuously improved.
3. Competence of each implementer related to responding to complaints through service is considered to be a source of low handling - so services do not reach prime, while programs have been rolled out to resolve each category of problems.
4. Furthermore in entering the industrial era 4.0. the absence of a leading sector-based application service in the city of Medan whose integrity includes all the interests of the community.

6. Suggestion

Based on the explanation above, the implementation of application policies still needs to be improved because the implementation has not been maximized This is because, some people in Medan City consider that the MRK application policy still lacks in terms of the system, which is difficult to carry out the exact location of the problem, the socialization for the MRK application is still minimal so that not all citizens know how to operate the MRK application which in the end the community is not using the MRK application properly.

References

[1] Amin M, Muchtar M A, Siregar I and Nasution T H 2018 Utilization of Cloud Computing Service for Smart City Development of Medan City *MATEC Web Conf.*
[2] Sung T K 2018 Industry 4.0: A Korea perspective *Technol. Forecast. Soc. Change*
[3] Rahmat R F, Purnamawati S, Saito S, Ichwan M F, Lubis T M 2019 Android-based automatic detection and measurement system of highway billboard for tax calculation in Indonesia *International Journal of Electrical Enginnering and Computer Science* **14** (2), pp. 877-886.
[4] Muchtar M A, Syahputra M F, Syahputra N, Ashrafia S, Rahmat R F 2017 Augmented Reality for Searching Potential Assets in Medan using GPS based Tracking *Journal of Physics: Conference Series* **801**(1),012010.

[5] Sugiyono M 2014 Educational Research Methods Quantitative, Qualitative Approach and R&D *Bandung Alf.*
[6] KEMENPANRB 2017 Peraturan Menteri PEndayagunaan Aparatue Negara Dan Reformasi Birokrasi Republik Indonesia Nomor 14 Tahun 2017 *Pedoman Penyusunan Survei Kepuasan Masy. Unit Penyelenggara Pelayanan Publik*

Analysis of Post-Combustion Concrete Study with The Addition of Superplasticizer

Rahmi Karolina[1*], **Deni Malik**[1]

[1]Departmet of Civil Engineering, Universitas Sumatera Utara, Medan, Indonesia

*Email : rahmi.karolina@usu.ac.id

Abstract. This research explains that the use of superplasticizer admixture as an additional material in concrete mixes is expected to minimize concrete damage due to burning at high temperatures by increasing the quality of concrete in its mechanical properties which is compressive strength and physical properties which is concrete porosity. The variations in the addition of superplasticizer admixture used are 1%; 1.5%; and 2%, while variations in concrete combustion are 200°C; 500°C; and 800°C.The test results show a decrease in compressive strength and increase in concrete porosity due to combustion at high temperatures. The tendency of concrete compressive strength will decrease along with the increasing temperature. At the temperature of 200°C, a decrease of 7.14%; 40.57%; and 65.71% occurs, at the 500 °C, decreased by 2.74%; 15.07%; and 65.75%, while at 800°C decreased by 13.79%; 14.47%; and 53.95% of the initial strength of concrete without combustion. The porosity of concrete will rise along with the increasing temperature and the percentage increase will be greater along with the combustion temperature. At the temperature of 200°C there is an increase in porosity of 7.41% - 11,853%, at 500°C increased by 13,334% - 17,779%, and at 800°C increased by 19.26% - 23.704%, from normal concrete porosity.

1. Introduction

Fires can cause damage to building structures, including concrete structures. When burning, concrete cannot produce fire but can absorb heat so that there will be excessive high temperatures which will result in changes in the concrete microstructure. At a certain temperature limit, heating will cause the stability of cement gel bonds in the concrete to be lost, expansion of gravel (aggregate), the release of cement bonds and expansion in granules, this will cause a decrease in the stability of the concrete unit itself so that the concrete strength drops. Changes or damage caused by fire are influenced by the height of the temperature, duration of combustion, type of materials forming the concrete mixture, and loading behavior.

The main problem faced in handling buildings after a fire is how to assess the strength of the rest of the building after a fire. Knowing the strength of the rest, we can make the most efficient remedial action to restore its original condition so that buildings that have experienced fires can be re-functioned. To date, buildings that have undergone fire are immediately demolished, even though there is a possibility that the burnt structural elements still have strength.

Based on the description above , it is necessary to do research to find out the relation between the increase in temperature height, combustion time, and type of concrete-forming materials against the changes in the properties and characteristics of concrete due to fire on building structural elements so the residual strength can be identified from concrete material. Thus, it can be determined whether or not the structure of the building will be used again.

2. Literature review

Concrete is a composite material (mixture) of several rock materials which are bonded by binding material. Concrete is formed from a mixture of aggregates (coarse and fine), cement, water with a certain ratio and can be added with certain mixtures if deemed necessary. Water and cement material are put together to form a cement paste which functions as a binding material while fine aggregate and coarse aggregate are as fillers. The strength, durability, and other properties of concrete depend on the nature of the basic ingredients, the comparison value of the ingredients, the method of stirring and how to work during pouring the concrete, the method of compaction, and the method of treatment during the hardening process (Gideon, 1993).

According to Tjokrodimuljo (2000) basically concrete is not expected to be able to withstand heat above 250°C. Concrete that is heated to more than 800°C will experience degradation in the form of a significant reduction in strength which may not return (recovery) after the cooling process. The high loss of strength and whether or not the strength of the material returns to its original state is determined by the type of materials used, the severity of the fire process, and the length of combustion. The high level of severity (temperature) and the length of combustion time causes a reduction in the compressive strength of concrete material especially the emergence of internal shear stress as a result of differences in thermal properties between cement and aggregate.

2.1. Effect of high temperatures on concrete

The increase in temperature due to fire causes the concrete material to change its properties. The temperature that can be achieved in a burning building room is ± 1000°C with a fire duration generally more than 1 hour. Most structural concrete can be classified into three types of aggregates, which are carbonate, silicate, and light-weight concrete. Carbonate aggregates include limestone and dolomite and are included in one group because these two substances undergo changes in chemical composition at temperatures between 700°C to 980°C. Silicate aggregates which include granite, quartzite and sandstone do not experience chemical changes at temperatures commonly encountered in fires (Norman Ray, 2009).

A phenomenon that can be seen in concrete exposed to extreme heat (fire) is the occurrence of sloughing off (peeling), hair cracks and wide cracks, and the color of concrete. From visual observations, the temperature experienced by concrete can be estimated . (Nugraha, P., 2007) is set 17 point Times Bold, flush left, unjustified. The first letter of the title should be capitalized with the rest in lower case. It should not be indented. Leave 28 mm of space above the title and 10 mm after the title.

3. Methodology

The method used in this research is an experimental study which generally includes:

1. Provision of concrete constituent materials.
2. Examination of materials.
3. *Mix Design*.
4. Making of samples.
5. Examination of slump value.
6. Concrete compressive strength test at 28 days curing.
7. Concrete porosity test at 28 days curing.
8. Concrete combustion test at 28 days curing.

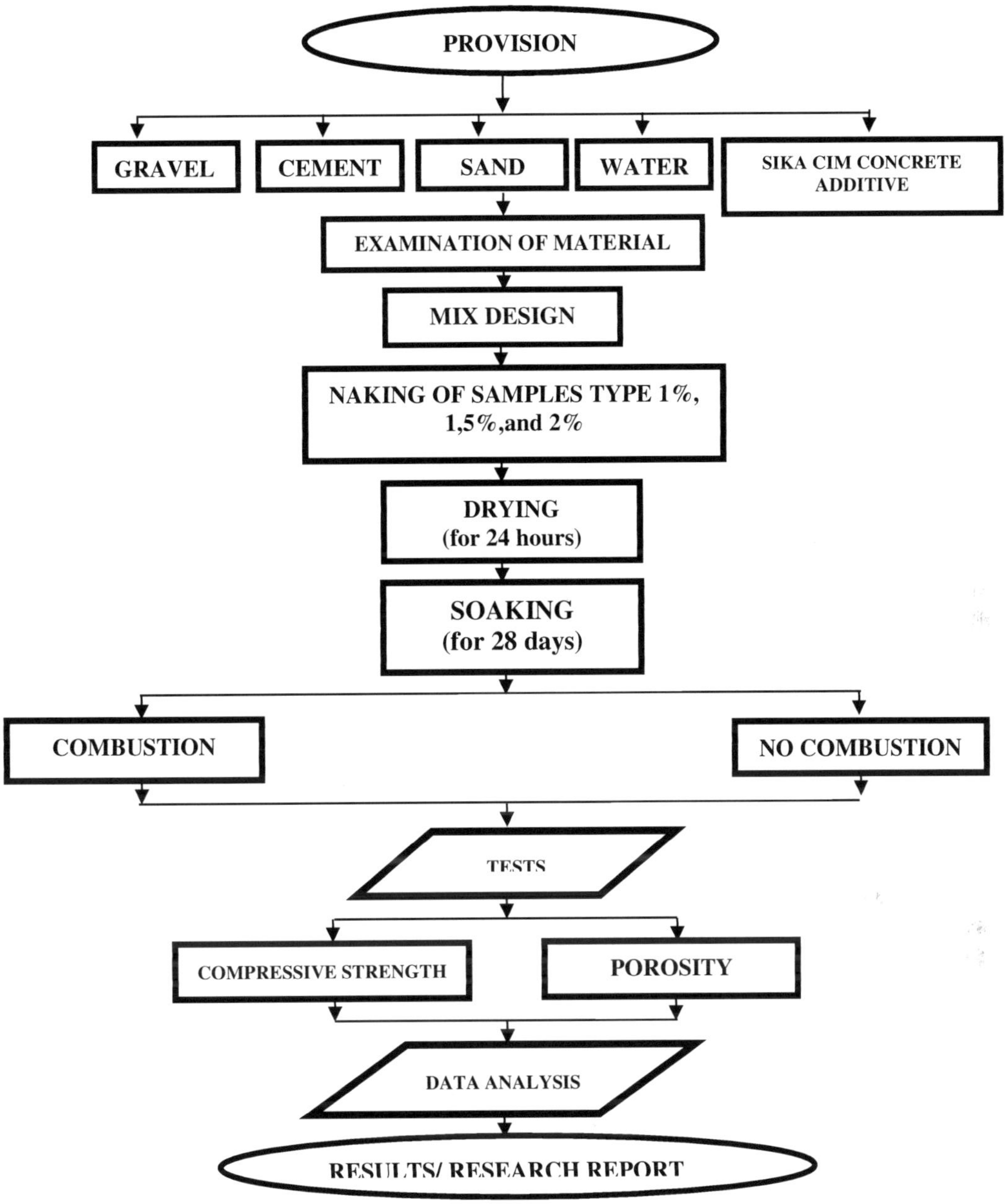

Figure 1. Research flowchart

3.1. Making of samples

The making of samples consist of four mixture variations for the trial, which is the normal mixture without substitute materials and the mixture with the addition of superplasticizer admixture by 1%; 1.5%; and 2% of cement use.

3.2. Concrete mixture planning

Concrete mixture planning is intended to determine the composition or proportion of concrete constituent materials. The proportion of the constituent materials of this concrete is determined through mix design. This is done so that the proportion of the mixture can meet the technical requirements economically. In determining the proportion of mixtures in this research, the method of the Department of Public Works is used which is based on SK SNI T-15-1990-03.

The basic criteria for concrete design by using the Department of Public Works method is compressive strength and the relation with cement water factor. The complete mix design calculation can be seen in the attachment. From the results of the mix design calculation, the comparison of the concrete mixture between cement: sand: gravel: water = 1.00: 1.85: 2.83: 0.53.

3.3. Concrete combustion process
In this research, the combustion test of the concrete mixture is carried out with the addition of 1% ; 1.5%; and 2% superplasticizer admixture compared to normal concrete. The combustion test, using a furnace machine, is carried out at the Mechanical Engineering Growth Centre Kopertis Laboratory of Region I Aceh-North Sumatra.

4. Results and discussion

4.1. Concrete compressive strength test
Concrete compressive strength test is done by using a Compressor Mechine. The compressive strength of concrete can be obtained using a formula. Data from the test results of the concrete compressive strength with the addition of superplasticizer admixture for each combustion temperature are as follows:

Table 1. Concrete with the addition of superplasticizer admixture without combustion

Sample	Compressive Load (ton)	Compressive Strength (kg/cm^2)
1% admixture concrete	70	311,111
1,5% admixture concrete	73	324,444
2% admixture concrete	87	386,667

Table 1 shows the compressive load data obtained from the test results for concrete with the addition of 1% ; 1.5%; and 2% superplasticizer admixture without combustion respectively, which are 70 tons, 73 tons and 87 tons. The three data show that the addition of superplasticizer admixture to the concrete mixture could increase the concrete compressive load.

Figure 2 shows the compressive load and compressive strength data for concrete with the addition of 1% superplasticizer admixture with combustion at temperatures of 200°C, 500°C, and 800°C, indicating that the increased maximum compressive load temperature that can be borne by concrete decreases its strength. This means that the concrete is getting weaker and the compressive strength of concrete has decreased by 7.14%, 40.57%; and 65.71% of the initial strength of concrete.

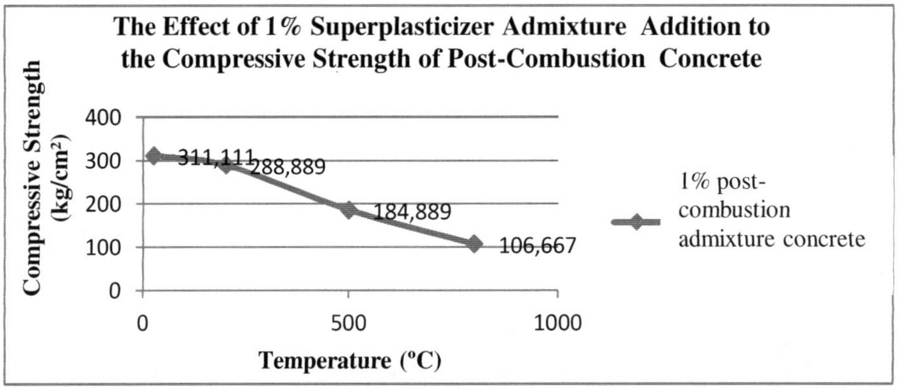

Figure 2. Graph of 1% post-combustion admixture concrete compressive strength to the temperature

Figure 3 shows the compressive load and compressive strength data for concrete with the addition of 1,5% superplasticizer admixture with combustion at temperatures of 200°C, 500°C, and 800°C, indicating that the increased maximum compressive load temperature that can be borne by concrete decreases its strength . This means that the concrete is getting weaker and the compressive strength of concrete has decreased by 2.74%; 15.07%; and 65.75% of the initial strength of concrete.

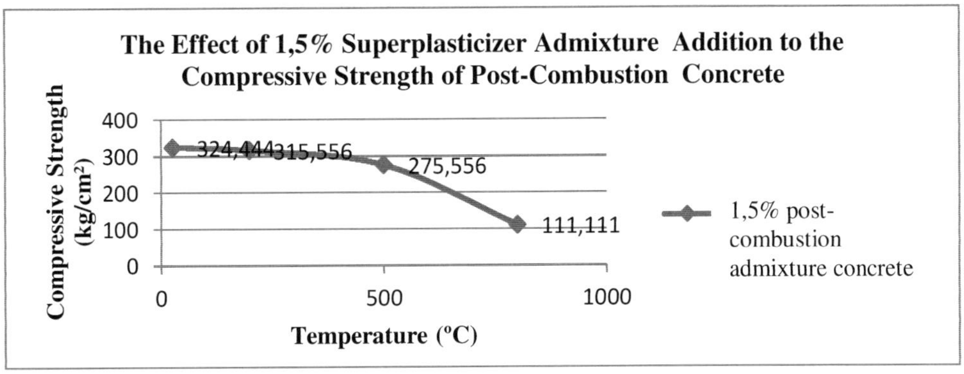

Figure 3. Graph of 1,5% post-combustion admixture concrete compressive strength to the temperature

Figure 4 shows the compressive load and compressive strength data for concrete with the addition of 2% superplasticizer admixture with combustion at temperatures of 200°C, 500°C, and 800°C, indicating that the increased maximum compressive load temperature that can be borne by concrete decreases its strength. This means that the concrete is getting weaker and the concrete compressive strength has decreased by 13.79%; 14.47%; and 53.95% of the initial strength of concrete.

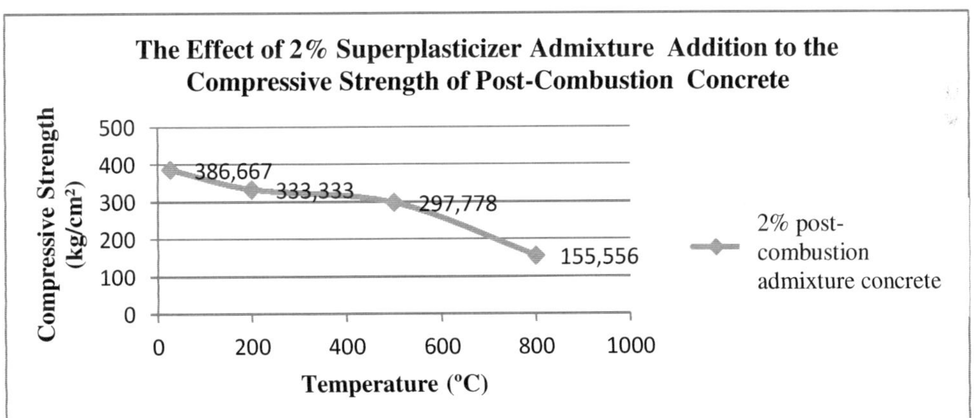

Figure 4. Graph of 2% post-combustion admixture concrete compressive strength to the temperature

Table 2. Data summary of post-combustion concrete compressive strength with addition superplasticizer admixture

Temperature (°C)	Compressive Strength (kg/cm²)		
	1% admixture concrete	1,5% admixture concrete	2% admixture concrete
27	311,111	324,444	386,667
200	288,889	315,556	333,333
500	184,889	275,556	297,778
800	106,667	111,111	155,556

Table 2 is a summary of the data between the increase in temperature and the compressive strength of concrete with each addition of superplasticizer admixture by 1%; 1.5%; and 2%. The table shows the tendency that the compressive strength of concrete will decrease along with the increasing temperature. At a temperature of 200 ° C decreased by of 7.14%; 40.57%; and 65.71%, at 500 ° C decreased by 2.74%; 15.07%; and 65.75%, at 800 ° C decreased by 13.79%; 14.47%; and 53.95% of the initial strength of concrete without combustion.

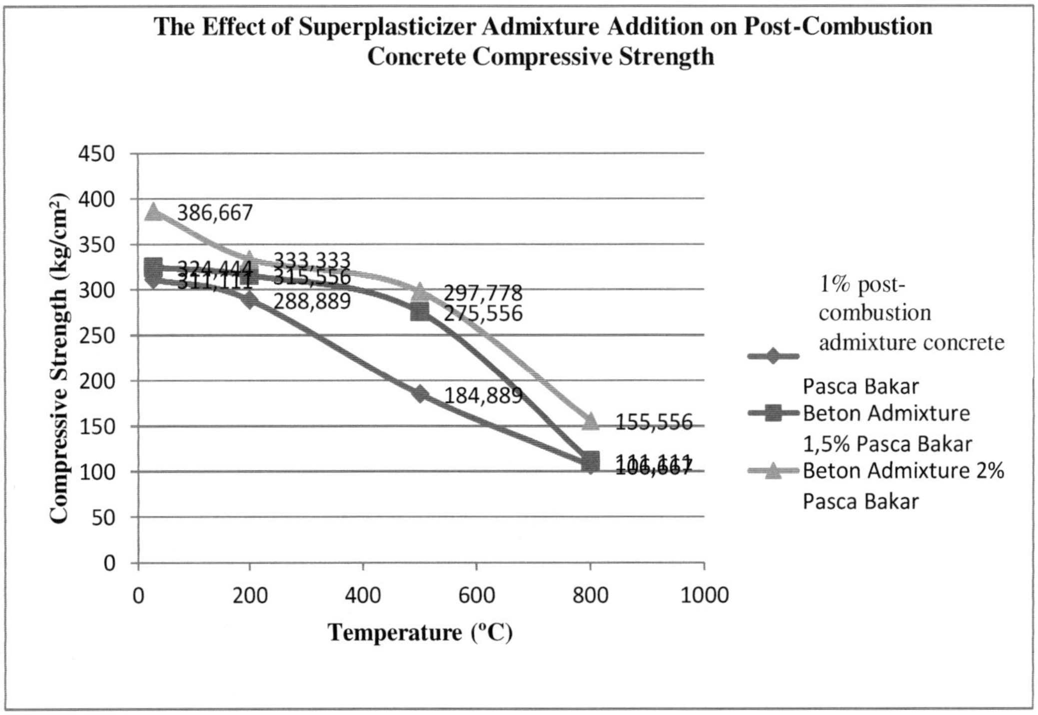

Figure 5. Graph of post-combustion admixture concrete compressive strength to the temperature

From Figure 5, it can be seen that the higher the temperature, the lower the compressive strength of the concrete. When the temperature is at 500°C with the addition of 1% superplasticizer admixture, the resulting compressive strength is 184.889 kg / cm^2, for the addition of 1.5% superplasticizer admixture, the compressive strength of the concrete is 275.556 kg / cm^2, and the addition of 2% superplasticizer admixture, the resulting compressive strength is 231.111 kg / cm^2. This result shows that the more concrete added with superplasticizer admixture , it will have a relatively high compressive strength. This happens because the concrete added by superplasticizer admixture has enough water to resist combustion heat even though water use is reduced by 10% -15% from the use of water in the making of normal concrete.

Water absorbed in the aggregate begins to evaporate, evaporation causes shrinkage of the cement paste. When the temperature is at 500°C to 800°C, a significant decreased of the compressive strength occurs, when the temperature of 800 ° C for concrete with the addition of 1% superplasticizer admixture, the resulting compressive strength is 106.667 kg / cm^2, for the addition of 1.5% superplasticizer admixture , the compressive strength is 111.111 kg / cm^2, and for the addition of 2% superplasticizer admixture, the compressive strength is 155.556 kg / cm^2.

This decrease in compressive strength is caused by the cement paste which has been hydrated, decomposes Ca (OH) $_2 \rightarrow CaO + H_2O$. CaO (lime) which is hygroscopic (absorbs water), while H_2O starts to evaporate at 100°C, causing dry and brittle concrete (Febrina, F., 2010).

In this case, it can be seen that the higher the combustion temperature, the lower the compressive strength of the concrete. When compared, the compressive strength value of the addition of superplasticizer admixture as much as 1%; 1.5%; and 2%, the resulting compressive strength of concrete decreases. For temperatures of 200 ° C, the decrease can reach 2.74% - 13.79%, for 500°C, it reaches 14.47% - 40.57%, for 800°C, it reaches 53.95% - 65.75% , where at 800°C with an additional admixture of at least 1% , it appears that the physical surface of the concrete has been peeled off and the inner structure is very fragile.

Lime from the combustion when added to water will expand and crack. Cement and water function as an adhesive and reinforce concrete. During the hydration process, the two most important components of the cement granules, C_2S and C_3S, react with H_20 and produce CSH and $(CaOH)_2$. C-S-H functions as a determinant substance of concrete hardness and aggregate binder. The hydration process is a process where the chemical composition of CaO cement is abbreviated C, SiO_2 is abbreviated S, Al_2O_3 is increased A, Fe_2O_3 is abbreviated F, reacts with water H2O abbreviated (H).

The effect of this hydration process is then to make micro and nano-sized crystals called gels and Ca (OH) 2 which will continue to grow to fill the crystal cavities where the cavities contain water and grow into solid crystals as time goes by, stodging into crystal spaces that are still empty.

The decrease in compressive strength is due to the elements decomposition of of C-S-H (Calcium Silica Hydrate) which decomposes into free lime CaO and SiO_2 which have no power at all. C-S-H is the main element that supports the strength of concrete, so a great reduction in C-S-H will greatly reduce the strength of concrete. When the temperature is raised to reach 800ºC, there is a carbonization process which is the formation of Calcium Carbonat ($CaCO_3$), whitish in color so that it changes the color of the concrete surface to become brighter. Besides, at this temperature, there is a drastic decrease in the attachment between rocks and cement paste characterized by cracks and fragility of concrete (concrete can be easily broken down by hand).

4.2. Concrete porosity test

Concrete porosity test is carried out using scales. Each concrete is weighed before and after experiencing the combustion process. The porosity of the concrete can be obtained by using a formula for concrete porosity.

Table 3. Post-combustion concrete porosity data with the addition of 1% superplasticizer admixture

Sample	Temperature (ºC)	Porosity (%)
1	27	5,925
2	200	17,778
3	500	23,704
4	800	29,629

Table 3 shows the average porosity data from the samples for concrete with combustion temperatures of 200 ° C, 500 ° C, and 800 ° C with the addition of 1% superplasticizer admixture. It is seen that the higher the temperature, the porosity of the concrete load increases. This means that the concrete mass is reduced and the concrete is more fragile and porous. This resulted in the compressive strength of the concrete decreasing.

Table 4. Post-combustion concrete porosity data with the addition of 1,5% superplasticizer admixture

Sample	Temperature (ºC)	Porosity (%)
1	27	2,962
2	200	16,296
3	500	22,222
4	800	28,148

Table 4 shows the average porosity data from the samples for concrete with combustion temperatures of 200 ° C, 500 ° C, and 800 ° C with the addition of 1,5% superplasticizer admixture . It is seen that the higher the temperature, the porosity of the concrete load increases. This means that the concrete mass is reduced and the concrete is more fragile and porous. This resulted in the compressive strength of the concrete decreasing.

Table 5. Post-combustion concrete porosity data with the addition of 2% superplasticizer admixture

Sample	Temperature (°C)	Porosity (%)
1	27	2,962
2	200	13,333
3	500	19,259
4	800	25,185

Table 5 shows the average porosity data from the samples for concrete with combustion temperatures of 200 ° C, 500 ° C, and 800 ° C with the addition of 2% superplasticizer admixture . It is seen that the higher the temperature, the porosity of the concrete load increases. This means that the concrete mass is reduced and the concrete is more fragile and porous. This resulted in the compressive strength of the concrete decreasing.

Figure 6 is a data summary between the increase in temperature and the porosity of the concrete with each addition of superplasticizer admixture by 1%; 1.5%; and 2%. The graph shows the tendency that the porosity of the concrete will rise along with the increase in temperature and the percentage increase will be greater along with the combustion temperature, at a temperature of 200 ° C there is an increase in porosity of 7.41% - 11.853%, at 500 ° C increasing to 13,334% - 17,779%, and at 800 ° C increasing to 19.26% - 23.704%, from normal concrete porosity.

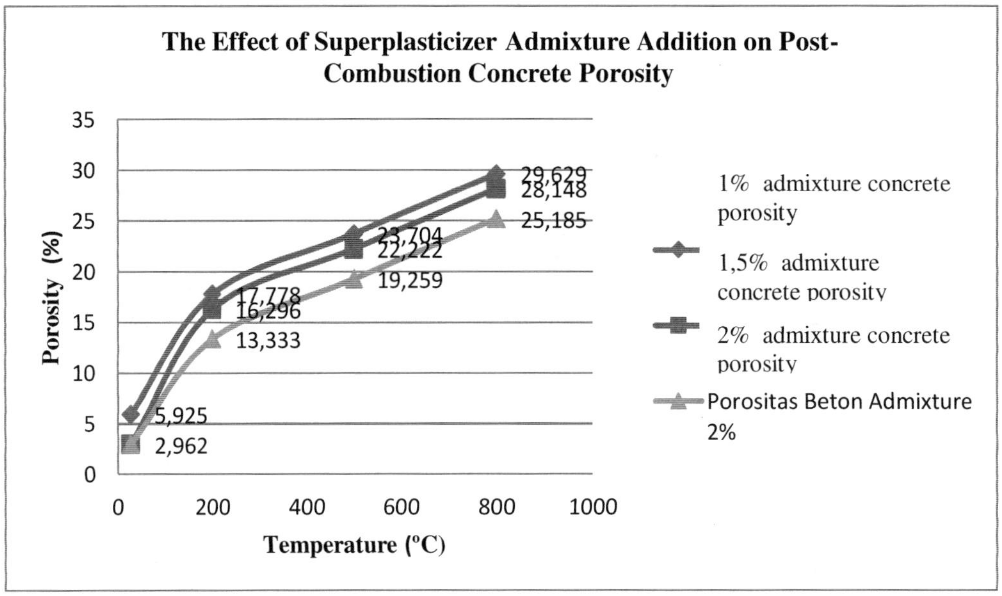

Figure 6. Graph of post-combustion admixture concrete porosity to the temperature

From the tests results above, it can be seen that the higher the combustion temperature, the greater the porosity of the concrete. This causes the concrete to become porous. This can be proven in Figure 6 where an increase in combustion temperature causes the number of concrete porosity to increase.

In the addition of superplasticizer admixture on post-combustion concrete, this is due to the fact that (H_2O) contained in the concrete will mostly evaporate. This condition is caused by the burnt concrete that will evaporate free water in large capillary pores, then followed by water in the gel pores that are smaller in size but more tightly bound by friction. The water molecules that will come out (migration) are blocked, then friction occurs with the pores of the concrete resulting micro cracks, so that the porosity of the concrete increases.

With increasing porosity, the compressive strength of the concrete will decrease and cause damage to the concrete structure. This is what to be overcome in the reason of adding superplasticizer admixture. With the addition of this type of admixture in the process of mixing fresh concrete allowing the concrete to be denser so that the density of concrete increases, indicating that the capillary pores in the concrete are so small that the porosity of the concrete due to combustion can be reduced which causes the concrete to still have sufficient residual strength. With the knowledge of residual strength, we can make the most efficient remedial measures to restore the condition to its original condition so that buildings that have experienced fires can be re-functioned. To date, buildings that have undergone fire are immediately crushed/destroyed (demolished action), even though the structural elements of the burnt building structure may still have strength.

4.3. Color and visual condition observation of concrete

Table 6. Color and visual condition observation of post-combustion concrete at certain temperatures

Temperature (°C)	Results of Visual Observation	
	Condition	Color of Concrete
200	Normal	Normal
500	Cracking	Grayish-brown
800	Crazing	Grayish-white

The first step in estimating the temperature experienced by concrete can be done through visual observation. The color of burnt concrete can indicate the level of fire, as presented in Table 4.6. The table shows the results of observing the difference in concrete color and the condition of the concrete surface after being burned. The color of the concrete that starts to brown to white indicates that the fire is quite severe. Aggregate or sand contains several iron compounds. Iron compounds produce cream, yellow, red, black and brown colors.

Liconite is a very common iron compound that produces cream, yellow and brown. While hematite will give a red color. Silica iron compounds give a green color, mangaan compounds produce brown and carbon compounds give blue, gray, green or brown. The discoloration of the concrete begin to appear when the temperature is at 500 ° C, which is grayish-brown in color. This occurs because of the presence of iron salt compounds in aggregates or concrete sand which causes the concrete to change color. If the temperature reaches 800 ° C, there is a carbonization process, which Calcium Carbonat (CaCO3) is formed that whitish in color so it changes the color of the concrete to become brighter. High temperatures in the concrete structure cause the concrete to experience fragility, and the visual signs that occur on the concrete are:

1. Discoloration of concrete
2. Transformation of concrete, such as:
   - Peeled off
   - Broken
   - Pore appearance

5. Conclusions

From the research data obtained and analysis of the data that has been done, it can be concluded that:

1. Fire phenomena in concrete structures will cause structural changes, including: cracks, damage / collapse, and changes in the color of the concrete surface. The color of the concrete will change along with the increasing temperature. The color change is caused by aggregates or sand which contains several iron compounds which can also cause corrosion.

2. Concrete will experience a decrease in strength along with the rising temperatures. In this case, it can be stated that temperature is very important in determining the mechanical properties of concrete.
3. The addition of superplasticizer admixture to the concrete mixture with 1% ; 1.5%; and 2% substitution of cement use can increase compressive strength and minimize damage to concrete after fire. This is because concrete with the addition of admixture has enough water to resist combustion heat even though the use of water is reduced by 10% -15% from the use of water in the manufacture of normal concrete.
4. The higher the combustion temperature, the greater the porosity of the concrete.
5. The addition of superplasticizer admixture to the concrete mixture with 1% ; 1.5%; and 2% substitution of cement use can reduce porosity in post-combustion concrete. Concrete compressive strength is inversely proportional to porosity. The smaller the porosity of the concrete, the greater the value of concrete compressive strength, this is because water (H2O) contained in the concrete will mostly evaporate. Increased porosity causes the compressive strength of the concrete to drop and cause damage to the concrete structure.

6. Suggestions
Some suggestions for further research are as follows:

1. It is expected that the next researcher can test the mechanical properties or physical properties of other post-combustion concrete such as: elasticity, tensile strength, or other post-combustion concrete properties that have never been studied before.
2. The next researcher is expected when testing porosity can measure samples after combustion to determine the impact of combustion on concrete porosity.
3. Tests for reinforced concrete can be carried out that focusing on reinforcement test.
4. The next researcher can test the mechanical or physical properties for other concrete qualities.

The next researcher conducts validation by examining the duration and temperature variations that have not been studied before.should be avoided whenever possible. If required they should be used only for brief notes that do not fit conveniently into the text.

7. Reference

[1] ASTM, Annual Books of ASTM Standards, 1991,*Concretes and Aggregates*, Vol.04.02 Construction, Philadelphia-USA: ASTM,1991,PA19103-1187.
[2] Sutapa, Dede A A, 2011 Porositas, Kuat Tekan dan Kuat Tarik Belah Beton dengan Agregat Kasar Batu Pecah Pasca Dibakar, *Jurnal Ilmiah Teknik Sipil* **15** (1) Universitas Udayana-Denpasar.
[3] Bayuasri, Trisni, Indarto H., & Antonius 2006 Perubahan Perilaku Mekanis Beton Akibat Temperatur Tinggi, *Jurnal PILAR* **15** (2), Universitas Diponegoro-Semarang.
[4] Pujianto A,PutroY S, Retno T and Ariska O 2006 *Beton Mutu Tinggi dengan Admixture Superplasticizer dan Aditif Silicafume*,Jurnal Teknik Sipil, Universitas Muhammadiyah Yogyakarta-Yogyakarta.
[5] Mulyono T, 2003,*Teknologi Beton*,Penerbit Andi, Yogyakarta.
[6] Murdock LJ, and Brook K M 1986,*Bahan dan Praktek Beton*, Penerbit: Erlangga, Jakarta.
[7] Nugraha P and Antoni, 2007,*Teknologi Beton*, Penerbit Andi, Yogyakarta.
[8] Sagel R, Kole P, and Gideon K, 1993,*Pedoman Pengerjaan Beton*, Penerbit: Erlangga, Jakarta.

IOP Conf. Series: Materials Science and Engineering **648** (2019) 012023 doi:10.1088/1757-899X/648/1/012023

Autonomous Surface Vehicle (ASV) Obstacle Avoidance Using Fuzzy Kohonen Network (FKN)

R Passarella[1], A Zarkasi[1], H Maghfur[1], Sutarno[1], K Exsaudi[1], APP Prasetyo[1], H Veny [2]

[1]Department of Computer Engineering, Faculty of Computer Science, Universitas Sriwijaya, Palembang, Indonesia

[2] Faculty of Chemical Engineering, Universiti Teknologi MARA (UiTM), Shahalam, Malaysia

E-mail: passarella.rossi@gmail.com

Abstract. Autonomous Surface Vehicle (ASV) is robot boats which can navigate autonomously to avoid obstacles in their path direction. The ASV has designed to detect and measure the distance of the position of the obstacle. The Fuzzy Kohonen Network (FKN) method is applying to the ASV as its brain to divine the manoeuvre what should do. The FKN is getting the information (Crips) from two sonar sensors, where are located in front of the ASV. In this experiment the FKN has four (4) pattern scenario which is three pattern normal condition, and one pattern danger condition. The three normal conditions define as if no obstacle detecting by two sensors, or either one of the sensor has detected an obstacle. The last condition pattern (danger) will occur when each sensor has detected the obstacle. These pattern condition is coding in C# and embedded into ATMega328. The range sensor is setting for 50-210 cm with the error rate is less than 1%. The manoeuvre of this ASV is providing two DC motors by controlling the PWM value. As the results of the experiment,The manoeuvre is quicker and smooth without a crash to obstacles.

1. Introduction

The navigation system is a system that determines the direction and purpose to accomplish. This system nowadays is familiar to use in the area of the Autonomous Navigation System (ANS)[1][2][3] . The implementation of this system can see in several vehicles, such as Unmanned Aerial Vehicles (UAV)[4][5] and Unmanned Surface Vehicles (USV)[6][7]. The ANS has integrated sensors and microcontroller or processor that enable autonomous navigation, path-planning and perception. The manoeuvre of this ANS is capable of sensing the environment and move with no human interruption.

The part of ANS is an Autonomous Surface Vehicle (ASV). The ASV is robot boats which can navigate autonomously to avoid obstacles in their path direction. The ASV for this research has a design with two DC motor as propulsion and microcontroller ATMega328 (Figure 1). Besides that, the ASV has two sonar sensors (MB7060 XL-Max Sonar-WR1). These sensors have a function to detect and measure the distance of the obstacle in front of the ASV.

The data measured from the sensors then sent to Arduino UNO rev3 with ATMega328 as its CHIP. Inside this module, the algorithm of Fuzzy Kohonen Network (FKN) is coding and embedding into it. The FKN method chosen due to the ability to recognize patterns of the environment and also has a fast decision response movement [8]. The FKN algorithm is still a new method for the minimum resource of the brain system. The FKN algorithm has embedded in the particle swarm robot reported by [9] to achieve real-time reactive formation control for swarm robots. Based on this, the experiment is conducting to implement FKN algorithm into the ASV.

Figure 1. The Autonomous Surface Vehicle

2. Design and Method

In this section, the methodology of the built ASV is dividing into several steps. The first step is designing the architecture of the system. The architecture of the system is representing by the block diagram, in general, the block diagram has input, process and output. The input block is consist of the two sonar sensor, the process block has a microcontroller with an FKN algorithm, and the output block has a motor driver and DC motor. The illustration of this system is showing in `Figure 2`.

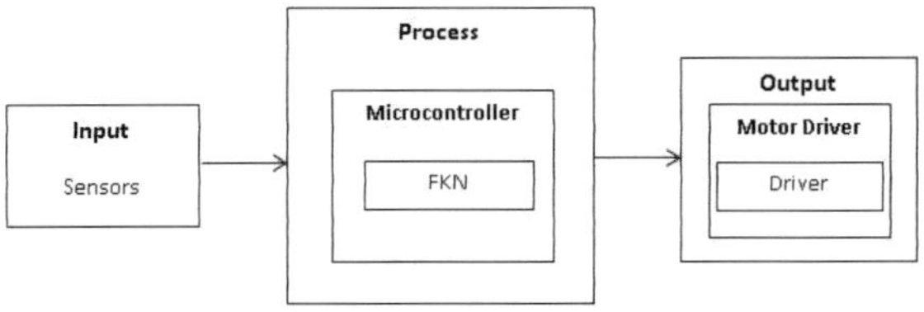

Figure 2. The block diagram of the ASV system

The second steps are calibrating the sonar sensor. The sonar sensor needs to calibrating to make the capability of measuring is precise. To do this, the calibrating is setting for measuring the distance object/obstacle in the range of 50 -210 cm. The procedure of calibrating is putting the object in the various distance such as 50 cm, 70 cm, 90 cm, 110 cm, 130 cm, 150 cm, 170 cm and 210 cm. The sensor detects the object and sent the data in the form of pulse width to the microcontroller (Arduino Uno rev3), this pulse width is getting from TX (transmitter) and receiving by RX of the sensor. The illustration of the minimum system calibrating is showing in `Figure 3`.

The output of the sensor then sent to the microcontroller for converting into digital form by using Analog Digital Converter (ADC) with a data width of 10 bit. The ratio between real distance and value ADC needs to find in the form of calibrating data by using the equation (1). From the `eq(1)`, the ratio between real distance and value ADC is 1.338.

$$q = \frac{bit}{md} \tag{1}$$

where :

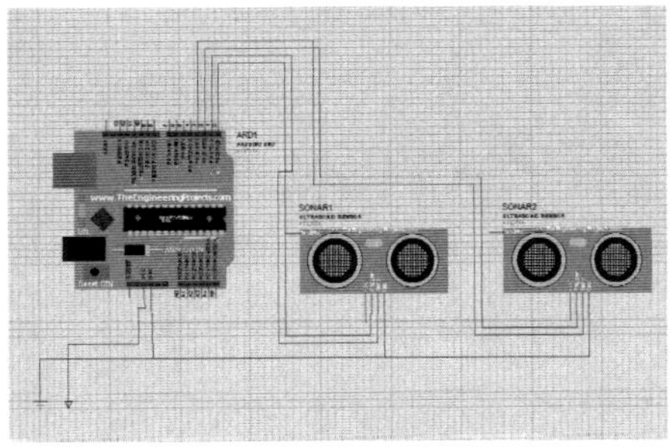

Figure 3. The ilustration of calibrating the distance object by sonar sensor

- q = the rasio between real distance and value ADC;
- md = maximum distance of the sensor(765);
- a = bit(1024);

To find the estimation distance of the object, the equation (2) is applying.

$$x = \frac{valueADC}{q} \tag{2}$$

where x = the estimation distance between sensor and object (cm).

As a result of calibrating sonar sensor is showing in `Table 1`. from `Table 1` is sighting that the sonar sensor is precise for the distance between 50 cm - 210 cm.

Table 1. Result of calibrating sonar sensor.

Measuring distance	Value ADC(S1)	Value ADC(S2)	Est Distance(S1)	Est Distance(S2)
50 cm	66	67	49 cm	50 cm
70 cm	93	93	70 cm	70 cm
90 cm	120	120	90 cm	90 cm
110 cm	147	147	110 cm	110 cm
130 cm	174	174	130 cm	130 cm
150 cm	200	200	150 cm	150 cm
170 cm	225	225	168 cm	168 cm
210 cm	278	277	208 cm	207 cm

The third step is developing the algorithm of FKN. The FKN is a part of Neuro-Fuzzy Method, which is the result of integration between Fuzzy Method and Neuron Network (Kohonen)[10]. The steps for learning the process in the fuzzy Kohonen network describing as:

- Step 1: Quantization of Input Value (Fuzzy Logic Section);
- Step 2: Calculate Euclidean Distance (Kohonen Section);
- Step 3: Calculate the Degree of Membership (Fuzzy Logic Section);
- Step 4: Output Processing (Fuzzy Logic Section).

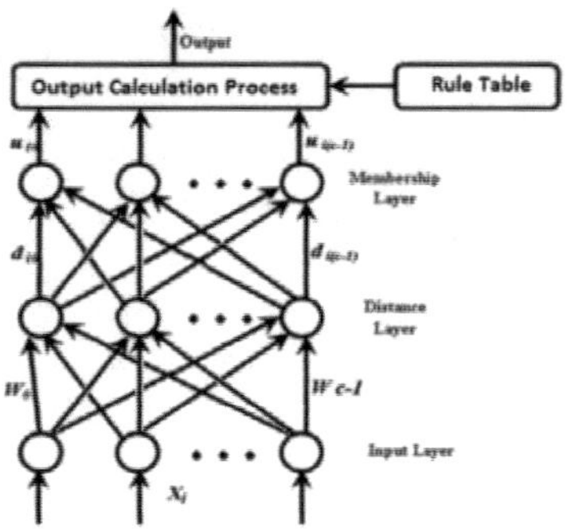

Figure 4. FKN Structure.

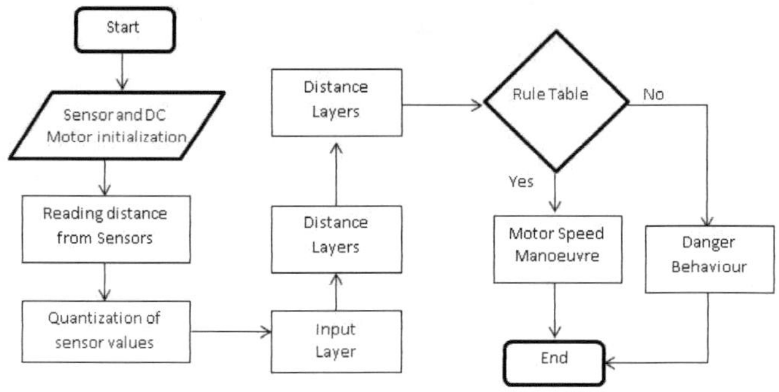

Figure 5. Flowchart Autonomous Surface Vehicle (ASV)

Inside the FKN methods, the process of quantitation of value ADC by using the eq(3).

$$X_i = \begin{cases} 1 & \text{for } 50\text{cm} < Y_i \leq 104\text{cm} \\ 2 & \text{for } 105\text{cm} < Y_i \leq 159\text{cm} \\ 3 & \text{for } Y_i \geq 160\text{cm} \end{cases} \quad (3)$$

where: i = number of sonar sensor.

Euclidean distance is the distance between the similarity between the input value pattern and the embedded rule pattern. The calculation process carried out using the Euclidean distance formula (d_{ij}) found in equation (4)

$$d_{ij} = \sqrt{\sum \left\langle \|x_{i(n)} - w_{j(n)}\| \right\rangle^2} \quad (4)$$

where :

- d = euclidean distance;
- n = Number of Sensors used;
- w = Value of implanted weight;

The calculating membership degree (μ_{ij}) is applying the equation (5). After getting the value of μ_{ij}, the value of the highest membership degree will be multiplied by the reference value of the membership degree pattern so that it gets the output value (equation (6)).

$$\mu_{ij} = \begin{cases} 1 & d_{ij} = d_{min} \\ (d_{max} \text{ - } d_{ij} \ / \ d_{max} \text{ - } d_{ij}) & d_{max} \leq d_{ij} < d_{max} \\ 0 & d_{ij} \geq d_{max} \end{cases} \quad (5)$$

where :

- u = degree of membership;
- dmin = smallest Euclidean distance = 0;
- dmax = The biggest euclidean distance;

$$Output = \mu_{ij} * Out_{ref} \quad (6)$$

The rules base is developing according to the `Figure 6` . For the last condition pattern (danger) will occur when each sensor has detected the obstacle, so the motor dc reversing until the condition of the environment is back to each of the rules base.

Pattern (X_i)	Pattern Linguistic		Image pattern condition	Value pattern Prototype		Output (% PWM)	
	S_1	S_2		S_1	S_2	M_1	M_2
X_1	Near	Far	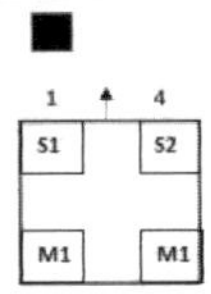	1	3	90	40
X_2	Far	Near	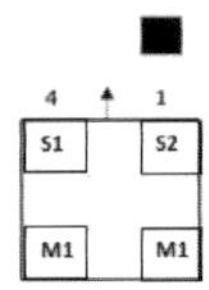	3	1	40	90
X_3	Far	Far	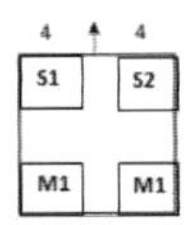	3	3	90	90

Figure 6. The Rules of Base instilled in the FKN

3. Results

In this test, the ASV is placing in the water (swimming pool) to monitor the movement that carried out in a particular environment to find out how the ASV moves through specified pattern recognition. The pattern tested is a pattern commonly used for testing boat robots. The manoeuvre robot in water environments is quite different from the manoeuver robot in the ground. The manoeuver robot in the ground environment is more responsive and quick compared to the water environment. This difficult must be considered by a design engineer to make a simple discussion on its brain system to reduce time-consuming before taking manoeuvre. As this is the case, the FKN algorithm suitable to implement in boat robot (ASV).

3.1. Testing of movements in the environment without obstacles

In this condition, the ASV has no obstacles blocking in front of it. The data extract from Arduino Uno Rev 3 by using cable data to monitor the FKM algorithm in 700 hundredths of seconds (hs).This value of 700 hs is equal to 7 seconds. The scenario is conducting in an environment without obstacles, as shown in Figure 7. The ASV is moving forward with a time that set, and the real condition is showing in Figure 8. The results of this experiment are showing in Table 2.

3.2. Testing of movements in the environment with obstacles on the right

For this testing, the ASV is placing in an environment where obstacle puts in front of the right sensor (Figure 9). Based on the human logical perception of this scenario, the ASV should manoeuvre to left sides. According to the rules base (Figure 3) that coding in FKN algorithm, the pattern is matched with the pattern X_2, with the output of propulsion left is 40% PWM and 90% PWM for the right. The slow capture with interval collecting data is showing in Figure 10. However, from Table 3 as the data interval collecting is showing the condition of transition percentage of PWM in time 152,456 and 532. This transition percentage PWM happens because the value of quantitation of value ADC by eq(3) shows the sensor (2) has a value of 2. Nonetheless, pattern recognition is correct (X_2).

IOP Conf. Series: Materials Science and Engineering **648** (2019) 012023 doi:10.1088/1757-899X/648/1/012023

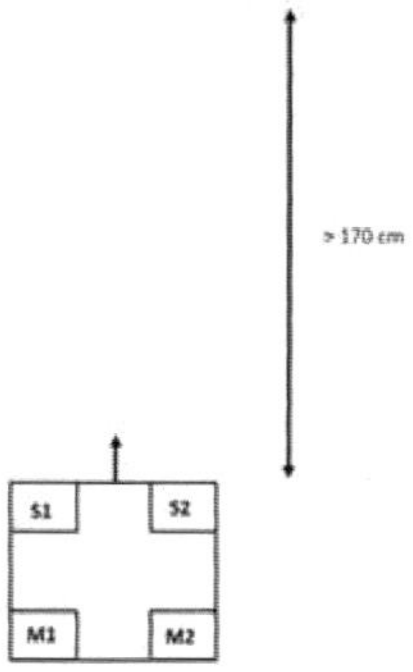

Figure 7. The scenario that conduct for experiment without obstacles

Figure 8. The real condition of ASV experiment without obstacles

Table 2. Results of movements ASV in the environment without obstacles

No	Time (hs)	Sensor(1)	Sensor(2)	% PWM left	% PWM Right	pattern recognition
1	0	3	3	90	90	X_3
2	76	3	3	90	90	X_3
3	152	3	3	90	90	X_3
4	228	3	3	90	90	X_3
5	304	3	3	90	90	X_3
6	380	3	3	90	90	X_3
7	456	3	3	90	90	X_3
8	532	3	3	90	90	X_3
9	608	3	3	90	90	X_3
10	684	3	3	90	90	X_3

Table 3. Results of movements ASV in the environment with obstacles in right side

No	Time (hs)	Sensor(1)	Sensor(2)	% PWM left	% PWM Right	pattern recognition
1	0	3	3	90	90	X_3
2	76	3	3	90	90	X_3
3	152	3	2	25.86	58.18	X_2
4	228	3	1	40	90	X_2
5	304	3	1	40	90	X_2
6	380	3	1	40	90	X_2
7	456	3	2	25.86	58.18	X_2
8	532	3	2	25.86	58.18	X_2
9	608	3	3	90	90	X_3
10	684	3	3	90	90	X_3

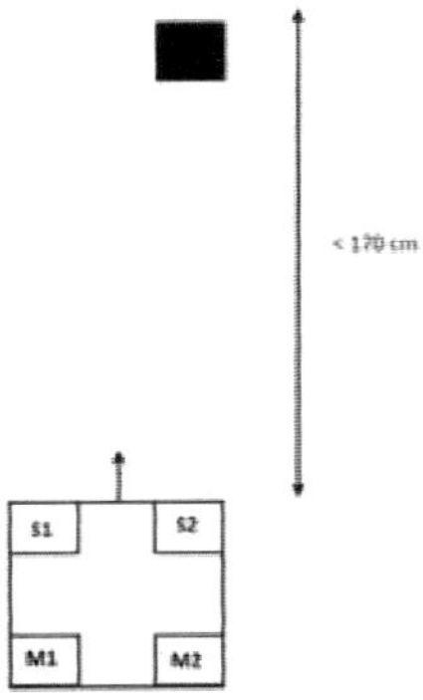

Figure 9. The scenario that conduct for experiment with obstacles in right side

Figure 10. The real condition of ASV experiment with obstacles in right side

3.3. Testing of movements in the environment with obstacles on the left

The next test is the manoeuvre to the right side to verify the rule base of this action (X1), to do this, the obstacle is placing in front of the left sensor (170 cm) as showing in `Figure 11`, with the perception that the ASV will move to the right side. From the experiment, the captured movement between time interval collecting data shows the ASV manoeuvre to the right with a smoothly without stop or delay `Figure 12`. For the detail data collecting, `Table 4` is presenting. The transition in percentage PWM is occurring in time 152, 228 and 532. However, pattern recognition is still correct (X_1).

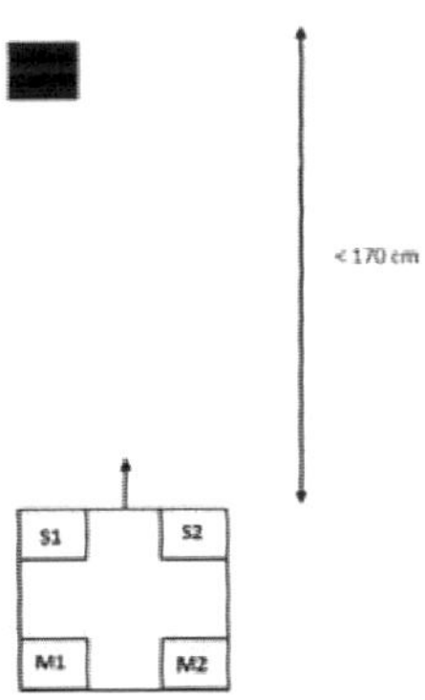

Figure 11. The scenario that conduct for experiment with obstacles in left side

Figure 12. The real condition of ASV experiment with obstacles in left side

Table 4. Results of movements ASV in the environment with obstacles in left side

No	Time (hs)	Sensor(1)	Sensor(2)	% PWM left	% PWM Right	pattern recognition
1	0	3	3	90	90	X_3
2	76	3	3	90	90	X_3
3	152	2	3	58.18	25.86	X_1
4	228	2	3	58.18	25.86	X_1
5	304	1	3	90	40	X_1
6	380	1	3	90	40	X_1
7	456	1	3	90	40	X_1
8	532	2	3	58.18	25.86	X_1
9	608	3	3	90	90	X_3
10	684	3	3	90	90	X_3

4. Conclusion

Based on the result from the experiments, the ASV with FKM Algorithm is showing successful implement as autonomous navigation thought three patterns recognition. The manoeuvre is quicker and smooth without a crash to obstacles. This manoeuvre is similar to the ground robot experiment with FKN that has been conducted by [8][9]. The manoeuvre robot in the water environment is a different treatment compared to the ground environment. This result shows that the ASV with FKN algorithm has the same behaviour with a ground robot.

References

[1] Courbon, J., Mezouar, Y. and Martinet, P., 2009. Autonomous navigation of vehicles from a visual memory using a generic camera model. IEEE Transactions on Intelligent Transportation Systems, 10(3), pp.392-402.
[2] Huntsberger, T., Aghazarian, H., Howard, A. and Trotz, D.C., 2011. Stereo vision–based navigation for autonomous surface vessels. Journal of Field Robotics, 28(1), pp.3-18.
[3] Galyean, T.A., 1995, April. Guided navigation of virtual environments. In Proceedings of the 1995 symposium on Interactive 3D graphics (pp. 103-ff). ACM.
[4] Duan, G.J. and Zhang, P.F., 2014. Research on Application of UAV for Maritime Supervision. Journal of Shipping and Ocean Engineering, 4, pp.322-326.
[5] Ahilan, T., Adityan, V.A. and Kailash, S., 2015. Efficient Utilization of Unmanned Aerial Vehicle (UAV) for Fishing through Surveillance for Fishermen. World Academy of Science, Engineering and Technology, International Journal of Mechanical, Aerospace, Industrial, Mechatronic and Manufacturing Engineering, 9(8), pp.1468-1471.
[6] Yang Fang, Muye Pang, Biao Wang, 2017. A course control system of unmanned surface vehicle (USV) using back-propagation neural network (BPNN) and artificial bee colony (ABC) algorithm.Procedia Computer Science. Vol 111, pp 361-366.
[7] Blaich, M., Wirtensohn, S., Oswald, M., Hamburger, O. and Reuter, J., 2013. Design of a twin hull based USV with enhanced maneuverability. IFAC Proceedings Volumes, 46(33), pp.1-6.
[8] Prasetyo, A.P., Astrowulan, K. and Fatoni, A., 2016. Implementation of Navigation Target Seeker Mobile Robot Based on Pattern Recognition with Fuzzy Kohonen Network (FKN) Methods. IPTEK Journal of Proceedings Series, 2(1).
[9] Nurmaini, S., Tutuko, B. and Aditya, A., 2015. Pattern Recognition Approach for Formation Control for Swarm Robotics Using Fuzzy-Kohonen Networks. Proceeding of the Electrical Engineering Computer Science and Informatics, 2(1), pp.143-144.
[10] Song, K. T., Huang, S. Y. (2004, September). Mobile robot navigation using sonar direction weights. In Proceedings of the 2004 IEEE International Conference on Control Applications, 2004. (Vol. 2, pp. 1073-1078). IEEE.

IOP Conf. Series: Materials Science and Engineering **648** (2019) 012024 doi:10.1088/1757-899X/648/1/012024

Separation of Basic Words in Angkola Batak Text Documents using Enhanced Confix Stripping Stemmer Case: Mandailing Ethnic

M A Muchtar[1*], I Jaya[1], Manguji Nababan[2], U Andayani[1], Lisa Noprianti Siregar[1], Erna B Nababan[1], Opim S Sitompul[1]

[1]Department of Information Technology, Faculty of Computer Science and Information Technology, Universitas Sumatera Utara, Medan, Indonesia
[2]Center of Documentation and Assessment of Batak Culture, Universitas HKBP Nommensen, Medan, Indonesia

*Correspondence Email: anggi.muchtar@usu.ac.id

Abstract. Stemming is the process of separating essential words from an affixed word. Stemming works by eliminating morphological variations which attached to a word by removing affixes in a word with a dictionary as a reference to the stemming process. One effective algorithm for resolving stemming in Indonesian is an Enhanced Confix Stripping Stemmer algorithm. In this case, the improved Confix stripping stemmer algorithm is implemented for the separation of essential words in the Angola-Mandailing Batak language document that has different phonologies and morphologies with Indonesian. The documents used are Latin character documents in the Angkola-Mandailing language that goes through the stages of filtering, folding cases, and tokenization before stemming. The stemming process is done by removing particles, ownership, suffixes, and prefixes. After testing this research, it was concluded that this algorithm was able to separate essential words in the Angkola-Mandailing Batak language with accuracy 87,05%.

1. Introduction

Indonesia is a country consisting of many tribes and languages. Based on the 2010 Central Bureau of Statistics census results, there are 1340 tribes and 2500 regional languages in Indonesia. The Batak tribe is one of the three tribes with the largest population in Indonesia. The Batak tribe itself is divided into five sub-groups, namely: Toba Batak, Karo Batak, Pakpak Batak, Simalungun Batak, and Angkola-Mandailing Batak. The Angkola-Mandailing Batak Population is a community living in the Mandailing Natal area, South Tapanuli, Padang Lawas, and Padang Sidimpuan. The language used in people's daily lives in this region is the Angkola Batak language - Mandailing. Angkola Batak Language - Mandailing itself has its own rules in pronunciation, sentence structure, and grammar. Grammar itself includes essential words and affixed words in a sentence.

One process for transforming essential words from words in a document is the process of stemming [2]. Stemming is a process or way of finding the essential words of Scripture. Stemming itself serves to eliminate the morphological variations that are attached to a word by removing the affixes to the word. The stemming process for each language is different because each style has different grammar rules in the use of affixed words [6]. Example sentence in Indonesian sentence "Saya berlari ke sekolah". The word that needs to be stemmed in the sentence is the word "berlari" because in the word "berlari", there is a ber-/ affix. The stemming process is done to get the word "lari" from the word "berlari". If the phrase "Saya berlari ke sekolah" is translated into the Angkola Batak language - Mandailing becomes the phrase "au marlojong tu sikola". The word that needs to be stemmed in the sentence is the word "marlojong" because the word contains the prefix mar- /. The stemming process is done to separate the word "lojong" from the word "marlojong."

One algorithm for stemming is an enhanced Confix stripping stemmer (ECS) algorithm. The stemming algorithm that was used for the first time to stimulate Indonesian was the Nazief-Adriani algorithm [1]. The enhanced Confix stripping stemmer (ECS) algorithm is an algorithm developed by Mahendra to correct the lack of the Confix Stripping Stemmer algorithm [6]. This algorithm previously used in Tahitoe research [8].Furthermore, to solve the stemming of Indonesian absorption words [4] or the improvement of summarization by using TextTeaser[9] .

Additionally, Maulidi completed the stemming of the Madura language by implementing a modified, enhanced Confix stripping stemmer (ECS) algorithm [7]. In previous research, stemming was done for Indonesian. Research for regional languages is still very little. Research for deriving local languages was previously conducted for Madurese language stemming. Based on this, the author is compelled to implement an enhanced Confix stripping stemmer (ECS) algorithm for stemming processes in the Angkola Batak language - Mandailing in obtaining the exact base words of a sentence that is affixed by doing the right prefix and suffix. The proper stemming process for the separation of essential words from the sentences in a text document is Angkola-Mandailing Batak Language as in Indonesian.

In 2012, Ayyidar et al. implemented an enhanced Confix stripping stemmer for absorption in Indonesian [4]. Enhanced Confix stripping stemmer (ECS) algorithms can well complete the stemming process for absorption words by adding additive rules but require a longer time due to the addition of stemming rules to overcome absorption words in Indonesian. Furthermore, a study was conducted to compare the stemming process using the Enhanced Confix Stripping Stemmer algorithm and the Porter stemmer algorithm. The study was conducted by testing 300 documents with the results: Enhanced Confix Stripping Stemmer produced 81% accuracy in 11.8311 seconds and a stemmer porter with 61% accuracy in 0.0291 seconds.

Furthermore, in 2013, Iqramitha Annifa applied the enhanced Confix stripping stemmer algorithm to the stemming process for the dictionary of Indonesian language cues with results [5]. The enhanced Confix stripping stemmer algorithm can be applied in a sign language learning system. This algorithm is able to chop words with a combination of Prefix 1 + Prefix 2 + Basic Words + Suffix 3 + Suffix 2 + Suffix 1 where suffix 3 = suffix, suffix 2 = pronoun belongs, and suffix 1 = particle. However, the stemming process cannot be carried out because in Indonesian sign language there are no rules that explain the insertion while for the homonym word in sign language there is no difference in form, like snake can with what I can do. Subsequent research in 2016, conducted a study to complete the stemming process in the Madurese language by modifying the rule base on Enhanced Confix Stripping Stemmer according to the morphology of the Madura language and getting good results [3].

2. Methodology

The system was built to complete the stemming process of the Angkola batak language document - Mandailing that had never been made before. The system will process stemming from a document and then decompose it into word words which are then stemmed using an enhanced Confix stripping stemmer (ECS) algorithm. The general architecture carried out in this study is shown in Figure 1.
The description are as follows:

2.1 Input

In the first stage, the system is inputting documents with the extension .docx. As for documents in the form of folk tales or oral transcripts of the traditional ceremony of the Angkola Batak language - Mandailing.

2.2 Pre-processing

At this stage, the process of changing the form of unstructured data into structured data is carried out according to system requirements. There are three stages to be carried out, namely:

- Filtering

This process aims to take the alphabet character of the document. In this process, the numbers, punctuation, and characters are removed from the letters of the alphabet.

- Case Folding

Case folding is the process of converting all letters in a document to lowercase. Only the letters 'a' up to 'z' are accepted. Non-letter characters are omitted.

- Tokenization

This process aims to separate each word that composes a document into words per word by marking a space character. Each word, the tokenization result will be saved in the array.

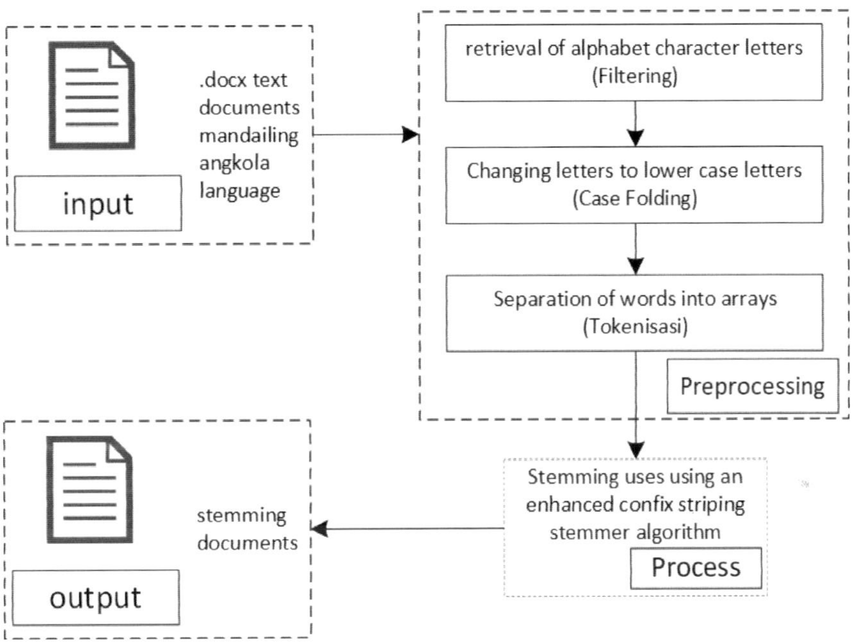

Figure 1. General Architecture

2.3 Processing

At this stage, a stemming process is carried out with an enhanced Confix stripping stemmer algorithm. One by one, the word will be checked, whether it is in the dictionary or not. If the word is included in the dictionary of the basic words, the stemming process for the word will not be confirmed. Then proceed to the next word, if the word is not in the basic word dictionary, a stemming process will be carried out with an enhanced Confix stripping stemmer algorithm. The process of stemming this algorithm has the following stages, namely:

a. Check the dictionary. If the word is in the dictionary, the basic word has been found.
b. Remove inflectional suffixes (particles) (-do, -ma, -pe)
c. Remove possessive pronoun (ownership) (you, me, -na)
d. Check the dictionary
e. Remove the derivation prefix (prefix) (mar- /, ma--, tar- /, pa-, di- /, paN- /, par- /, saN- /, um- /, si- /)
f. Delete infix
g. Check the dictionary
h. Remove the derivation suffixes (khiran) (- an, -on, -kon, -hon, -i)

i. Check the dictionary
j. Replenishment returns

The stemming process for the Angkola-Mandailing Batak language refers to the rule table 1:

Table 1. Rules for the Angkola-Mandailing Batak Language prefix.

<table>
<tr><th>Rule</th><th>Word Format</th><th>Syllabification</th></tr>
<tr><td>1</td><td>marV</td><td>mar-V</td></tr>
<tr><td>2</td><td>m-arC</td><td>mar-C</td></tr>
<tr><td>3</td><td>mam{p|b}</td><td>mar-{p|b}</td></tr>
<tr><td>4</td><td>man{j|d|c}</td><td>man-{j|d|c}</td></tr>
<tr><td>5</td><td>mang{k|g}</td><td>mang-{k|g}</td></tr>
<tr><td>6</td><td>mangV</td><td>mang-V</td></tr>
<tr><td>7</td><td>manga{1|r}</td><td>manga-{1|r}</td></tr>
<tr><td>8</td><td>many{s}</td><td>ma-ny{s}</td></tr>
<tr><td>9</td><td>man{t}</td><td>ma-{t}</td></tr>
<tr><td>10</td><td>tarV</td><td>tar-V</td></tr>
<tr><td>11</td><td>tarC</td><td>tar-C</td></tr>
<tr><td>12</td><td>diV</td><td>di-V</td></tr>
<tr><td>13</td><td>diC</td><td>di-C</td></tr>
<tr><td>14</td><td>paC</td><td>pa-C</td></tr>
<tr><td>15</td><td>paV</td><td>pa-V</td></tr>
<tr><td>16</td><td>pam{b|p}</td><td>pam-{b|p}</td></tr>
<tr><td>17</td><td>pang{g}</td><td>pang-{g}</td></tr>
<tr><td>18</td><td>pan{c|d|j|t}</td><td>pan-{c|d|j|t}</td></tr>
<tr><td>19</td><td>parV</td><td>par-V</td></tr>
<tr><td>20</td><td>parC</td><td>par-C</td></tr>
<tr><td>21</td><td>saV</td><td>sa-V</td></tr>
<tr><td>22</td><td>saC</td><td>sa-C</td></tr>
<tr><td>23</td><td>siC</td><td>si-C</td></tr>
<tr><td>24</td><td>sal{1}</td><td>sal-{1}</td></tr>
<tr><td>25</td><td>pang{h}</td><td>pa-ng{h}</td></tr>
</table>

2.4 Output
After all the words in the document have been processed, the results of the stemming process will be displayed in the form of files with the extension .docx that can be stored on the user's computer. All words in the original document will be displayed, and the words that are stemmed will be separated by prefix, essential words, and endings with space characters.

3. Results and Discussion
System testing is done to find out whether the system has been built functions properly and runs as desired. This test will explain how much the accuracy of the implementation of Enhanced Confix Stripping Stemmer is applied to the separation of essential words in text documents Angkola Batak Language - Mandailing. The transliteration system interface can be seen in Figure 2.

On the transliteration page, there is the Select File button to enter the text document file in the Angkola-Mandailing Batak Language. The user will wait a few seconds until the document entered is displayed by the system. After the document is displayed, then the user selects the Stemming and Waiting button until the system displayed the stemming results. The last step is to choose the Save button to save the transliteration results in the form of files with the extension .doc. This stemming process can be seen in Figures 3, 4, and 5.

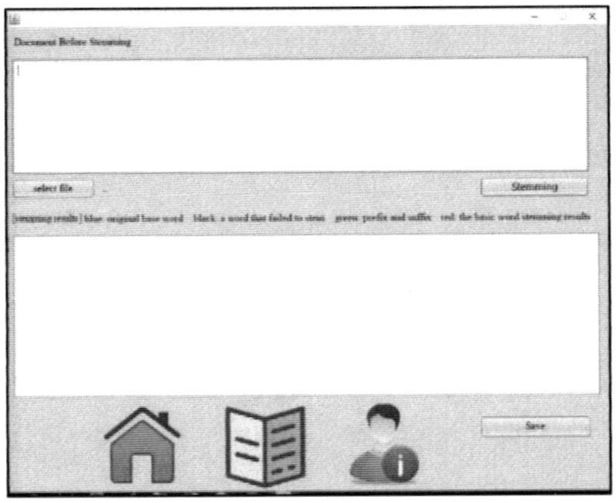

Figure 2. Transliteration system interface.

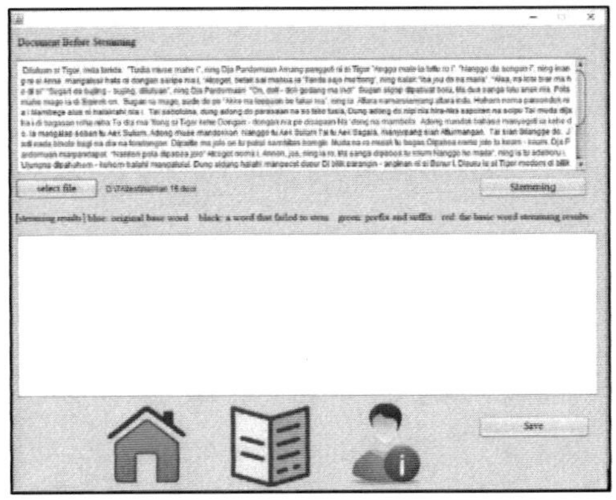

Figure 3. Input document interface appearance.

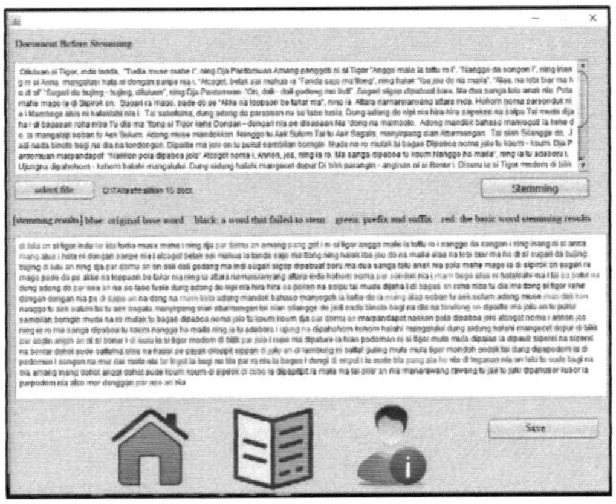

Figure 4. Stemming results interface.

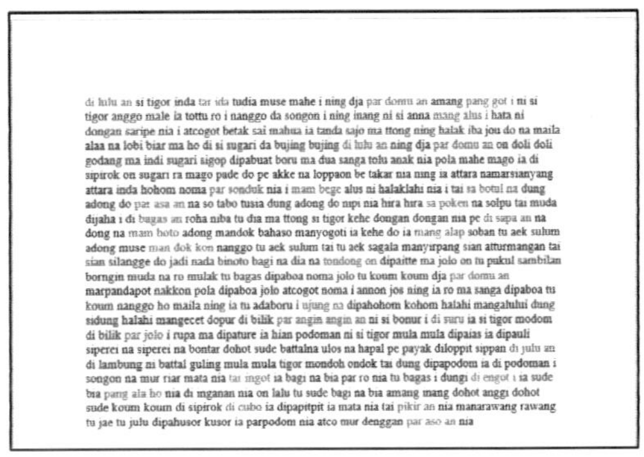

Figure 5. Stemming results interface with the .doc extension.

The test results on this system will use the following formula 1:

$$\text{Accuration} = \frac{\text{Total Successful Stemmed Words}}{\text{Total Stemmed Words}} \times 100\% \quad (1)$$

The calculation of the accuracy of the test results for each script document can be seen in table 2.

Table 2. Result of system testing.

Input Documents	**Words in Documents**	**Stemmed Words**	**Successes Stemmed words**	**Acurration**
Halilian 1	246	24	22	91%
Halilian 2	195	22	19	86%
Halilian 3	225	38	35	92%
Halilian 4	228	26	22	84%
Halilian 6	230	34	30	88%
Halilian 7	238	26	20	76%
Halilian 12	474	55	41	74%

Halilian 11	483	57	51	98%
Si Lian 2	225	22	20	90%
Mangupa	585	90	78	86%
Markobar 4	202	35	29	82%
Markobar 2	262	43	39	82%
Markobar 5	122	25	23	92%
Markobar 3	127	28	25	89%
Halilian 13	434	46	40	86%
Halilian 14	424	64	56	87%
Halilian 15	421	54	48	88%
Hari Minggu	188	19	17	89%
Si Lian 3	267	40	35	87%
Si Lian 1	211	34	31	91%
Total		782	681	87,08%

4. Conclusion

From the results of testing on the stemming system of the Angkola-Mandailing Batak language text document, outcomes are obtained as follows.

- Stemming the Angkola-Mandailing Batak language with the enhanced Confix stripping stemmer algorithm for the Angkola-Mandailing Batak language obtained an accuracy of 87.08%.
- The system cannot resolve to stem for decays, which are given the prefix t and p.
- The system cannot resolve to stem for words with the prefix manga- / and panga- /.
- Suggestions that can be given by the author for the development of further research are as follows:
- Further research is expected to be able to overcome the shortcomings in this study, namely the prefix that decays and the prefix manga / / and panga /.
- Further research is expected to be able to overcome more than one word of additive.
- The next research is expected to be able to overcome the addition of letter characters to words ending in vowels and affixed with the word ownership.
- Further research is expected to be able to process the Angkola-Mandailing Batak language with other algorithms to improve accuracy.
- Stemming data can be developed to do translations into other languages.

Acknowledgments

This research was supported by the Universitas Sumatera Utara. All lecturers and staff members of the Information Technology Study Program. Mr. Manguji Nababan as lecturer and Pusat Dokumentasi Pengkajian Kebudayaan Batak (PD & PKB) Universitas Nommensen Medan who was willing to provide research data to the Author.

References

[1] Adriani, M., Asian, J., Nazief, B., Tahaghoghi, S. M., & Williams, H. E. (2007). Stemming Indonesian: A confix-stripping approach. *ACM Transactions on Asian Language Information Processing (TALIP)* Hal 1-33.

[2] Agusta L. 2009. Perbandingan Algoritma Stemming Porter dengan Algoritma Nazief &Adriani untuk Stemming Dokumen Teks Bahasa Indonesia. *Proceeding Konferensi Nasional Sistem dan Informatika 2009*, pp. 09-036.

[3] Alif M, Solihin F,and Husni H. (2013). Perbandingan Metode Enhanced Confix Stripping dan Porter Stemmer Untuk Stemming Konten Bahasa Indonesia. *Jurnal Buana Informatika* Vol. 4 No.2

[4] Ayyidar, N. H., Syukur, A., & Subagyo, H. Algoritma Stemming Untuk Kata Serapan Bahasa Indonesia. Jurnal Teknologi Informasi, Volume 8 Nomor 1, April 2012, ISSN 1414-9999

[5] Iqramita, Annifa. 2013. Implementasi Algoritma Enhanced Confix Stripping Stemmer Pada Kamus Sistem Bahasa Isyarat Indonesia. Skripsi. Universitas Sumatera Utara.

[6] Mahendra, I. P. A. K., Arifin, A. Z., dan Ciptaningtyas, H. T. 2008. Penggunaan Algoritma Semut dan Confix Stripping Stemmer Untuk Klasifikasi Dokumen Berita Berbahasa Indonesia. Skripsi. Institut Teknologi Sepuluh Nopember.

[7] Maulidi, Rakhmad. 2016. Stemmer Untuk Bahasa Madura Dengan Modifikasi Metode Enhanced Confix Stripping Stemmer. Prosiding Seminar Nasional FDI 2016 Hal 12–15. ISSN. 2460-5271

[8] Tahitoe, A. D. dan Purwitasari, D. 2010. Implementasi Modifikasi Enhaced Confix Stripping Stemmer Untuk Bahasa Indonesia Dengan Metode Corpus Based Stemming. Tesis. Institut Teknologi Sepuluh Nopember

[9] Gunawan D, Pasaribu A, Rahmat R F, Budiarto R 2017 Automatic Text Summarization for Indonesian Language using TextTeaser *IOP Conference Series : Material Science and Engineering* **109**(1) 012048

IOP Conf. Series: Materials Science and Engineering **648** (2019) 012025 doi:10.1088/1757-899X/648/1/012025

Implementation of Porter Stemmer Algorithm to Obtain Basic Words in Toba Batak Language Documents with the Two-Level Morphological Method

M A Muchtar[1*], Erna B Nababan[1], Manguji Nababan[2], UAndayani[1], Tuti Simanjuntak[1], Opim S Sitompul[1]

[1]Department of Information Technology, Faculty of Computer Science and Information Technology, Universitas Sumatera Utara, Medan, Indonesia
[2]Pusat Dokumentasi dan Pengkajian Kebudayaan Batak, Universitas HKBP Nommensen, Medan, Indonesia

*Correspondence Email: anggi.muchtar@usu.ac.id

Abstract. Stemming is the process of deciphering words by carrying out affixes to words to get the right base words. In its development stemming has been developed for various languages, including Porter Stemmer. In this study, the stem was used to add Toba Batak language. Decapitation of affixes in the Toba Batak language is difficult because there are essential words that contain affixes and there are essential words that decay when affixed, namely the essential words beginning with the letter's 't', 's' and 'p'. To solve this problem by offering the implementation of a porter stemmer algorithm using the two-level morphological method. This stemmer method uses a dictionary of essential words. Each step in the stemmer process will always check the lexicon of the essential words so that the essential words that contain affixes will not be monitored, and if the word is not in the dictionary the basic word will not be stemming so when there are a name people or place names that contain affixes will not be implemented. The use of two levels of morphology was applied to the essential words beginning with the letter's 't', 's' and 'p' to get the right essential words. From the results of testing on 20 Toba Batak language documents, the Porter stemmer algorithm with two levels of morphology can produce an accuracy rate of 87.503% with an average speed of 0.000460798 seconds per word.

1. Introduction

Indonesia is a nation known for having many tribes. Each tribe has a different language and culture. For example, the Batak tribe. The Batak people themselves still consist of five sub-groups, namely Angkola-Mandailing, Toba, Simalungun, Pakpak-Dairi, and Karo [1]. The five Batak tribes have different languages as well, so the authors only focus on the Toba Batak language. The Toba Batak language is included in an ancient Malay language family that has its grammar system and the meaning of the word [2]. The Toba Batak language and Indonesian language come from the same family so that Indonesian grammar can be used as a reference tool. However, in the Toba Batak language, some words are not equivalent to Indonesian so that they need to be translated themselves. The Toba Batak language also has its phonetics, so the writing and pronunciation are different [3].

Stemming is a process for determining the essential words of a word by removing all affixes which consist of prefixes (prefixes), insertions (infixes), suffixes (suffixes) and combinations of prefixes and suffixes (confixes) in derivative words. One of the algorithms used for stemming is the Porter stemmer algorithm. The morphology of the Indonesian language is not much different from the morphology of the Toba Batak language, which consists of prefixes, inserts, suffixes, proprietary pronouns, and particles [3]. The affix syllabification in the Toba Batak language are difficult because there is an affix which turns out to be part of the essential words themselves. For example, the word " manakko" 'mencuri'. If the word is stemmed with one level of morphology, the basic word is "akko" which should be "takko". In order to get an accurate root word, it must be solved with two levels of morphology.

In this study, the system to be built is a system for determining essential words in Toba Batak language documents. The input of this system is the text of the Toba Batak language in a document that will be processed using the Porter stemmer algorithm to produce output in the form of input documents before the system and after the order. Words that cannot be applied using a single morphological level will be used using two levels of morphology to obtain more accurate results [4].

Previous studies that have been conducted include [5], the approach taken is two levels of morphology that are implemented with xfst and lexc. This approach has been able to handle non-concatenative reduplication and morphology.

Subsequent research [6], modifications made to the rules of the stemmer porter by combining them with the regulations on Nazief & Adriani's algorithm can improve the accuracy of the stem. The complete dictionary used for stemming processes affects the accuracy of the stem. Increasing the efficiency of the stem in many cases of search decreases the value of precision but does not affect the recall value.

Subsequent research using the Porter stemmer algorithm is [4]. By applying the 2-level morphology rule, it can increase the ability of stemmer to get the right base words on the base words beginning with the letter's "k", "t", "s" and "p" with 100% accuracy. The prefix and suffix combination rules can also increase the stemmer's ability to determine the suffix attached to a word especially for the "-kan" and "-a" suffixes with an accuracy rate of 96.6%. But the phrase homonym, homophone, homograph, and polysemy still cause problems. Besides, a method is also needed to identify the base words on repeated words correctly.

Furthermore, [7] with an accuracy level of 0.66 while the average processing time for each word is 0.000000160 seconds. The results of stemming accuracy are very dependent on the collection of original word dictionaries related to affixes — the more complete the group, the better the accuracy of the stemming results.

Next is [8], where the Porter stemmer algorithm determines the essential words in the inputted document and helps speed up the winnowing algorithm in determining the fingerprint value of a text so that the results are in the form of similarity between two coded documents.

The next study [9] is the utilization of the Porter stemmer algorithm for Indonesian in the process of classifying book types. The text pre-processing stage that is carried out on the tested document documents makes it easier for the stemming process. Stemming algorithms with stemmer porters for Indonesian that use rule-based algorithms significantly help speed up the classification process because the word generated from the test documents is the proper base word to facilitate matching with keywords and calculating the probability of occurrence. The prototype built with the application of porter stemmer rules for Indonesian can be used to help speed up and simplify the process of classifying book types. Other study shows how to do the next stage of text processing by summarizing the paragraph [10] which can be applied too later on in this Batak documents.

2. Methodology

The method proposed for determining the basics in the Toba Batak language document consists of several steps. The steps are the collection of Toba Batak language documents that will be used as test data; tokenization is to divide sentence by sentence into words per word (token) and syllabification (stemming). After the steps are taken, you will get the essential words in the document. Each of these stages will be explained in more detail in the next section. The general architecture that describes each step of the method used in this study is shown in Figure 1.

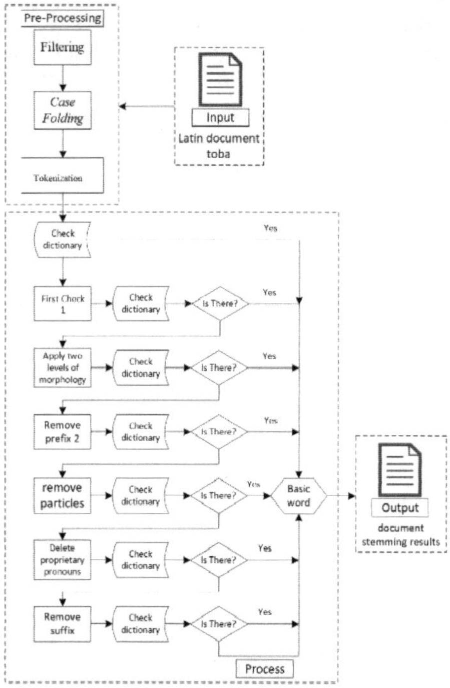

Figure 1. General Architecture.

The description are as follows:

a. Input

The input used in this system is a text file for Toba Batak Latin language documents with an extension of .docx. The document is inputted to be processed by the system.

b. Preprocessing

1. Filtering

Filtering is the process of removing punctuation marks and characters that are considered invalid, such as numbers and symbols. For example, the sentence "You are on Sipahutar taon 1995." will be filtered into "You are on Sipahutar taon".

2. Case Folding

Case-folding is the process of converting an entire text of a document into a standard form, namely lowercase letters. The purpose of this case-folding is to equalize the case in the document because in a document it is not all capital letters or all lowercase letters. For example, from the results of filtering, "You always on Sipahutar taon". Be "au on the taon line".

3. Tokenization

Tokenisation is the process of cutting a document into parts called tokens. After the filtering process and folding case are completed, the tokenization will be carried out to produce words per word that have been separated. For example, the sentence "Halaki do na manakko abit i." Is typified into [halaki] [do] [na] [manakko] [abit] [i].

c. Process Stemming

The steps taken in the stemming process are as follows.

1. Check the dictionary, if the word is in the dictionary, then the word is a basic word, and if it is not there then going to stage 2.
2. Check prefix 1 (ma-, mang-, mar-, pan-, par-, tar-), if there is no step 4, if there is the first prefix, then check the dictionary. If the word is in the dictionary, then the word is the original word, if it does not go to stage 3.
3. Apply two levels of morphology then check the glossary, if any, the word is the base word, if it does not go to stage 4.
4. Check the second prefix (si-, na-, ni-, di-, hu-, ha-, tu-, um-, pasi-, hina-), if there is no continue to stage 5, if there is delete prefix 2 then check the dictionary, if the word is in the dictionary, then the word is the primary word, if it does not go to step 5.
5. Check the particle, if there is no further step 6 if there is delete the particle then check the dictionary if the word is in the dictionary then the word is the essential word. If there is no further, go to step 6.
6. Check the proprietary pronouns, if there is no progress to stage 7, if there are delete pronouns then check the dictionary. If the word is in the dictionary, the word is the essential word. If there is no continuing to stage 7.
7. Check suffix, if it does not exist, the word is considered as the base word and returns the original word. If there is, delete the suffix then check the dictionary. If the word is in the dictionary, the word is a base word; if it does not exist, the word is considered as the base word and returns the original word.

Examples of sentences from the results of the previous process, namely [halaki] [manakko] [abit] [i] will be changed to [halak] [i] [man] [takko] [abit] [i]. The word "halaki" goes through process 1-6 and enters the 7th process which is removing the ending "-i". The word "manakko" goes through the 1-2 process and enters the third process, namely removing the "man-" prefix by applying two levels of morphology. While the word "abit" only enters the first process.

d. Output

The output of the design of this system is stemming results in the form of essential words from the inputted document. After the stemming process is complete, the stemming results in the form of essential words can be saved with extension of Microsoft Word documents.

3. Results and Discussion

When the system is completed, the order will be tested. The testing of this system aims to determine whether the system is functioning correctly and measure the accuracy of the Porter stemmer algorithm with the two-level morphological method implemented in the Toba Batak language stemming system. The accuracy of the results of testing this system will be calculated using the following formula (1).

$$\text{Accuration} = \frac{\text{Total Successful Stemmed Words}}{\text{Total Stemmed Words}} \times 100\% \quad (1)$$

The results of system testing can be seen in Table 1.

Table 1. Result of a testing system.

Input Document	**Amount of words in Document**	**Amount of Stemmed words**	**Amount of Successful Stemmed Words**	**Accurate (%)**
Doc 1	268	60	56	93,33
Doc 2	288	59	56	94,91
Doc 3	378	60	50	83,33
Doc 4	386	81	71	87,65
Doc 5	175	43	39	90,69
Doc 6	232	60	54	93,33
Doc 7	777	208	191	91,82
Doc 8	209	52	47	90,38
Doc 9	663	181	159	87,84
Doc 10	846	225	188	73,72
Doc 11	230	83	79	95,18
Doc 12	208	78	74	94,87
Doc 13	156	77	70	90,90
Doc 14	349	88	77	87,5
Doc 15	252	70	57	81,42
Doc 16	205	79	72	91,13
Doc 17	218	63	58	92,06
Doc 18	196	71	67	94,36
Doc 19	211	65	62	95,38
Doc 20	241	76	61	80,26
The Average of Accuration				89,503

Based on the results of testing, as found in Table 1, several affixes cannot be implemented by the Porter Stem algorithm with the two-level morphological method. The stemming failure is caused by several things, such as those found in Table 2.

Table 2. Stemming error.

No.	**Error**	**Example**	**Description**
1	The basic word contains more than two prefixes	Mar-si-pa-sangap-an, Mar-si-pa-sombu-an, Mar-si-pa-ture-an, Mar-na-man-sohot, etc.	The porter stemmer algorithm used in this study can only delete one and two prefixes
2	Essential words contain more than one suffix	Ihut-hon-on, Togi-hon-on, Hudus-hon-on, etc.	The porter stemmer algorithm used in this study can only delete one suffix
3	Essential words ending in vowels and affixed with pronouns belonging to hu and ta	Rohanghu, Rohanta, Pisonghu, Pisonta, etc.	The essential words ending in vowels and spelled with the pronouns belonging to -hu and -ta there are additional letters ‘ng 'and ‘n'.
4	Essential words containing prefixes and affixed endings,	Panganan (pangan), Pamatangna (pamatang), Haemuna (hae),	The first process carried out in the Porter stemmer algorithm used in this study is to delete the prefix so

	particles and proprietary pronouns		that the prefix attached to the word will be removed first then eliminate the particles, proprietary pronouns, and suffixes
5	Essential words that are affixed with inserts	Tinogu (togu) Jumadi (jadi) Harungkut (hungkut) Halaput (haput)	The porter stemmer algorithm used in this study has not been able to remove inserts

4. Conclusion

There are several conclusions:

- The porter stemmer algorithm with two morphological levels can be applied to the Toba Batak language with an accuracy rate of 89.503%.
- By using two standards of morphology to the essential words that decay if given a prefix, namely the critical words are beginning with the letter's 't', 's' and 'p', get the right essential words.
- Words that are systemized but the results of stemming are not in the dictionary will be returned to the original word. For example, a person's name and place name that is not in the dictionary.
- The more complete the word data in the dictionary, the stemming results will be more accurate because if the word is not in the dictionary, then the word will not be programmed.

Further research**:**
The development of further research are as follows:
1. It is hoped that further research can overcome the shortcomings of this stemming system, such as:
   a. To be able to be stemming on affixed words that have more than two prefixes and more than one suffix.
   b. To be able to be stemming on the essential words ending in vowels and spiked with pronouns belonging to hu
   c. To be able to be stemming on essential words that contain prefixes and affixed suffixes, particles, and proprietary pronouns.
   d. To be able to do stemming on crucial words that are attached with inserts
2. The porter stemmer algorithm can be combined with other stemming algorithms to improve stemming accuracy results.
3. Stemming results can be used to do translations into other languages
4. This Toba Batak language stemming system can be developed in the mobile version.

Acknowledgments

This research was supported by the Universitas Sumatera Utara. The study program of Information Technology Lecturer and Staff member. Mr. Manguji Nababan as lecturer and Pusat Dokumentasi Pengkajian Kebudayaan Batak (PD & PKB) Universitas Nommensen Medan who was willing to provide research data to the Author.

References

[1] Sibarani, Robert. 1997. *Sintaksis Bahasa Batak Toba*. Medan: USU Press.
[2] Malau, G. G. 2000. *Aneka Ragam Budaya Batak*. Yayasan Bina Nusantara: Jakarta.
[3] Sinaga, A. B. 2002. *Tata Bahasa Batak Toba: Meresapkan Jiwa dan Darah Batak*. Bina Media: Medan.

[4] Wiguna, P. B. S. & Bimo, S. H. 2013. Peningkatan Algoritma Porter Stemmer Bahasa Indonesia berdasarkan Metode Morfologi dengan Mengaplikasikan 2 Tingkat Morfologi dan Aturan Kombinasi Awalan dan Akhiran. *JNTETI*, Vol. **2**, No. 2.
[5] Pisceldo, F., Mahendra, R. & Manurung, R. 2008. A Two-Level Morphological Analyser for The Indonesian Language. *Australian Language Technology Association Workshop*, Vol. **6**.
[6] Budiasih, N. N., Wirayuda, T. A. B. & Dayawati, R. N. 2009. Analisis Dan Implementasi Stemming Teks Berbahasa Indonesia Dengan Menggunakan Porter Stemmer. Tugas Akhir Teknik Informatika, Fakultas Teknik Informatika, Universitas Telkom: Bandung.
[7] Zaman, B. 2014. Modifikasi Algoritma Porter Untuk Stemming Pada Kata Bahasa Indonesia. *Seminar Nasional Teknologi Informasi dan Komunikasi:* Yogyakarta.
[8] Ferari, L.A. 2014. Implementasi Algoritma Winnowing dan Porter Stemmer Mendeteksi Kemiripan Dua Dokumen Berbasis Web. Tugas Akhir Ilmu Komputer, Fakultas Ilmu Komputer dan Teknologi Informasi, Universitas Sumatera Utara: Medan.
[9] Indriyono, B. V., Utami, E. & Sunyoto, A. 2015. Pemanfaatan Algoritma Porter Stemmer Untuk Bahasa Indonesia Dalam Proses Klasifikasi Jenis Buku. *Jurnal Buana Informatika*, Vol. **6**, No. 4.
[10] Gunawan D, Pasaribu A, Rahmat R F, Budiarto R 2017 Automatic Text Summarization for Indonesian Language using TextTeaser *IOP Conference Series : Material Science and Engineering* **109**(1) 012048

Interactive Learning Media of the Solar System for Children using Kinect

R F Rahmat[1*], N Amalia[1], S Faza[2], M A Muchtar[1], R Budiarto[3]

[1]Department of Information Technology, Faculty of Computer Science and Information Technology, Universitas Sumatera Utara, Medan, Indonesia
[2]Politeknik Negeri Medan, Medan, Indonesia
[3]Albaha University, Saudi Arabia

*Corrospondence Email: romi.fadillah@usu.ac.id

Abstract. Children education adheres to the emphasis of playing while learning or learning while playing. Playing using the body muscles can stimulate the body senses of a child and assist the child in exploring the surroundings. Children discover and learn new things or skills through playing a game. In creating an interactive learning experience to stimulate the brain of a child, a solar system is included in the formal education curriculum packed in an interactive media using Kinect where the children will play an active role in studying the solar system. The Kinect sensor detects 23 joints of a child to be used to move the 3-dimensional astronaut object directly on the body (coordinates mapper) and uses the background removal on the image to separate the background and the child pixels, so the child seems to be beside the planet. Based on the test results using the respondents of 50 children, it was shown that the system effectiveness level for the children is 86%.

1. Introduction

The solar system consists of the Sun, eight official planets, at least three "dwarf planets," more than 130 satellites on the planet, many comets and asteroids, and interplanetary media. The inner solar system contains the Sun, Mercury, Venus, Earth, and Mars. The main asteroid belt is located between the orbits of Mars and Jupiter. The outer planets in the solar system are Jupiter, Saturn, Uranus, and Neptune. The first thing to note is that the solar system is mostly empty space. The planets are atomic compared to the space of the solar system. It is even too large to be measured on a scale in relation to the orbit size.

The golden age is the reason for the importance of early childhood education. Beeker, in his book, classified the golden age as the children in the age of 3-6 years. This group of children usually attends pre-school or kindergarten. In the book, child education must be prepared in a planned and holistic manner so, in the golden age, the development of the children can be complete to develop and showcase various potentials.

An interactive media for the introduction to solar system using Kinect is required to reach the peak point of the children potential in the golden age. The media provides the children with a learning-by-doing method to learn actively.

A research was conducted to create a robot that can be controlled using Kinect sensors so the robot can move according to user movement. The robot can follow the user and calculate the distance between it and the user using the depth value of the coordinates mapping component and the Kinect component. The limit ranges of the robot movement are 2-4 meters for forward motion, 1.5-2 meters for stopping and 0.5-1.5 meters for backward movement [1].

Another study was performed to create a virtual fitting room system that can change the texture and color of the users using Kinect 2. The proposed method of the study had been jointly developed focusing on virtual fitting room applications. This research replaced the shirt texture with a new costume texture from an image. This was made using the camera of Microsoft Kinect 2 to obtain the depth image of the scene. In the study, the shirt segmentation was carried out using Kinect depth information to effectively eliminate the background and manually mark some boundaries of the garment and body. The texture coordinates were calculated using the information of the real object coordinates that matches with the pixel coordinates in the image. The system created a realistic perspective by only considering the real world coordinates of X and Y when making the maps. The effect and result look very realistic. The lighting estimation was achieved by equalizing the lighting of the original image. The estimated lighting was used to shade the texture color of the new image in order to acquire more realistic results where some parts of the surface are in shadow while the other parts are in bright [2].

Another study was conducted on Kinect project mapping using the combination of augmented reality and the sensor and projector of the Kinect. In the research, a framework was developed to calculate the calibration data and achieve dynamic mapping using Microsoft Kinect. The ultimate goal was for the graphical applications to be able to utilize data derived from the Kinect-Calibration projector and apply it to a real environment which results in the visualizations of Augmented Reality from the captured scenes in the runtime [3].

A research on virtual fitting room using Kinect sensor was conducted in 2012. The system allowed the users to choose their clothes in the form of a sleeveless dress and save the result. The input image of the system is a dataset consisting of dress images that have been separated from the background. Kinect recorded the user movements and identified the head, shoulder center, shoulder left, shoulder right and hip center. The processed image will be mapped out, so the user seemed to wear a different outfit [4]. Several interactive application has been developed before such as Augmented reality (AR) for Hijaiyah Learning [5], AR for children with special needs [6], activity learning using traditional game in Android [7]

The solar system is a collection of celestial bodies that are bound by gravitational forces to form a system. A system is required to simplify the learning process that can be used as an interactive learning media about the Solar System where children can play a direct role through actions and visuals in real-time (learning by doing).

2. Methodology

The proposed method of this study consists of several processes. The initial step is to connect the system with a Kinect sensor. There are several processes in the system; the first is to detect the presence of a user within the distance of Kinect scope, then transform the image of the user into a skeleton form using SDK function in the Kinect SDK and background removal to detect the joints of the user. After the joints are detected, the GameObject (Astronaut) is mapped (Coordinates Mapper) to the joints, and the motions are adjusted to the joint motion (Body Joint Orientation). The general architecture of this study is shown in Figure 1.

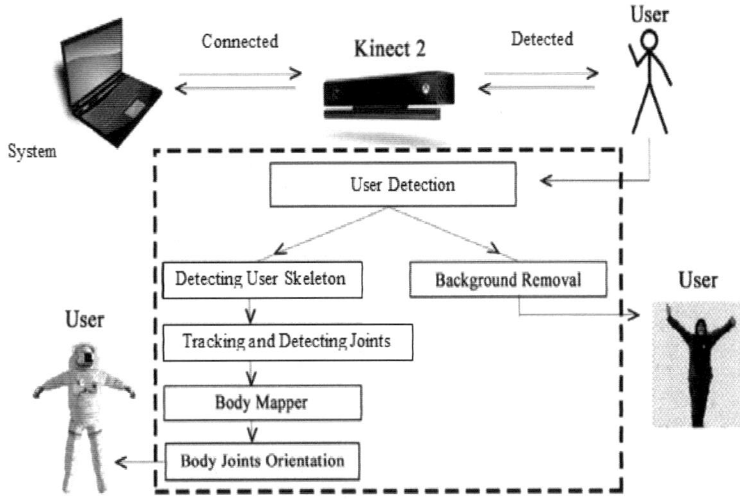

Figure 1. General Architecture.

2.1. Research Data

The system collected the data of the joints in the Kinect to use the coordinates mapper. Of all the 25 detected joints, the system selected 23 joints to be applied. The list of joints is shown in Table 1.

Table 1. List of joints

Joint Order	Joint Name	Joints Type
0	Spine base	Spine Base
3	Head	Head
4	Shoulder left	Shoulder Left
5	Elbow left	Elbow Left
6	Wrist left	Wrist Left
7	Hand left	Hand Left
8	Shoulder right	Shoulder Right
9	Elbow right	Elbow Right
10	Wrist right	Wrist Right
11	Hand right	Hand Right
12	Hip left	Hip Left
13	Knee left	Knee Left
14	Ankle left	Ankle Left
15	Foot left	Foot Left
16	Hip Right	Hip Right
17	Knee right	Knee Right
18	Ankle right	Ankle Right
19	Foot right	Foot Right
20	Spine shoulder	Spine Shoulder
21	Hand tip left	Hand Tip Left
22	Thumb left	Thumb Left
23	Hand tip right	Hand Tip Right
24	Thumb right	Thumb Right

In addition to joints data, the system requires 3D astronaut and planetary objects that are given a texturing image of a planet. Planet texturing aims for a planet to look like its actual state. The image texture of the planetary can be seen in Table 2.

Table 2. List of the planet textures

Object	Texture
Sun	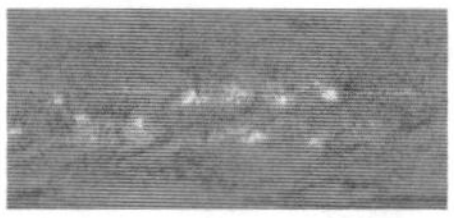
Mercury	
Venus	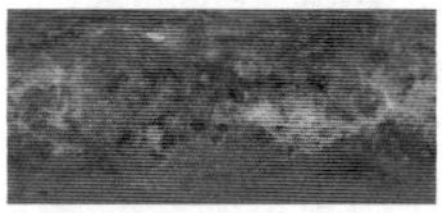
Earth	
Moon	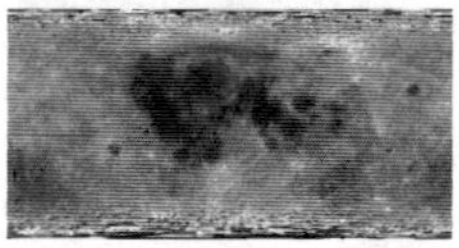
Mars	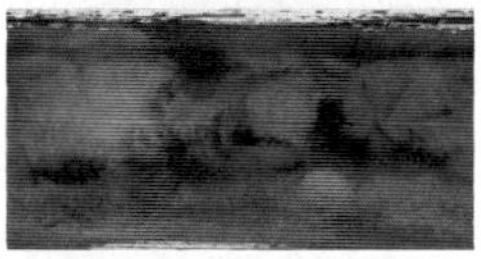
Jupiter	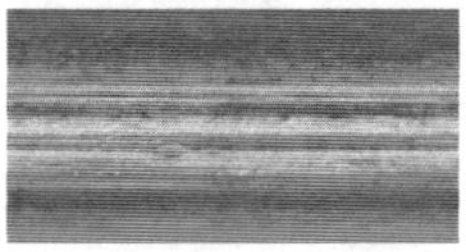
Saturn	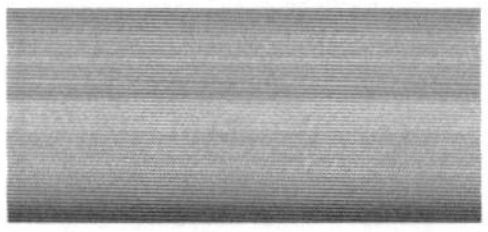
Ring of Saturn	
Neptune	

IOP Conf. Series: Materials Science and Engineering **648** (2019) 012026 doi:10.1088/1757-899X/648/1/012026

2.2. Detection of User Skeleton

The skeleton in the frame can have a tracking status of "Tracked" or "Position Only." The tracked framework provides detailed information about the position in the camera viewpoint consisting of twenty-five joints of the user's body. A framework with tracking status "Position Only" has information about the user's position, but there are no details about the joint.

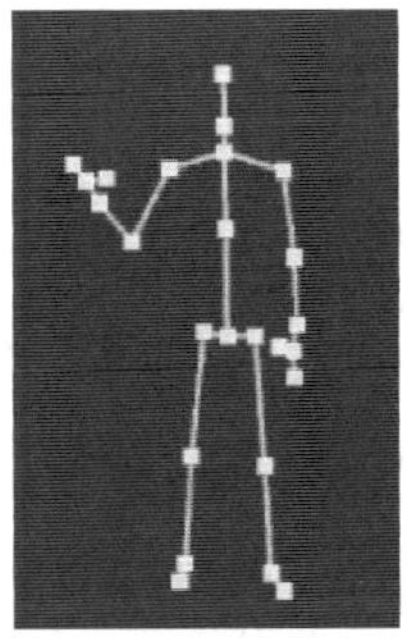

Figure 2. Skeleton and Joints Tracking.

The application can specify the framework to be tracked using the tracking ID as shown in the Active User Tracking section. For the tracked skeletons, a joint array provides the position of twenty-five human joints in space. In an example, the application can use the hand joint to guide the cursor on the screen or pull the position of the user body on the screen. The joint also has a tracking status of "Tracked" for a connection that is visible, "inferred" if the connection is not visible and Kinect will decide its position, and "non-tracked," for the bottom joint when the user is sitting down.

2.3. Background Removal

Background removal is a separation process between user pixels and background pixels. When we wipe out the background, it is necessary to keep the pixels representing the user and remove the other redundant pixels. The depth camera of Kinect sensor is utile in determining the user body. However, instead of the distance, it is compulsory to find the value of RGB. We need to determine the RGB value that matches with the depth value of the user. Using Kinect, every point in the space has the following information: Color value: Red (Red) + Green (Green) + Blue (Blue). Depth Value: Distance from the sensor.

The depth camera provides the depth values, and RGB camera produces the color value. CoordinateMapper is used to map these values. CoordinateMapper is a useful Kinect property that determines which color value suits each distance and vice versa. Note that the RGB frame (1920 × 1080) is wider than the frame depth (512 × 424). As a result, not every color pixel has the appropriate depth mapping. However, body tracking is mainly performed using depth sensors, so there is no need to worry about missing values. Background removal has a result like the green screen.

2.4. Coordinate Mapper and Body Joints Orientation

The Coordinate Mapper is to identify whether the dots of 3D space correspond to 2D color or depth space and vice versa. CoordinateMapper is part of the KinectSensor class. In the Kinect, there is a Camera Space referring to the 3D coordinate system used by Kinect. The coordinate system is defined as follows: Origin ($x = 0$, $y = 0$, $z = 0$) is located in the center of the IR sensor on the Kinect, X grows to the left of the sensor, Y grows up above the sensor (note that this direction is based on the sensor slope), Z grows in the direction of the sensor and 1 unit = 1 meter. Depth Space is a term used to describe 2D locations in depth images. Consider this as the location of row or column of a pixel where x is the column and y is the row. So $x = 0$, $y = 0$ corresponds to the top left corner of the image and $x = 511$, $y = 423$ (width-1, height-1) is the lower right corner of the image. In some cases, a z value is needed to map the inner space. In this case, it is sufficient only with the sample of the depth image in the row or column and uses that value (in millimeters) directly as z.

A common operation in depth image is to produce 3D points at the target point. In this case, it will not project from the depth space to the camera space. The operation uses one of the array-based functions (MapDepthPointsToCameraSpace) or the mapping table (GetDepthFrameToCameraSpaceTable) then the multiplication can be performed. After the skeleton and joints of the user are obtained, the next task is to map 3D objects into the joint coordinates. It is necessary to divide the 3D objects according to the types of mapped joints. The joints of the user can be accessed using BodyJoint handleft = body.Joints [JointType.HandLeft], and carried out to each joint to be accessed.

The Body Joints Orientation is applied for the mapped 3D objects to move according to the type of the mapped joint. Kinect reads the orientation value of the joints as a quaternion. A quaternion is a set of 4 values: X, Y, Z, and W. Kinect SDK encapsulate the quaternion into a structure called Vector4. This Vector4 needs to be converted into a set of 3 numerical values. The rotation of the joints around the axes X, Y, and Z can be calculated using quaternion orientation. The rotation around the X-axis is called Pitch, the rotation around the Y-axis is called Yaw, and the rotation around the Z-axis is called Roll. The body joints orientation can be summoned using the command of BodyJointOrientation handLeftRot = body.JointsOrientations[JointType.HandLeft].

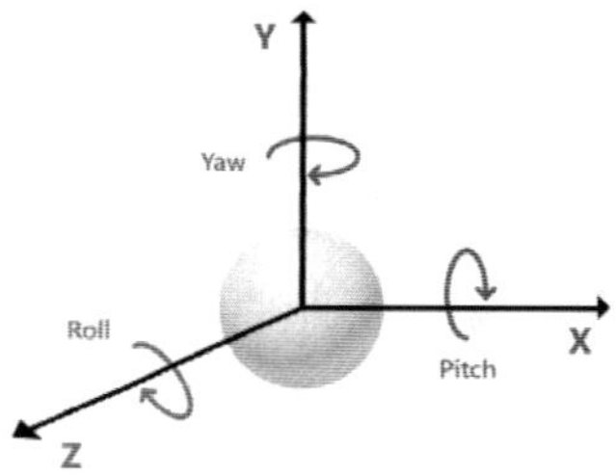

Figure 3. Joint Direction.

2.5. *Activity Diagram*

Figure 4 describes the process of system activity. First, the user runs the application and checks the connection of the Kinect sensor. If the Kinect is connected, the system displays the user interface, and If not connected, it will go back to the start screen. After the system connected, it will detect whether there is any user within the distance range of the Kinect. If no user is detected, the display remains unchanged. The system detects the user but only when the user raises both hands. These users will be tracked as the player. Every movement of the detected player will be tracked, and the player's hand will serve as the cursor to run the system. The cursor will move according to the hand movements, and the user can change which hand is made for the cursor. The user should put his current hand cursor behind and put the other hand upfront to change the hand for the cursor.

After the player is detected, the player appears in the system. The user can choose to appear as himself or an astronaut in 3D while playing in real-time. When a player uses the system, the system can detect and display up to 6 players on the system screen in background removal mode. Players can choose the selection of the solar system and planet to learn and see the information of the selected objected in detail. When a player is in the solar system screen, the player can see how each planet moves around the sun. The player then can move to the next view, which is the detailed view of the planet. The user is required to touch his head to see the details of the planet when in astronaut mode. The users can see the planet and the direction of its rotation up close. They can also see the information on each planet in the information bar which is presented in the form of text and sound. When the information bar shows, the voice information will also be heard by the user. Players can exit the system by selecting the exit button symbolized by an X. However, players can also return to the previous Solar System view with a menu symbolized by a triangle.

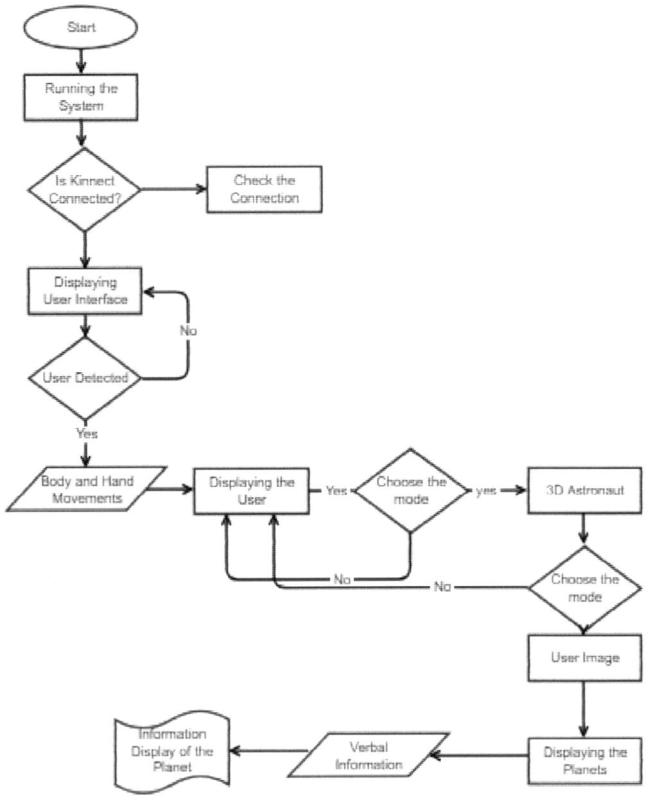

Figure 4. Activity Diagram.

3. Result and Analysis

3.1. System Requirements

Several specifications were required in building an interactive learning system using a Kinect sensor to support the application running smoothly. The minimum requirements are as follows:

1. Processor Intel(R) Core(TM) i3-2348M
2. 64-bit (x64) processor
3. Dual-core 3.1 GHz
4. Memory (RAM) of 4 GB
5. Kinect adapter
6. USB 3.0 host controller
7. The graphical adaptor of DX11:
   - Intel HD 4400 integrated display adapter
   - ATI Radeon HD 5400 series
   - ATI Radeon HD 6570
   - ATI Radeon HD 7800 (256-bit GDDR5 2GB/1000Mhz)
   - NVidia Quadro 600
   - NVidia GeForce GT 640
   - NVidia GeForce GTX 660
   - NVidia Quadro K1000M
8. The operating system of Windows 8, Windows 8.1 or Windows 10

9. Visual Studio 2013 and Unity 5
10. Kinect SDK V 2

3.2. Implementation of System Design

The first page that appears is the opening screen. The camera runs from the direction of the sun to the last planet, Neptune. This view aims to introduce the whole planet. The opening display can be seen in Figure 5.

Figure 5. Welcome Page.

The second display of this application is to present the images of the orbital movements of the entire planet around the sun or the revolution. This has the purpose of introducing the direction of each planet in the revolution process. Figure 6 shows the orbit of the planet in the three-dimensional model in the astronaut mode, while Figure 7 shows the planet's orbit with the background removal mode or using the image of the player himself. For the background removal mode, the system can display up to 6 people, but only one person is tracked as a pointer.

Figure 6. Planet Orbit View in Three-Dimensional Object Mode.

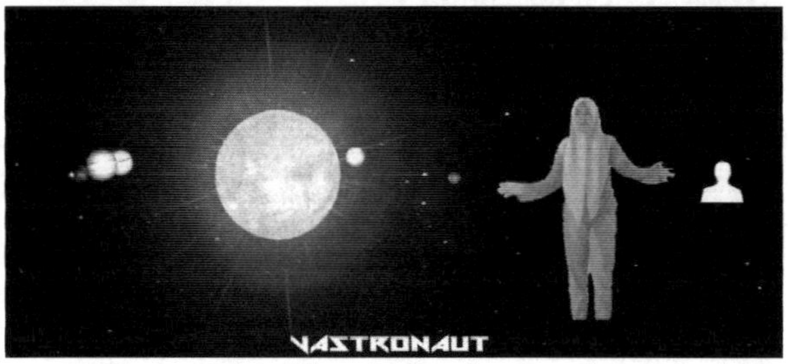

Figure 7. Planet Orbit View in Background Removal Mode.

The next one is a screen containing information about each planet and the sun. Figure 8 shows a planetary information display in three-dimensional Object mode, while Figure 9 presents the display in background removal mode. In this mode, Kinect can detect up to 6 users.

Figure 8. Information Display in three-dimensional object mode.

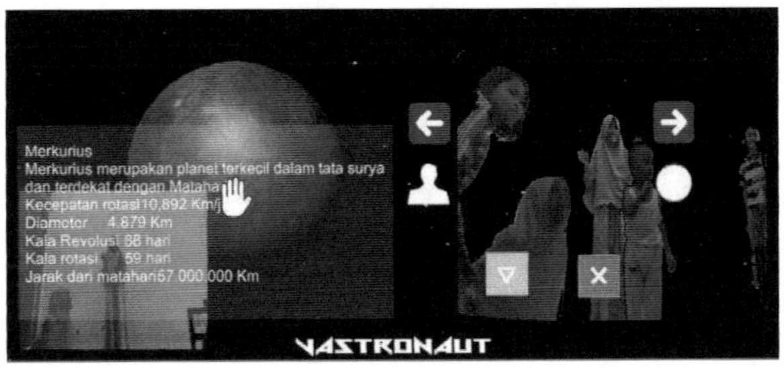

Figure 9. Information Display in Background Removal Mode.

3.3. Application Testing

Authors conducted a questionnaire on 50 respondents of all elementary school students in SDN 105287 with ages ranging from 6 to 9 years to obtain a more accurate result. The respondents were initially guided in the practice of using the system; then they would try the features in the system independently. After the respondents tried all the features available, they were invited to fill out a questionnaire containing a poll on the level of agreement between interaction and gesture. The statements in the poll are as follows:

1. The solar system is easy to use.
2. The solar system is very interactive.
3. Convenience in using the solar system application.
4. The body movements on the screen are synchronized with mine.
5. Able to see the image of oneself on the screen
6. Able to change the body to an astronaut on the screen.
7. Can use the hand as the pointer.
8. The design of the planets is intriguing.
9. Can remember the planet names in the solar system.

All of these statements are linked with a Likert scale to obtain the results. The scale marks the strongly agreed (SA) points as 5 and the strongly disagreed (SD) as 1. The full marks can be seen in Table 3.

Table 3. List of questionnaire value

Scale	Value
Strongly Disagree (SD)	1
Disagree (D)	2
Neutral (N)	3
Agree (A)	4
Strongly Agree (SA)	5

Based on the values in Table 3, the average value of each question can be calculated. The average value of each statement at the questionnaire for interaction and gesture can be seen in Figure 10.

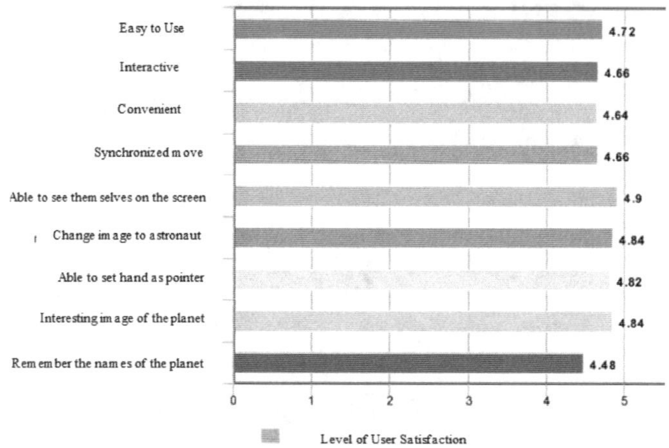

Figure 10. Graphic of User Satisfaction on Interaction and Gesture.

Furthermore, the system testing was carried out on the user height with a predetermined distance. The test result can be seen in Table 4.

Table 4. System testing based on height

User Height	Distance from the Kinect	Detection Time
98 cm	± 1 m	2.60 Seconds
102 cm	± 1.15 m	5.5 Seconds
106 cm	± 1.15 m	2.39 Seconds
111 cm	± 1.15 m	1.75 Seconds
123 cm	± 1.5 m	1.05 Seconds
137 cm	± 1.5 m	1.44 Seconds
142 cm	± 1.5 m	3.76 Seconds
160 cm	± 2 m	1.18 Seconds
166 cm	± 2 m	6.21 Seconds
180 cm	± 2.10 m	4.43 Seconds
Total Time		**30.31 Seconds**
Average Detection Time		**3.031 Seconds**

Based on Table 4, the test was conducted on users with different heights. The players were ranged in age of 6 to 25 years old. Every function in the system performed well. The distance between Kinect and user was adjusted to be able to show the entire body of the user on the screen. If the screen only displayed half of the user body, the leg will fall apart in the astronaut mode. However, it is not a problem in background removal mode. NPS (Net Promoter Score) method was used to measure the level of user satisfaction on the system. The method uses the scale of 0 - 10 where 0 - 6 indicates dissatisfaction and detractors, 7 - 8 represents the passives who are quite satisfied but still indifferent to the system, while 9 - 10 indicates the promoters, users who are satisfied with the system. NPS score can be obtained by subtracting the scores of satisfied users with the unsatisfied ones. Net Promoter Score is always illustrated in absolute number and not a percentage. Net Promoter Score can range from as low as -100 (if all the users are detractors) to as high as 100 (if all the users are Promoters). The NPS data of this study are shown in Table 5.

Table 5. NPS Score of The Users

Score	Number of Users
10	7
9	14
8	13
7	11
6	5
5	0
4	0
Total User	**50**

$$Percentage\ of\ the\ promoters = \frac{21}{50} \times 100$$
$$= 42\%$$

$$Percentage\ of\ the\ passives = \frac{24}{50} \times 100$$
$$= 48\%$$

$$Percentage\ of\ the\ detractors = \frac{5}{50} \times 100$$
$$= 10\%$$

$$NPS\ Score = Percentage\ of\ the\ promoters - Percentage\ of\ the\ detractors$$
$$= 42\% - 10\%$$
$$= 32\%$$

There are many reasons in the necessity to measure the system functions. The most common one is the need to communicate effectively with the interests of the evaluated system. Other is to meet the requirements of comparing the benefits of two or more systems and to identify any malfunction in the system. Of the total 50 users, 43 of them understood the solar system from the system trial. These users answered questions about the solar system correctly. The effectiveness of the system is shown in a percentage using the following equation.

$$Effectiveness = \frac{Successful\ task}{Total\ task}\ x\ 100\ \%$$
$$= \frac{43}{50}\ x\ 100\%$$
$$= 86\ \%$$

3.4. Kinect Limitations

During the period of building and testing the system, the authors noted some limitations in using Kinect sensors. First is that the Kinect sensor cannot function properly in a room with a full background. Kinect cannot detect users if the background has lots of patterns or other objects. Second, Kinect cannot detect users if the room is dark or in minimum light condition.

4. Conclusion and future research

Based on the test results, some conclusions can be drawn as follows:

1. The system can detect the users with a height of 98 cm - 180 cm in a predetermined distance.
2. Although the system is created for the children, it is also suitable for adults.
3. The system can detect up to 6 people but only those who raise their hands to be tracked and considered as users.

References

[1] Gahlot A, Agarwal P, Agarwal A, Singh V and Gautam A K 2016 Skeleton based human action recognition using Kinect *Intl J. of Computer Applications* 0975 – 8887.
[2] Traumann A, Anbarjafari G H and Escalera S 2015 A new retexturing method for a virtual fitting room using Kinect 2 camera *CVPR*.
[3] Motta T, Loaiza M and Raposo A 2014 Kinect projection mapping *SBC J. on Interactive Systems* vol. **5** 3.
[4] Lan Z 2012 *Augmented reality: virtual fitting room using Kinect* Singapore: University of Singapore.
[5] Rahmat R F, Akbar F, Syahputra M F, Budiman M A, Hizriadi A 2018 An Interactive Augmented Reality Implementation of Hijaiyah Alphabet for Children Education, *Journal of Physics: Conference Series* **978**(1) 012102
[6] Syaputra M F, Sari PP, Arisandi D, Abdullah D, Napitupulu D, Setiawan M I, Albra W, Asnawi, Andayani U 2018 Implementation of augmented reality to train focus on children with s pecial needs *Journal of Physics: Conference Series* **978**(1) 012109
[7] Rahmat R F, Ramadhan R, Arisandi D, Syahputra M F, Sheta O 2018 Rangku Alu – A Traditional East Nusa Tenggara Game in Android Platform *Journal of Physics: Conference Series* **978**(1) 012103

Identification of Melanoma through Dermoscopy Image using Learning Vector Quantization

S Purnamawati[1], R F Rahmat[1*], S Faza[2] and A J Lumbantobing[1]

[1]Department of Information Technology, Faculty of Computer Science and Information Technology, Universitas Sumatera Utara, Medan, Indonesia
[2]Politeknik Negeri Medan, Medan, Indonesia

*Email: romi.fadillah@usu.ac.id

Abstract. Melanoma is one of the rare and malignant types of skin cancer. There are numerous ways to detect the melanoma, and one of them is through doctor diagnosis. A doctor can detect melanoma after a thorough medical check-up. If the patient has symptoms of melanoma, a biopsy will be carried out. This process requires a long time making it inconvenient. Therefore, an approach of the image processing system is necessary to assist the experts in diagnosing melanoma. The process consists of image input using the dermoscopy image, a pre-processing process of grey-scaling and median filtering, and feature extraction using Grey-Level Co-Occurrence Matrix (GLCM). In the final step, a classification process will be performed using learning vector quantization. Based on the experimental test, the system generated an accuracy of 83.33% in identifying melanoma cancer.

1. Introduction

The skin is the outermost protective layer of the human body. The skin has many functions such as the protection of the body from dangerous things, the sense for touch, also has the role in excretion for regulating body temperature and other vital roles. Although the skin has an important role in the body, the skin can become infected. One of the severe diseases that attack the skin is melanoma skin cancer.

Melanoma skin cancer is a rare and terminal form of skin cancer. Melanoma is a malignant skin tumor originating from melanocytes. Approximately 3% of skin cancers are a melanoma [1]. Melanocytes are the skin cells that produce melanin dyes [2]. Melanin aims to absorb ultraviolet light and protect the skin from damage. Sunlight or ultraviolet light is the main causative factor [1]. If someone is exposed to sunlight for a long time, the risk of suffering from skin cancer and other skin diseases becomes higher.

There are many ways to diagnose melanoma. One of them is with the help of a doctor. Doctors can diagnose melanoma through a physical examination. If the patient has symptoms of melanoma, the doctor will refer the patient to see a dermatologist. The most common cases are in moles. Melanoma can develop from an existing or newly emerging mole [2]. Tissue moles that are considered suspicious will be removed surgically and observed whether they have become cancerous. This process is known as a biopsy. The abnormalities of the skin that are suspected as melanoma will be treated by extensive excision or amputation depending on tumor localization [1]. However, the weakness of biopsy is the need for a long preparation and healing time. This may result in cancer cells spread more widely.

As the advancement of technology, melanoma can be identified through imaging analysis. Based on the inputted image, the information will be obtained from the image after going through the image processing process.

Melanoma is one of the malignant types of cancer. In general, identifying melanoma through dermoscopy images is still performed manually by the pathologist. Therefore, a method is needed to identify melanoma automatically, so the examination results are faster than the manual diagnosis.

In this study, the authors limit the problem to prevent the problem scope from expanding in this study. The limitations of the problem, including:

1. The image data used is dermoscopy image.
2. The image is an image with the format .bmp
3. The dataset was collected from https://www.fc.up.pt/addi/ph2%20database.html
4. The classification is divided into two, namely melanoma and non-melanoma.

The purpose of this study was to identify melanomas through dermoscopy images using Learning Vector Quantization. The benefits of this research are:

1. Assist the doctor or the user to identify melanoma through dermoscopy image.
2. Reference for further research.

Numerous past researches related to this study had been performed. A study entitled "Advanced Earlier Melanoma Detection Algorithm Using Colour Correlogram" discussed a detection system using color correlogram. The authors implemented the algorithm for feature extraction and SFTA for texture analysis. A Bayesian classifier was applied in the classification process to cluster the result into three types, i.e. cancer, non-cancer and atypical [3].

Bhati et al. performed a research entitled “Early Stage Detection and Classification of Melanoma". The study implemented a thresholding and Otsu algorithm, while ABCD method was applied for feature extraction. In the classification process, TDS Calculation Classification identified the results as malignant or benign cancer [4], similar approach by Nababan et al, which is using Evolving Connectionist Systems as classifier [5].

Another study was conducted by Chatterjee et al. entitled "Mathematical Morphology aided Shape, Texture and Color Feature Extraction from Skin Lesion for Identification of Malignant Melanoma". In the segmentation stage, the authors implemented a mathematical morphology method, and GLCM was applied to extract the features based on shape, texture, and color. Support Vector Machine (SVM) would generate the classification results [6].

Suganya conducted another study about an automated computer-aided diagnosis of skin lesions detection and classification using dermoscopy images. In the research, k-means clustering was applied as the segmentation method. The classification process used a support vector machine (SVM) method to produce the results in two stages. The first stage is to classify the melanoma and non-melanoma while the second one is to classify the melanoma into melanoma or nevus and the non-melanoma into bcc or sk [7]. Other research using waveflet transform [8] and recurrent neural network [9]

2. Methodology

2.1. General Architecture

The proposed method to identify melanoma consists of several stages. These stages begin from acquiring image data consisting of melanoma images and non-melanoma images that will be used as training images and test images; then it will process in the pre-processing stage which consists of two processes, i.e., grey-scaling to uniform grey images, and median filtering to minimize the influence of small objects. The process continues to the feature extraction stage using Gray Level Co-occurrence Matrix (GLCM). The last step is the classification process using learning vector quantization method to identify whether the input is considered as melanoma or not. The general architecture of this study is shown in Figure 1.

2.1. Pre-Processing

This stage is one of the image processing stages aiming to produce a better image to be processed for the next stage. This pre-processing stage consists of grayscaling and median filtering.

Grayscaling aims to change the pixel intensity of the image to gray. The result will be used for the median filter process. Median filter process is to improve the image quality by minimizing the influence of small objects such as thin hair, scratches on the skin and air bubbles.

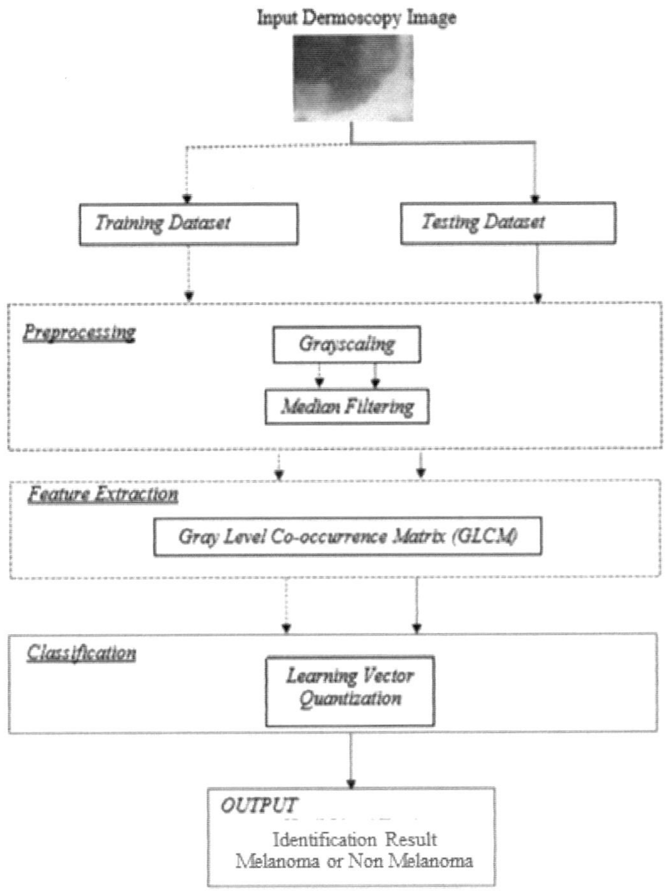

Figure 1. General Architecture.

2.2. Gray Level Co-occurrence Matrix (GLCM)

Gray Level Co-occurrence Matrix (GLCM) is one of the second-order texture analysis methods. This method was introduced by Haralick et al. in 1973. GLCM represents a neighboring two-pixel relationship where two related pixels have a certain gray intensity, distance, and direction between them. The distance and direction are expressed in pixels and angles respectively. The distance can be worth 1, 2, 3 and so on, while the direction can be set at 0 °, 45 °, 90 °, 135 ° and so on.

Only fourteen parameters are proposed to obtain the texture feature of GLCM. Some features to-be-used are:

1. *Contrast*

Contrast is the calculation of the difference in intensity between one pixel and adjacent pixels throughout the input image. Contrast can be calculated using the following formula:

$$Contrast = \sum_i \sum_j (i-j)^2 P_{i,j} \tag{1}$$

2. *Energy*

Energy or commonly called Angular Second Moment (ASM) is a measure of image homogeneity calculated in the following equation:

$$Energy = \sum_i \sum_j p_{i,j}^2 \tag{2}$$

3. *Entropy*

Entropy is a linear dependency measurement between gray-level values in an image using the following formula:

$$Entropy = \sum_i \sum_j P_{i,j} \log P_{i,j} \tag{3}$$

4. *Correlation*

Correlation is a measure of linear dependence between gray-level values in an image. Correlation is calculated as follows:

$$\text{Correlation} = \sum_{i=1}^{L} \sum_{j=1}^{L} \frac{(ij)GLCM(i,j) - \mu i \mu j}{\sigma i \sigma j} \tag{4}$$

5. *Homogeneity*

Homogeneity is used to measure the distribution proximity of each element in the GLCM matrix to the diagonal GLCM matrix. Homogeneity is calculated in the following way:

$$\text{Homogeneity} = \sum_{i,j} \frac{p(i,j)}{1+|i-j|} \tag{5}$$

6. *Shade*

Shade can be calculated using the following equation:

$$\text{Shade} = \sum_{i=0}^{G-1} \sum_{j=0}^{G-1} \{i + j - \mu x \mu y\}^3 X\ P(i,j) \tag{6}$$

2.3. Learning Vector Quantization

Learning Vector Quantization (LVQ) is one of the supervised neural network algorithms that uses competitive learning strategies known as winner-take-all networks [10]. A competitive layer will automatically learn to classify input vectors. Classes are generated based on the distance of the vector. If two vectors have close distances, they will be grouped into one same class. The model of LVQ algorithm can be seen in Figure 2.

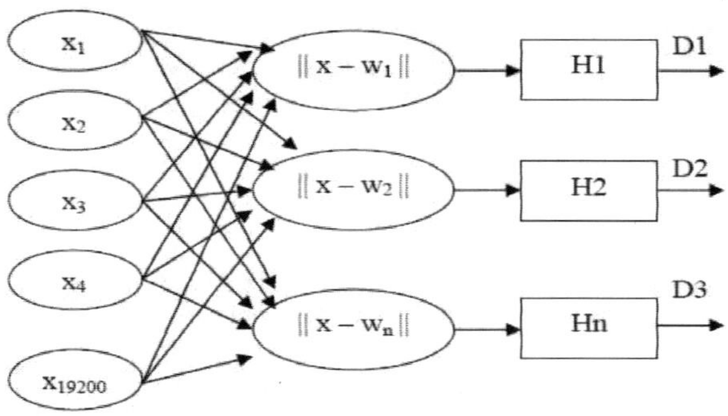

Figure 2. LVQ Architecture.

3. Result and Analysis

3.1. System Implementation and Design

The classification system of melanoma using the Learning Vector Quantization (LVQ) method requires numerous supporting hardware and software, including Intel® Core ™ i3-2332M CPU @ 2.20GHz, hard disk capacity of 500 GB, 4.00 GB RAM processor, the operating system of Windows 10 in 64 bit, and Matlab R2016a.

IOP Conf. Series: Materials Science and Engineering **648** (2019) 012027 doi:10.1088/1757-899X/648/1/012027

3.2. Data Training

The training process was conducted using 83 dermoscopy images consisting of 32 data of melanoma images and 51 data of non-melanoma images. The classification process implemented learning vector quantization neural networks. The parameters for the data training were learning rate (α) of 0.02, hidden size of 100, and epoch of 100.

3.3. Data Testing

In data testing process, the inputs were 30 images consisting of 10 images of melanoma cancer and 20 images of non-melanoma. The results of data testing can be seen in Table 1.

Table 1. Test results

No	Image Name	Actual Output	Desired Output	Status
1	1.bmp	Melanoma	Non-Melanoma	Failed
2	2.bmp	Melanoma	Non-Melanoma	Failed
3	3.bmp	Melanoma	Melanoma	Succeed
4	4..bmp	Melanoma	Non-Melanoma	Failed
5	5.bmp	Melanoma	Melanoma	Succeed
6	6.bmp	Melanoma	Non-Melanoma	Failed
7	7.bmp	Melanoma	Melanoma	Succeed
8	8.bmp	Melanoma	Melanoma	Succeed
9	9.bmp	Melanoma	Non-Melanoma	Failed
10	10.bmp	Melanoma	Melanoma	Succeed
11	11.bmp	Non-Melanoma	Non-Melanoma	Succeed
12	12.bmp	Non-Melanoma	Non-Melanoma	Succeed
13	13.bmp	Non-Melanoma	Non-Melanoma	Succeed
14	14.bmp	Non-Melanoma	Non-Melanoma	Succeed
15	15.bmp	Non-Melanoma	Non-Melanoma	Succeed
16	16.bmp	Non-Melanoma	Non-Melanoma	Succeed
17	17.bmp	Non-Melanoma	Non-Melanoma	Succeed
18	18.bmp	Non-Melanoma	Non-Melanoma	Succeed
19	19.bmp	Non-Melanoma	Non-Melanoma	Succeed
20	20.bmp	Non-Melanoma	Non-Melanoma	Succeed
21	21.bmp	Non-Melanoma	Non-Melanoma	Succeed
22	22.bmp	Non-Melanoma	Non-Melanoma	Succeed
23	23.bmp	Non-Melanoma	Non-Melanoma	Succeed
24	24.bmp	Non-Melanoma	Non-Melanoma	Succeed
25	25.bmp	Non-Melanoma	Non-Melanoma	Succeed
26	26.bmp	Non-Melanoma	Non-Melanoma	Succeed
27	27.bmp	Non-Melanoma	Non-Melanoma	Succeed
28	28.bmp	Non-Melanoma	Non-Melanoma	Succeed
29	29.bmp	Non-Melanoma	Non-Melanoma	Succeed
30	30.bmp	Non-Melanoma	Non-Melanoma	Succeed

Based on the test results of the melanoma identification system, the accuracy rate of the system is 83.33%.The accuracy value was obtained from the following equation.

$$Accuracy = \frac{Number\ of\ correctly\ identified\ data}{Number\ of\ data\ testing}$$

$$= \frac{25}{30}\ x\ 100\ \%$$

$$= 83.33\%$$

3.4. Analysis of Precision and Recall

Precision is the fraction of relevant instances among the retrieved instances, while recall is the fraction of relevant instances that have been retrieved over the total amount of relevant instances. In this study, precision and recall were used to measure the performance of learning vector quantization algorithms (LVQ) in identifying melanoma. The result of the values of precision and recall can be seen in Table 2.

Table 2. Precision and recall analysis

	Melanoma	**Non-Melanoma**
Relevant **(a)**	5	20
Irrelevant **(b)**	5	0
Total **(a+b)**	10	20
Missing **(c)**	5	0
Total **(a+c)**	10	20
Recall **[a/(a+c)] x100%**	50%	100%
Precision **a/(a+b)]x100%**	50%	100%
Average	**75%**	**75%**

Based on Table 2, the average value of precision and recall is 75%. Although the value of precision is the same as the recall value, the effectiveness of the information retrieval system in this study can be considered as effective. Thus it can be concluded that this system performs well.

4. Conclusion and future research

Based on the test result of the melanoma identification system using learning vector quantization algorithms, it can be drawn some conclusions as follows:

1. The Learning Vector Quantization (LVQ) method can properly identify melanomas through dermoscopy images and obtain an accuracy rate of 83.33%.
2. Based on the experimental results for the selection of the parameters, the learning rate of 0.02 is the best value in identifying melanoma in data testing.
3. The shape and size of the image can affect the accuracy. The similarity of the images of melanoma and non-melanoma may cause a failure in the identification process.

References

[1] Tambunan G W 1991 *Diagnosis and treatment of ten most common cancer in Indonesia* Jakarta: EGC Publisher.

[2] Supriyanto W 2010 *Threat of early-stage cancer and its treatments* Yogyakarta: Cahaya Ilmu Publisher.

[3] Soumya S R, Neethu S, Niju T S, Renjini A and Aneesh R P 2016 Advanced earlier melanoma detection slgorithm using colour correlogram *Intl Conf. on Communication Systems and Networks (ComNet)*.

[4] Bhati P and Singhal M 2015 Early stage detection and classification of melanoma *Intl Conf. on*

Communication, Control and Intelligent Systems (CCIS).

[5] Nabbaban E B, Iqbal M, Rahmat R F 2017 Breast cancer identification on digital mammogram using Evolving Connectionnist Systems *International Conference on Informatics and Computing* pp 132-136.

[6] Chatterjee S, Dey D and Munshi S 2015 Mathematical morphology aided shape, texture and color feature extraction from skin lesion for identification of malignant melanoma *Intl Conf. on Condition Assessment Techniques in Electrical Systems (CATCON)*.

[7] Suganya R 2016 An automated computer-aided diagnosis of skin lesions detection and classification for dermoscopy images *5th Intl Conf. On Recent Trends In Information Technology*.

[8] Damayana I,Atmaja R D and Fauzi H 2016 Detection of melanoma skin cancer using the wavelet transform *e-Procs. Eng.* **3** 3.

[9] Setianingrum R 2014 *Classification of colorectal cancer stages using a recurrent neural network model* Yogyakarta: Universitas Negeri Yogyakarta.

[10] Rahmat R F, Pulungan A F, Faza S, Budiarto R 2017 Image Classification of Ribbed Smoke Sheet using Learning Vector Quantization *Journal of Physics: Conference Series* **801**(1) 012050.

Contribution of Solar Photovoltaic on the Reduction of Greenhouse Gas Emission in the Indonesia Power Sector

Wilson Nababan[1], Hendrik V. Sihombing[1,2], Hideki Kawai[3], and Himsar Ambarita[1,2*]

[1]Mechanical Engineering, Universitas Sumatera Utara, Jl. Almamater Kampus USU, Medan 20155, Indonesia
[2]Sustainable Energy Research Centre, Universitas Sumatera Utara, Jl. Almamater Kampus USU, Medan 20155, Indonesia
[3]Mechanical Engineering Department, Muroran Institute of Technology, Mizumoto-cho 27-1 Muroran-shi Hokkaido 050-8585, Japan

[*]Email: himsar@usu.ac.id

Abstract. Indonesia has released the target of reducing Greenhouse gas (GHG) emissions. The strategy includes enhancement renewable energy utilization. One of the most potential renewable energy resources in Indonesia is solar energy. State Electricity Company of Indonesia has planned to develop solar Photo voltaic (PV) power plan in the document planning 2019-2028. The objective of this study is to explore the contribution of solar PV to the reduction GHG of Indonesia. The potency of solar irradiance and solar PV output in several big cities in Indonesia are discussed. The results show that by using the PV technology of Crystalline silicon with mounting position building integrated and the slope and azimuth is 0°, the average solar PV output in Indonesia is 1361.2 kWh/kWp. The planned solar PV installed capacity by the end of 2028 is 985.5 MW. The electricity resulted by the system in a year will be 1341.47 GWh or it is only 0.31% of national electricity consumption. The contribution of solar PV to the GHG emission reduction is 1.341 million tons CO_{2eq} in the year 2028.

1. Introduction

Indonesia has released its pledge on climate change mitigation by submitting the Intended National Determined Contribution (INDC) to the UNFCC [1]. In the INDC document it is stated that Indonesia is committed to reducing greenhouse gas (GHG) emissions by 29% compared to the business as usual (BAU) scenario by 2030. The target can be up to 41% by using international support. The target of reduction divided into five source categories, they are (*i*) Industrial Process and Product Use, (*ii*) Agriculture, (*iii*) Land-use, Land-use Change and Forestry, (*iv*) Energy, and (*v*) Waste. For Indonesia case, several studies on the GHG emission reduction have been found in literature [2,3,4].

In energy sector, the source of GHG emissions is all activities that burn fossil fuel. The activities include transportation, electricity production, and machineries. The strategy of reduction GHG emission is sustainable energy policy. The sustainable energy policy has two pillars, energy efficiency and renewable energy utilization. Renewable energy sources include wind energy, solar energy, hydropower, geothermal, etc. This work focus on the utilization of solar energy to producing electricity by using solar cells (Photovoltaic or PV). Mousa et al. [5] reported a comparative energy and GHG assessment of industrial rooftop-integrated PV and solar thermal collector. Kirshner et al. [6] examined the emergence and adoption of solar PV in the Mozambique and South Africa. Duan et al. [7] explored the potential of GHG mitigation and temperature benefits by substituting PV-generated power for coal fired electricity. The results showed that PV solar promises to dominate GHG mitigation, with the highest contribution reaching 64.67% in the year 2050. Pillot et al. [8] investigated the historical interactions between energy and society in both global and Sub-Saharan contexts and how renewable energy power has ended up being considered accordingly. Zurita et al. [9] reported their review on the solar PV development in Chile. In the year 2017, the net PV installed capacity has reached 1802 MW. This results in 3% of electricity. Moldal and Islam [10] reported the impacts of CO_2 emission constraints on the penetration of solar PV in the Bangladesh power sector. It was stated that the total primary energy requirement would be reduced in the range of 4.5 – 22.37% and the primary energy supply system would be diversified. Solar PV plays an important role in achieving reasonable energy security. The above reviewed studies show that solar PV has a great contribution in reducing GHG emissions. However, many barriers do exist and different for every country.

In order to meet the climate change pledge, Indonesia has released several targets on utilization renewable energy to produce electricity. In the power sector the target is announced in the Electricity Supply Business Plan (known as RUPTL). The objective of this study is to investigate the future contribution of solar PV development in the Indonesia power sector. The results are expected to supply the necessary information to meet Indonesia INDC targets.

2. Method and Theoretical approach

In the first section of this study global solar irradiation will be discussed. The development of renewable energy, include solar energy, in the Indonesia power sector will be discussed. Solar irradiation of clear sky condition on a horizontal surface on the earth can be calculated using the below equation.

$$G = G_{on} \cos\theta_z \left[\tau_b + 0,271 - 0,294\tau_b\right] \quad (1)$$

where G_{on} is extraterrestrial irradiation, θ_z is zenith angle, and τ_b is the atmospheric transmittance for beam radiation. All of these parameters and the methodology to calculate the clear sky radiation can be found elsewhere [11]. The solar irradiation that reaches the surface of solar cell is the converted into electricity.

The conversion of solar irradiation into electricity is estimated using method proposed by Huld et al. [12]. The method is explained in the followings. The instantaneous power (Watt) of the solar PV can be calculated by the below equation.

$$P(G',T') = G'\left(P_{STC,m} + k_1 \ln(G') + k_2 \ln(G')^2 + k_3T' + k_4T'\ln(G') + k_5T'\ln(G')^2 + k_6T'^2\right) \quad (2)$$

where G' is the normalized in-plane irradiation and T' is module temperature relative to standard test condition, they are defined in equation (3) and equation (4), respectively.

$$G' = \frac{G}{G_{STC}} \quad (3)$$

$$T' = T_{mod} - T_{STC} \quad (4)$$

here, *STC* is standard test condition and *mod* stands for module. By using the above equations, the energy resulted from the solar PV can be estimated. In this study, an online software named as Photovoltaic Geographical Information System (PV-GIS) [13] is employed to estimate energy resulted by the solar PV system.

3. Results and Discussions

The results will be presented in four sub sections. In the first section Global Solar irradiation will be discussed and followed by solar energy development in Indonesia in the second section. In the last two sections the solar electricity produced and emission reductions are discussed.

3.1. *Global Solar Irradiance*

The solar irradiance on the horizontal surface in Indonesia islands is shown in Fig 1. The solar resource data obtained from the Global Solar Atlas, owned by the World Bank Group and provided by SOLARGIS [14]. The figure shows that daily solar irradiation in Indonesia varies from 3.2 kWh/m^2 to 6.2 kWh/m^2. In a year it can be 1300 kWh/m^2 to 2200 kWh/m^2. The lowest of solar irradiation captured in the middle part of Irian Jaya (eastern part of Indonesia). On the other hand, the highest is shown by the West Timur and Sumbawa islands (in the figure shown by red areas). The solar irradiation in Indonesia islands, in comparison with Middle East and North Africa, can be categorized as medium. In the Middle East countries, the solar irradiance can be up to 7 kWh/m^2 per day or yearly total up to 2500 kWh/m^2.

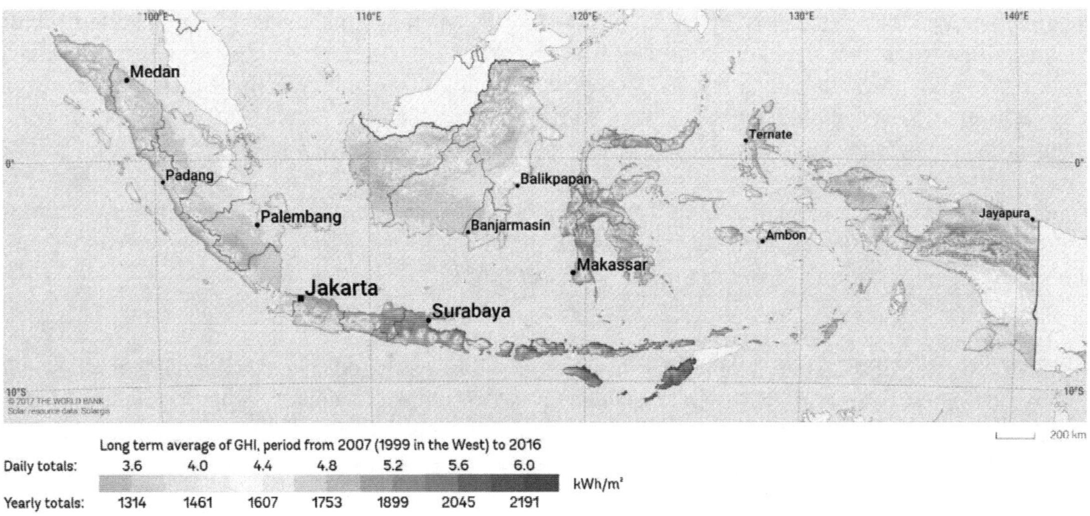

Figure 1. Global Horizontal Irradiance in Indonesia [14]

3.2. *Solar energy development in Indonesia*

The planned renewable energy development in Indonesia during 2019-2028 are shown in Table 1. In the table the development is shown in capacity [MW].

Table 1 Renewable energy development

Year	Technology							Total
	Geothermal [MW]	Hydro [MW]	Mini Hydro [MW]	Solar [MW]	Wind [MW]	Biomass [MW]	Ocean [MW]	
2019	190	154	140	63	0	12	0	559
2020	151	326	238	78	0	139	0	932
2021	147	755	479	219	30	60	7	1697
2022	455	0	200	129	360	357	0	1501
2023	245	182	168	160	260	50	0	1065
2024	415	1484	232	4	50	103	0	2288
2025	2759	3047	27	250	150	19	0	6252
2026	45	129	20	0	0	5	0	199
2027	145	466	20	2	0	15	0	648
2028	55	1467	10	2	5	35	0	2416
Total	4607	8010	1534	907	855	795	7	16715

In Indonesia the electricity distribution is monopolized by State Electricity Company (named as PLN). The PLN generates the majority of the Indonesia electric power. Even though, the solar irradiance in Indonesia is categorized medium it is sufficient to be converted into electricity power. In the year 2019, PLN released the Electricity Supply Business Plan 2019-2018. In the document the list of renewable energy development capacities is mentioned and showed in the Table 1. In the table of the renewable developments are planned based on the potency of the renewable. In total during 2019 to 2028, a total of 16.7 GW renewable energy power plant will be developed. The highest capacity development is Hydropower. The total capacity, together with Mini Hydro, is 9.544 GW. The second is Geothermal power plant with a total capacity development of 4.6 GW. The third is solar power plant with a capacity of 907 MW. It is bigger than Wind power and Biomass. In addition, a small portion of ocean power is also planned. In term of capacity, the solar power has a portion of 5.4% from all renewables.

3.3. PV electricity potency

The renewable energy development shown in Table 1 is only capacity. The emissions reduction can be estimated by using electric energy resulted by the system. In order to estimate the electricity produced from the solar PV, the PV-GIS software is employed. As shown in Fig 1, the solar irradiance in Indonesia varied in every location. For instance, the potency of electricity resulted by solar PV 1 kWp in Medan city is shown in Table 2. The table shows average daily and monthly energies produced by the system. It can be seen that the maximum and minimum average monthly electricity is 120 kWh (March) and 91.7 (December), respectively.

Table 2 Characteristics solar energy and PV output in Medan city

Month	Ave. daily energy produced [kWh]	Ave monthly energy produced [kWh]	Ave. daily Irradiation received [kWh/m^2]	Ave. monthly Irradiation received [kWh/m^2]	Standard deviation of monthly energy [kWh]
Jan	3.39	105	4.68	145	9.84
Feb	3.70	104	5.15	144	7.02
Mar	3.88	120	5.48	170	10.3
Apr	3.71	111	5.24	157	9.1
May	3.49	108	4.96	154	4.41
Jun	3.52	106	5.02	151	7.8
Jul	3.54	110	5.04	156	7.31
Aug	3.56	110	5.03	156	8.08
Sep	3.45	103	4.86	146	6.84
Oct	3.32	103	4.64	144	6.81
Nov	3.11	93.2	4.32	130	5.1
Dec	2.96	91.7	4.07	126	5.41
Year	3.47	105	4.87	148	3.5

The different results are shown by the solar PV if it is installed in Jakarta. The typical average monthly solar electricity resulted by the system at several cities in Indonesia are shown in Fig 2. The cities are Medan, Jakarta, Surabaya, Samarinda, and Sumbawa. The figure shows that the characteristics of each city is different. The total produced electricity if the solar PV installed in Medan is 1264.9 kWh/kWp and in Jakarta is 1314.4 kWh/kWp. The energy in Surabaya, Samarinda and in Sumbawa is 1439.9 kWh/kWp, 1216.2 kWh/kWp and 1523 kWh/kWp, respectively. In order to estimate the energy produced by the solar power plant capacity planned in Table 1, an average electricity produced must be estimated. The total produced electricity in a year from all cities will be averaged by considering the total energy consummed in every city in comparison to Indonesia. By using this defenition, the average number is 1361.2 kWh/kWp. This number will be used to estimate electricty production from solar PV.

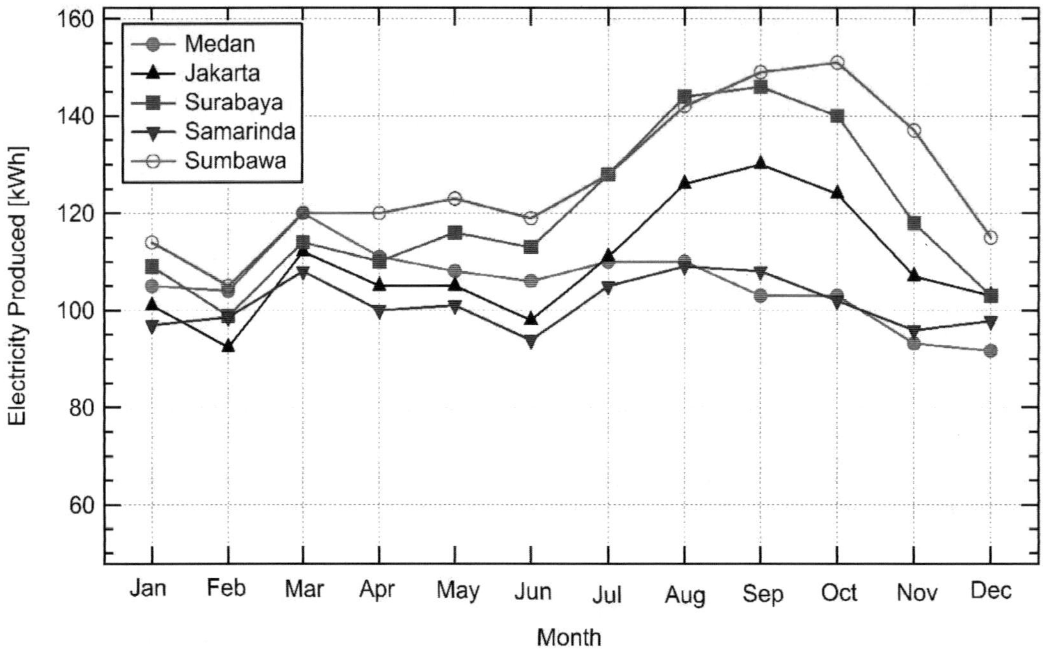

Figure 2. Average monthly energy produced from the PV system in several cities

3.4. Solar Electricity and Emission Reduction

By using average electric energy produced by PV in Indonesia, the solar electricity produced from the planned solar PV is calculated and presented in Table 3.

Table 3. Installed solar PV and Solar Energy produced by the PV

Year	Installed Capacity per Year [MW]	Total Installed Capacity [MW]	Solar Electricity Produced [GWh]	National Electricity Consumption [GWh]	Solar Energy Share
Baseline	78.5	78.5	106.86	232,296	0.05%
2019	63	141.5	192.61	245,397	0.08%
2020	78	219.5	298.79	261,450	0.11%
2021	219	438.5	596.89	279,353	0.21%
2022	129	567.5	772.49	299,510	0.26%
2023	160	727.5	990.28	320,004	0.31%
2024	4	731.5	995.73	339,923	0.29%
2025	250	981.5	1336.03	360,939	0.37%
2026	0	981.5	1336.03	383,310	0.35%
2027	2	983.5	1338.75	407,239	0.33%
2028	2	985.5	1341.47	432,713	0.31%

In the calculation several estimations have been made. The PV technology is Crystalline silicon, the mounting position is building integrated, and the slope and azimuth is 0°, respectively. The system loss is 14% and it includes losses in cables, power inverters, dirt on the modules, etc. The baseline of the installed solar capacity is 78.5 MW. In the year 2019, the electricity produced by the solar PV is 192.61 GWh. In the same year, the total electricity consumption in Indonesia is 245,397 GWh. This means, there only 0.08% of the energy produced from solar PV. In the year 2020, the energy produced by solar

PV, total energy consumed, and percent of the solar PV is 298.79 GWh, 261,450 GWh, and 0.11%, respectively. The data shown in Table 3 shows that the maximum contribution of the solar PV is only 0.37%, it is reached in the year 2025. As a note, the Government of Indonesia has a target of energy share of 23% renewables by 2025. The renewables include solar PV. Solar PV is expected to support this target. However, the contribution of the planned solar PV is still very low or less than 1%.

The emission factor in Indonesia power sector in the year 2019 is 0.817 ton CO_2/MWh. This emission factor will increase to 0.945 CO_2/MWh in the year 2028. By using these factors the contribution of solar PV to Indonesia emission reduction is 106.85 kilo tons CO_{2eq} in the year 2019. The contribution will be 1.341 million tons CO_{2eq} in the year 2028.

4. Conclusions

In this study, the contribution of solar PV development in the power sector to the reduction of GHG emission has been examined. The conclusions are as follows. The development of solar PV shows a significant in the capacity number. However, in term of energy resulted the contribution is very low. During the year 2019 to 2028, the highest contribution of electricity resulted by solar PV system is only 0.37% in comparison with national electricity consumption.

References

[1] https://www4.unfccc.int/sites/submissions/indc/Submission%20Pages/submissions.aspx (Acessed on May 13, 2019)
[2] Ambarita H 2017 *IOP Conference Series: Materials Science and Engineering* **237** 012013
[3] Ambarita H 2018 *IOP Conference Series: Earth and Environmental Science* **126** 012021
[4] Ambarita H, Soeharwinto S, Ginting N, Basyuni M and Zen Z 2018 *IOP Conference Series: Earth and Environmental Science* **126** 012114
[5] Mousa OB, Kara S, Taylor RA 2019 *Applied Energy* **241** 113-123
[6] Kirshner J, Baker L, Smith A, Bulkeley H 2019 Geoforum (Inpress Corrected Proof)
[7] Duan H-B, Zhang G-P, Zhu L, Fan Y, Wang S-Y 2016 *Renewable and Sustainable Energy Reviews* **53** 1076-1085
[8] Pillot B, Muselli M, Poggi P, Dias JB 2019 *Energy Policy* **127** 113-124
[9] Zurita A, Castillejo-Cuberos A, Garcia M, Mata-Torres C, Simsek Y, Garcia R, Antonanzas-Torres F, Escoba RA 2018 *Renewable and Sustainable Energy Reviews* **92** 701-727
[10] Mondal MAH and Islam AKMS 2012 *Renewable Energy* **43** 418-422
[11] Ambarita H and Kawai H 2016 *Case Studies in Thermal Engineering* **7** 36 – 46
[12] Huld T, Friesen G, Sckoczek A, Kenny RP, Sample T, Field M, Dunlop ED 2011 *Solar Energy Materials & Solar Cells* **95** 3359-3369
[13] http://re.jrc.ec.europa.eu/pvg_tools/en/tools.html#PVP (Acessed on May 15, 2019)
[14] https://globalsolaratlas.info/ (Acessed on May 15, 2019)

Acknowledgement
This work is a part of research project funded by Ministry of Research and Higher Education of Republic of Indonesia. The scheme of the research project is "Hibah Kompetensi DRPM" Year of 2019.

Effect of internal heat exchanger to the performance of vapor compression cycle using refrigerant R32

H V Sihombing[1], A H Nasution[1], and H Ambarita[1]*

[1]Sustainable Energy Research Centre, Universitas Sumatera Utara, Jl. Almamater Kampus USU, Medan 20155 Indonesia

*Corresponding Email: himsar@usu.ac.id

Abstract. Vapor compression refrigeration cycle generally is used in the Air-Conditioning (AC) system. In order to increase the energy efficiency, the Coefficient of Performance (COP) of the cycle need to be enhanced. Many innovations can be used, one is to utilize internal heat exchanger (IHX). In this work, the effect of IHX in the vapor compression cycle with refrigerant R32 is investigated and compared with refrigerant R22. The analyses are carried out at temperature evaporation of -5^{0}C, 0^{0}C, and 5^{0}C by using software *ASPEN PLUS*. The results show that effect of IHX in the cycle with refrigerant R22 increases the COP from 22.78% to 25.61%. On the other hand, in the cycle with refrigerant R32 the COP increases from 33.79% to 36.80%. The average increase for the cycle with refrigerant R22 and refrigerant R32 is 24.14% and 37.04%, respectively. This fact reveals that IHX in the cycle with R32 is more effective that the cycle with refrigerant R22.

1. Introduction

Vapor compression refrigeration cycles are widely used in air-conditioning cycle and also in industrial refrigeration applications. Since, the cycle consumes significant energy, the enhancement of the performance of vapor compression has come under scrutiny. There are many strategies to improve the performance of vapor compression cycle have been found in literatures such as refrigerant replacement, cycle modification, optimum temperature and pressure [1-3], etc. The performance of a single stage vapor compression cycle will down at high heat released and low evaporation temperatures. In general, this can be avoided by employing modification cycle. One of the strategies is to employ inter-cooling in order to decrease the specific volumes and the discharge temperatures, which as a result decreases the power input to the compressor [4].

The performance of a single stage using internal heat exchanger vapor compression cycles have been studied and analyzed by several researchers. Nakagawa et al. [5] and Liopis et al. [6] have studied internal heat exchanger used in CO_2 cooling cycle. Djuanda [7] reported an analysis of the internal heat exchanger in air conditioning cycle used refrigerant R410a. The used internal heat exchanger will increase the COP 9.02 %. Zhenying et al. [8] studied the effect an internal heat exchanger on the trans-critical carbon dioxide refrigeration cycle with expander. The results showed that internal heat exchanger increases specific cooling capacity and COP 17%. Wantha [9] investigated the characteristics of tube-in-tube internal heat exchanger for refrigerant R1234yf and R134a. It was shown that the heat transfer coefficient for R 1234yf is lower than R134a by 11-17%. Zhang et al. [10] reported the analysis and optimization a single stage and doubles stage vapor compression cycle with an internal heat exchanger. The work showed that double stage compression and the internal heat exchanger have a significant effect on the optimum COP.

The above studies reveal that study on the internal heat exchanger installed in a vapor compression cycle is an interesting topic to be explored. In particular, it shows different effect at different refrigerant. The present work focuses on the investigation performance of single stage vapor compression refrigeration cycle with internal heat exchanger. The effect of the internal heat exchanger will be investigated in different refrigerant. The used refrigerants are is R22 and R32. These refrigerants are typically used in the air-conditioning applications. In addition, the refrigerant R32 is a new refrigerant that can be used to replace R22. The objective of this study is to examine the maximum coefficient of performance (COP) of the cycle at several different working pressures. The results are expected to supply the necessary information on development highly efficient vapor compression refrigeration cycle.

2. Solution Method

In the present work, the vapor compression cycle is simulated using Aspen plus commercial software. The studied vapor compression cycle is shown in Figure 1. Aspen plus is a computer-aided code which uses the underlying physical relationships (e.g., material and energy balance a thermodynamics equilibrium, rate equations) to predict process performance (e.g., stream properties, operating conditions, and equilibrium sizes). Process analysis may involve the use of experimental means to predict and validate performance. Then in the simulation process, we need to give the initial process initial process input and flow sheet are required to predict process output.

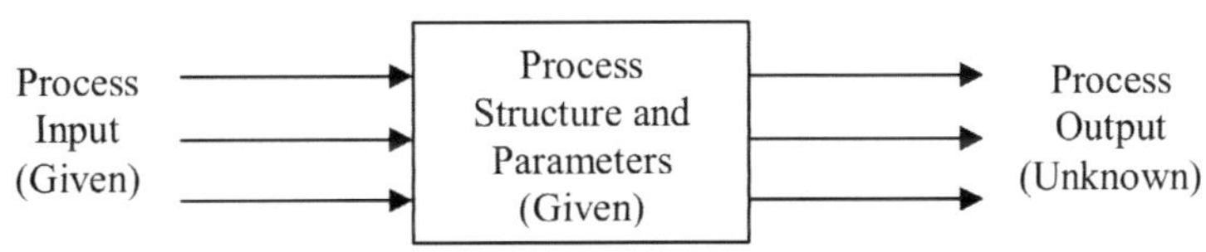

Figure 1. Process simulation

The schematic diagram of single-stage using internal heat exchanger vapour compression refrigeration cycle is shown in Figure 2. In the figure, the *p-h* diagram is also presented. It can be seen that the component of the cycle consists of one compressor, condenser, internal heat exchanger, and evaporator.

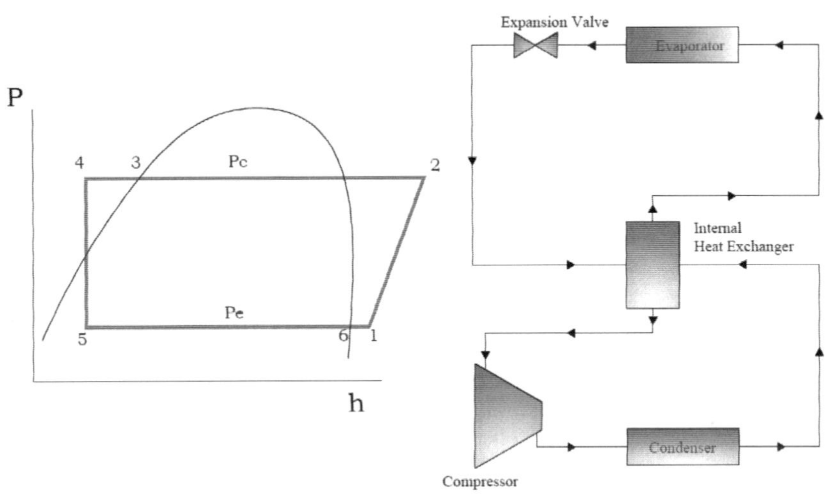

Figure 2. The component and *P-h* diagrams of the vapor compression cycle

IOP Conf. Series: Materials Science and Engineering **648** (2019) 012029 doi:10.1088/1757-899X/648/1/012029

The governing equations of the cycles are developed as follows. The power of compressors ($\dot{W}_c$) is calculated by

$$\dot{W}_c = \dot{m}\left(h_2 - h_1\right) \quad (1)$$

Here, $\dot{m}$ [kg/s], h_1, and h_2 is the mass flow rate, enthalpy of the refrigerant at point 1 and 2, respectively. After leaving the compressor, the refrigerant release the heat to the ambient in the condenser. The heat release by the cycle to ambient is given by equation (2) and calculated as follows.

$$\dot{Q}_c = \dot{m}\left(h_2 - h_3\right) \quad (2)$$

Where, h_2 and h_3 is enthalpy of the refrigerant at the inlet and the exit of the condenser, respectively. The heat absorbed by the refrigerant in the evaporator, $\dot{Q}_e$ is given by:

$$\dot{Q}_e = \dot{m}\left(h_6 - h_5\right) \quad (3)$$

Here, h_5 and h_6 is the enthalpy of the refrigerant at the inlet and the exit of the evaporator. Refrigeration effect (ER) of the cycle is calculated by equation (4).

$$ER = \left(h_6 - h_5\right) \quad (4)$$

The coefficient of performance of the cycle (COP) is given by

$$COP = \frac{ER}{\dot{W}_c} \quad (5)$$

The above parameters will be examined in the results.

3. Results and Discussions

The analyses were carried out using ASPEN process simulation software. The flow diagram of the cycle analysed is shown in Figure 3. The figure shows that an internal heat exchanger is installed in the cycle. In this process each individual component is simulated and then connected as a closed loop.

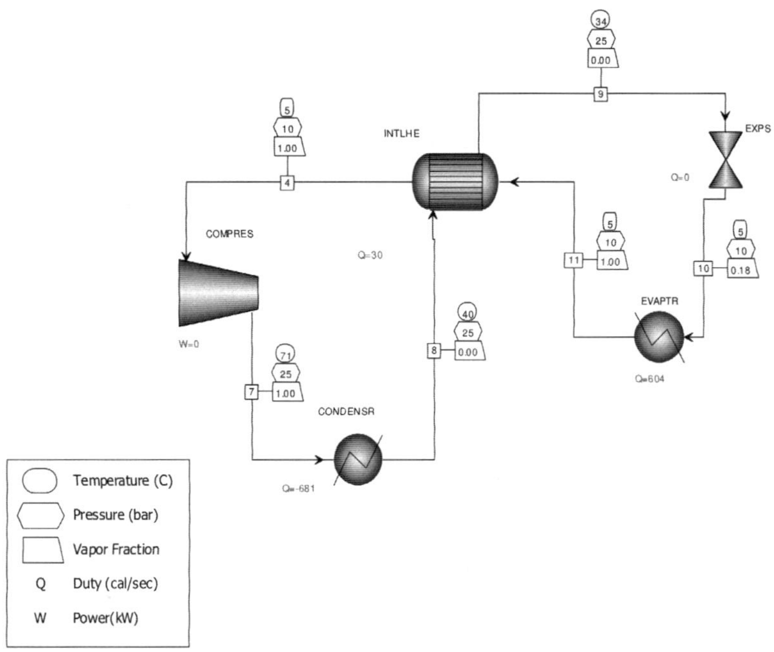

Figure 3. Flow sheet of refrigeration cycle using internal heat exchanger

As mentioned in the previous section, the analysis is divided into two cases, case 1 is the cycle with refrigerant R22 and case 2 is with refrigerant R32. In order to vary the operating conditions, three different conditions have been simulated. The first condition is when the vapor compression cycle is operated at temperature evaporation of -5^{0}C and temperature condensation of 30^{0}C. The second condition is at temperature evaporation of 0^{0}C and temperature condensation of 35^{0}C, and the third condition is at temperature evaporation of 5^{0}C and temperature condensation of 40^{0}C. For all cases and conditions, the cooling load is assumed to be constant at 1000 Ton.

The simulation results will be discussed in three sub section. They are refrigeration effect, compressor power, and coefficient of performance. The results for each sub section are explained in the followings.

3.1. Refrigeration Effect

Figure 4 shows the refrigeration effects of each cycle. In the figure, the refrigeration effect from the evaporator using internal heat exchanger is shown by red line. The refrigeration effect from the evaporator without internal heat exchanger (IHX), known as standard cycle, is shown by blue line.

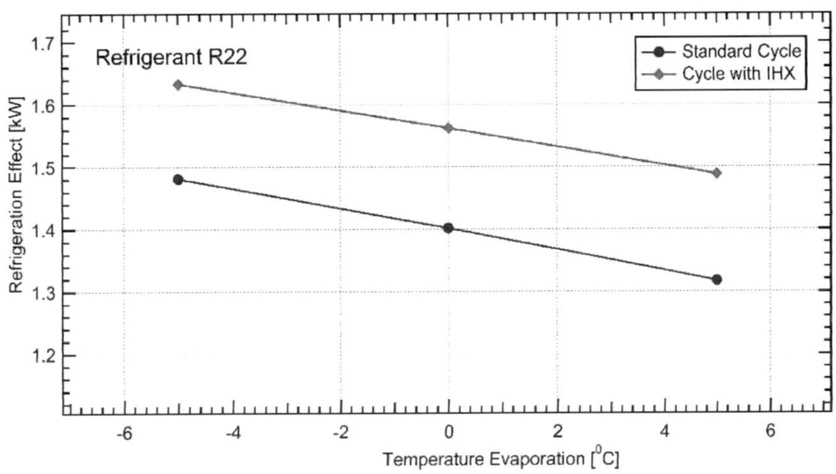

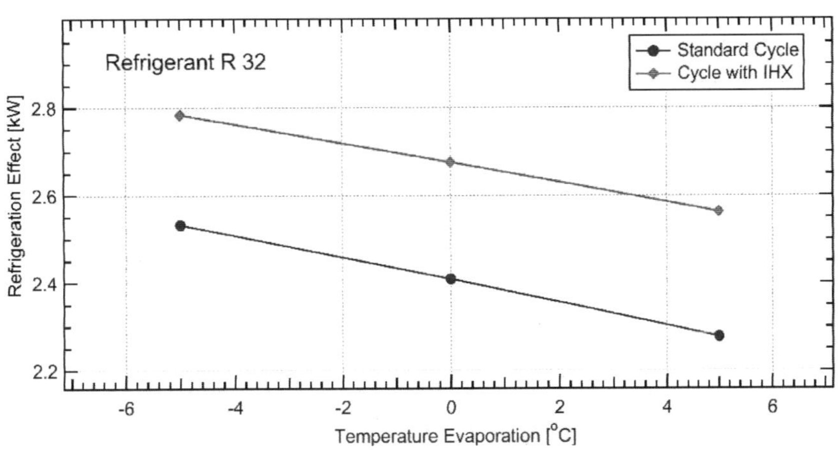

Figure 4. Refrigeration effect for Refrigerant R22 and Refrigerant R32

It can be seen that, as expected, the refrigerant effect of the cycle with internal heat exchanger is hinger that the standard cycle. This occurs for all conditions. This fact suggest that internal heat exchanger do increase the refrigeration effect of the cycle. The same trend is also shown by the cycle

IOP Conf. Series: Materials Science and Engineering **648** (2019) 012029 doi:10.1088/1757-899X/648/1/012029

with refrigerant R32. The figure will be used to examine the effect of IHX to the cycle with refrigerant R22 in comparison with refrigerant R32. For refrigerant R22, at temperature evaporation of -5°C the RE of the standard cycle and cycle with IHX is 1.481 kJ/kg and 1.633 kJ/kg, respectively. Thus, IHX increases the RE about 10.27%. At evaporation temperature of 0°C and 5°C, the RE increases about 11.46% and 12.95%, respectively. On the other hand, at the similar evaporation temperatures with refrigerant R32, the RE increases about 9.94%, 11.09%, and 12.53%, respectively.

3.2. Compressor Power

Figure 5 shows the compressor power of the cycle for all cases. The compressor power almost constant for all temperature evaporations. The compressor power of the cycle with IHX is relatively lower than the standard one. It can be said that IHX decreases the compressor power. The similar trend is shown for both refrigerants. Fore refrigerant R22, at temperature evaporation of -5^{0}C, the compressor power of the standard cycle and cycle with IHX is 0.41 kJ/kg and 0.37 kJ/kg, respectively. IHX decreases the compressor power of 10.19%. On the other hand, for refrigerant R32 at the same temperature evaporation, the compressor power of the standard cycle and cycle with IHX is 0.52 kJ/kg and 0.43 kJ/kg, respectively. Here, IHX decreases the compressor power of 17.8%. This fact suggests that IHX works more effective in the cycle with refrigerant R32 in comparison with the cycle with refrigerant R22.

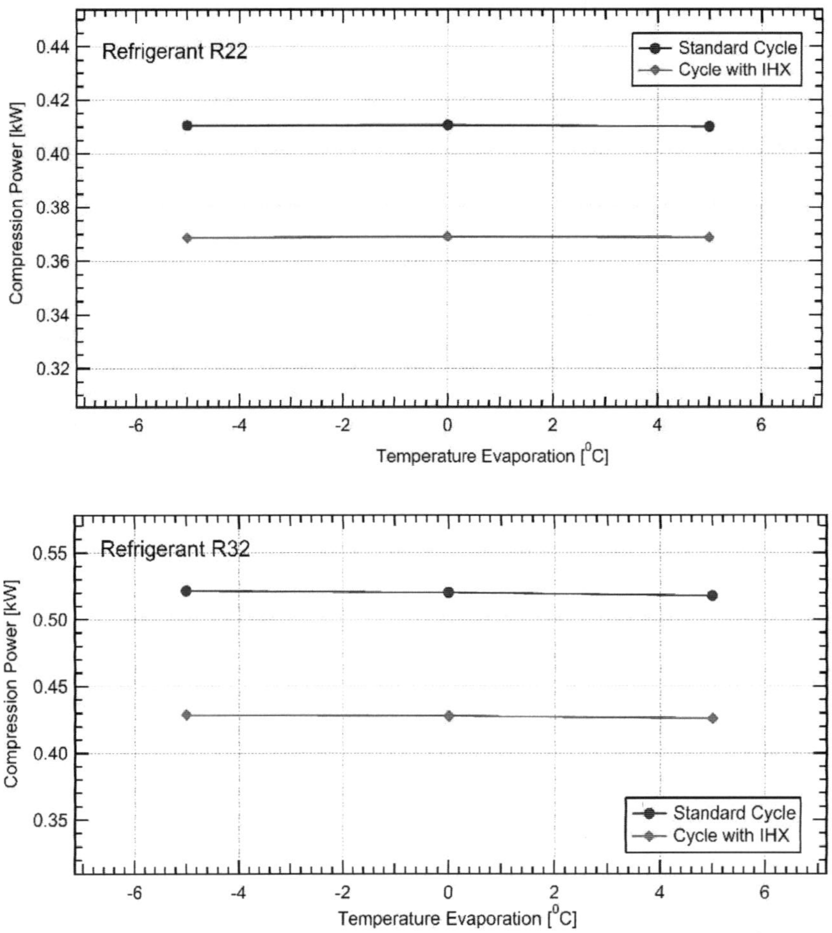

Figure 5. Compressor power for Refrigerant R22 and Refrigerant R32

3.3. Coefficient of Performance (COP)

Finally, the examination will be carried out using COP. Figure 6 shows the performance of the cycle for all cases. The figure shows that, in general, the COP of the cycle increases with decreasing temperature evaporation. It can be seen that installing IHX will increase COP significantly. For refrigerant R22 the COP of the standard cycle at temperature evaporation of -5^{0}C, 0^{0}C, and 5^{0}C is 3.6, 3.4, and 3.2, respectively. While for the cycle with IHX, the COP at the similar temperature evaporation is 4.42, 4.23, and 4.03, respectively. These fats reveal that IHX increases the COP about 22.78%, 24.02%, and 25.6%, respectively. The average increase of the COP is 24.14%.

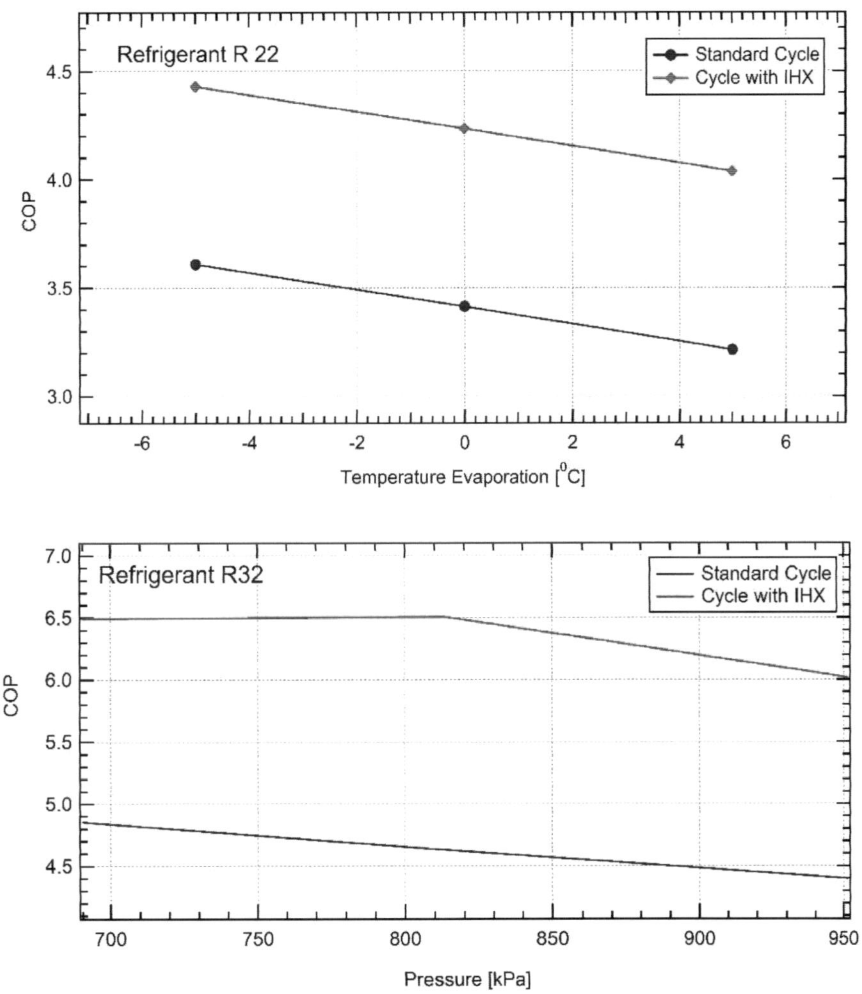

Figure 6. Coefficient of Performances for Refrigerant R22 and Refrigerant R32

Table 1. The effect of IHX to the Coefficient of the Performance

Temperature evaporation	Refrigerant R22			Refrigerant R32		
[^{0}C]	Standard	IXH	Difference	Standard	IXH	Difference
-5^{0}C	3.61	4.43	22.78%	4.85	6.49	33.79%
0^{0}C	3.41	4.23	24.02%	4.63	6.51	40.54%
5^{0}C	3.21	4.04	25.61%	4.40	6.02	36.80%
		Average	24.14%		Average	37.04%

In order to examine the effect of the IXH to the Coefficient of the Performance to the cycle with different refrigerant, the COP for all cases are shown in Table 1. The table shows that the effect of IHX in the cycle with refrigerant R22 increases the COP from 22.78% to 25.61%. On the other hand, in the cycle with refrigerant R32 the COP increases from 33.79% to 36.80%. The average increase for the cycle with refrigerant R22 and refrigerant R32 is 24.14% and 37.04%, respectively. This fact reveals that IHX in the cycle with R32 is more effective that the cycle with refrigerant R22.

4. Conclusions

The effect of the Internal Heat Exchanger in a vapor refrigeration cycle has been investigated numerically. The conclusions of this study are as follows. By installing the Internal Heat Exchanger, both refrigerants show the similar trend. The refrigerant effect increases and the compressor power decreases. As a result, the coefficient of the performance increases. The increasing of the COP in the cycle with refrigerant R32 is higher than refrigerant R22. Thus, IHX is more effective in the cycle with refrigerant R32 than R22.

References

[1] Ambarita H, Nasution AH, and Sihombing HV 2018 *IOP Conf. Series: Material Science and Engineering* **420** 012038.
[2] Ambarita H, Nasution A H, Siahaan N M and Kawai H 2016 *Case Studies in Thermal Engieering* **8** 105-114.
[3] Ambarita H, Nasution DM, Gunawan S and Nasution AH 2017 *IOP Conf. Series: Material Science and Engineering* **180** 012027
[4] Purohit N, Gupta DK, and Dasgupta MS 2015 *Energy Procedia* **90** 171-178.
[5] Llopis, R, Sanz-Kock C, Cabello R, Sánchez D, Torrella E 2016 *Applied Thermal Engineering* **103** 1077-1086
[6] Nakagawa M, Marasigan AR, Matsukawa T 2011 *International Journal of Refrigeration* **34** 1577–1586.
[7] Djuanda D 2018 *AIP Conference Proceedings* **1984** 020008.
[8] Zhang Z, Tian L, Chen Y, and Tong L 2014 *Entropy* **16** 5919-5934.
[9] Wantha C 2019 *Applied Thermal Engineering* **157** 113747.
[10] Zhang Z, Hou Y, and Kulacki FA 2018 *Applied Thermal Engineering* **140** 147-157.
[11] Arora R C 2010 *Refrigeration and Air Conditioning*, PHI, New Delhi

Acknowledgement

This work is a part of research project funded by Ministry of Research and Higher Education of Republic of Indonesia. The scheme of the research project is "Hibah Desentralisasi PDUPT DRPM" Year of 2019.

Effect of Chord Length on the Performance of H-Darrieus Wind Turbine with NACA 4415 Airfoil

Indro Pramono[1], Himsar Ambarita[1,2,*], and Koki Kishinami[3]

[1]Mechanical Engineering Department, Universitas Sumatera Utara, Jl. Almamater, Kampus USU, Medan 20155, Indonesia
[2]Sustainable Energy Research Centre, Universitas Sumatera Utara, Jl. Almamater, Kampus USU, Medan 20155, Indonesia
[3]Mechanical Engineering Department, Muroran Institute of Technology, 27-1 Mizumoto, Muroran 8585, Hokkaido, Japan

*Email: himsar@usu.ac.id

Abstract. The effect of chord length to the performance of an H-Darrieus Vertical Axis Wind Turbine (VAWT) was investigated numerically. The turbine blade made of NACA 4155 airfoil. A set of governing equations has been developed. In order to incorporate the turbulence effect, $k-epsilon$ model was adopted. A Computational Fluid Dynamic code has been used to solve the problem. The non-dimensional chord length of 0.2, 0.3, 0.4, and 0.5 are investigated. The contour velocity in the computational domain was plotted and discussed. By using the average velocity of the wind leaving the turbine, power coefficient is estimated. The results show that the power coefficient for non-dimensional chord length of 0.2, 0.3, 0.4, and 0.5 is 0.51, 0.57, 0.59, and 0.59, respectively. The conclusion here is that non-dimensional chord length strongly affects the power. Increasing the chord length will increase the coefficient power. However, there exists an optimum chord length it is 0.4. Thus, it is suggested to use the non-dimensional chord length of 0.4.

1. Introduction

There is a growing commitment to increase utilization of renewable energy sources to reduce the Greenhouse gas emission worldwide. One of the potential renewable energy resources is wind power. A wind turbine is an engineering innovation that can be used to convert the kinetic energy of wind into useful mechanic energy. There are many types of wind turbine can be found in the field. Those can be segregated into Vertical Axis Wind Turbines (VAWTs) and Horizontal Axis Wind Turbines (HAWTs). Mollerstrom [1] has reported a historical review of VAWTs rated 100 kW and above. The performances of VAWTs projects has been reviewed. It was stated that there are many barriers on development VAWTs mainly on metal fatigue and durability.

In this work, we focus on the exploring characteristics of H-Darrieus wind turbine. Several works related to investigation of H-Darrieus wind turbine have been found in literature. Hashem and Mohamed [2] investigated the effect of airfoil shape and wind-lens type to improve the performance of H-Darrieus wind turbine. It was stated that S1046 airfoil showed the highest output performance. Ali et al. [3] explored the effects of instantaneous tangential velocity on the aerodynamic performance of an H-Darrieus wind turbine. CFD simulations were used for five different values of tip speed ratios i.e. 1, 2, 3, 4, and 5. The results showed that instantaneous tangential velocity has significant effect to the

IOP Conf. Series: Materials Science and Engineering **648** (2019) 012030 doi:10.1088/1757-899X/648/1/012030

performance depend on the speed tip ratio. Abdalrahman et al. [4] studied the effect of blade pitching on performance of straight-bladed Darrieus vertical axis wind turbine (H-type VAWT) in term of power output. The investigation was performed numerically using CFD commercial code on 2D variable. The results showed that for H-type VAWT, compared to a conventional controller, a multilayer perceptron artificial neural network method results in superior power output. Bianchini et al. [5] proposed a new design criterion for H-Darrieus turbines based on the energy-yield maximization. The proposed design criterion has shown interesting prospects in terms of energy production improving. Lanzafame et al. [6] explored a strategy to develop a two-dimensional CFD model of H-Darrieus VAWT. The model has been compared with experimental work and show a good agreement. Mohamed et al. [7] employed CFD commercial code to analyze H-Darrieus VAWT at low speed. In the study, blade pitch angle has been investigated and the results indicated that the zero-pitch angle gives best performance. Asr et al. [8] investigated the start-up characteristics of H-Darrieus VAWT comprising NACA 4-digits series blade airfoils. The optimum start-up characteristics were observed with the use of a medium-thickness cambered airfoil, NACA 2418, put to use with an outward pitch angle of 1.5°.

The above literatures show that the performance of an H-Darrieus VAWT strongly depend on many parameters. CFD commercial code is widely used to explore the effect of design parameter to the performance. As a note, our research group is developing an aerator for agricultural uses. The aerator is powered by H-Darrieus VAWT comprising NACA 4155 airfoil. In this paper the effect of chord length will be examined numerically. The objective is to explore the effect of the chord length to the performance of the turbine.

2. Mathematical Model and Method

As mentioned above, in this study the developed of the H-Darrieus VAWT will be used as model to be investigated. The developed H-Darrieus VAWT is shown in Figure 1(a).

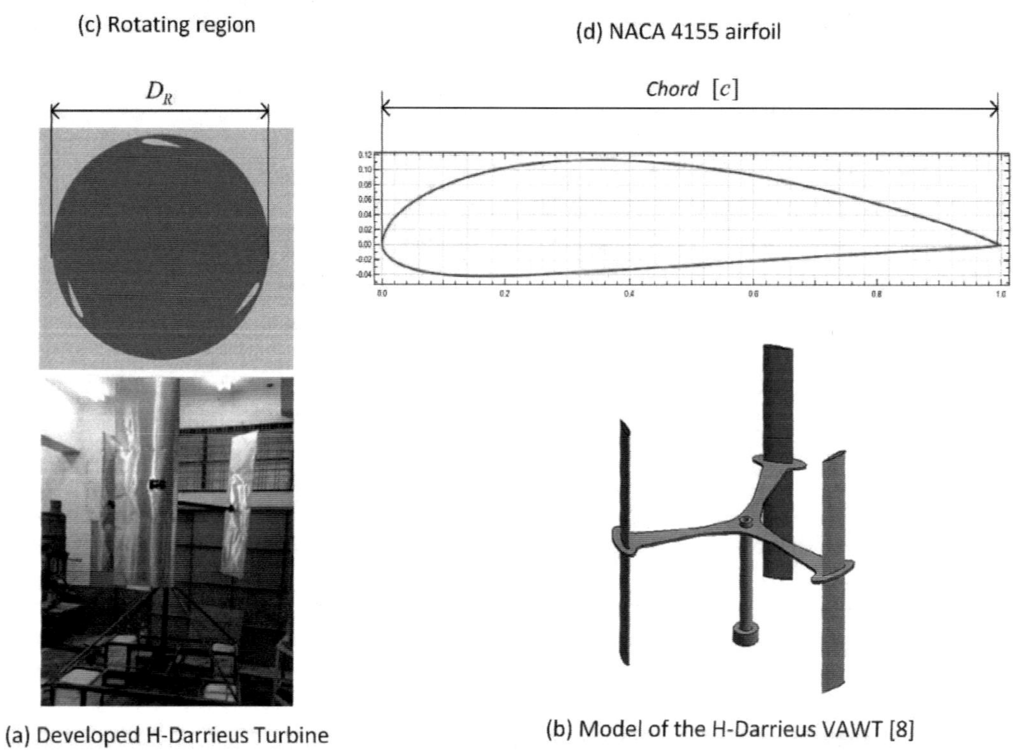

Figure 1. The developed H-Darrieus Wid Turbine and Models

It is used as an energy converter to power the aerator blade. Figure 2(b) shows the schematic of the H-Darrieus VAWT. The position of the air foil in the rotating region is shown in Figure 1(c). The blade of the turbine is developed by using NACA 4155 as shown in the Figure 4(d). The design parameter that is explored here is the chord length of the air foil. The non-dimensional chord length is defined as comparison of cord length to diameter of the rotating region. It is calculated by the below equation.

$$\bar{c} = \frac{c}{D_R} \tag{1}$$

In this work, Computational Fluid Dynamics commercial code is used to perform the analyses. Two-dimensional model was used and the turbulent flow is considered. The mass conservation equation and momentum equation can be written as follows [9].

$$\frac{\partial \rho}{\partial t} + \frac{\partial}{\partial x_i}(\rho u_i) = 0 \tag{2}$$

$$\rho \frac{Du_i}{Dt} = -\frac{\partial p}{\partial x_i} + \frac{\partial}{\partial x_j}\left[\mu\left(\frac{\partial u_i}{\partial x_j} + \frac{\partial u_j}{\partial x_i} - \frac{2}{3}\delta_{ij}\frac{\partial u_i}{\partial x_i}\right)\right] + \frac{\partial}{\partial x_j}\left(-\rho\overline{u_i' u_j'}\right) \tag{3}$$

The last term of equation (3) is named as the Reynolds-stresses and calculated by

$$-\rho\overline{v_i' v_j'} = -\rho\left(\overline{v_i' v_j'} - \bar{v}_i \bar{v}_j\right) \tag{4}$$

The Boussinesq hypothesis is used to combine the stresses with the mean velocity gradients. As a result, the Reynolds-stresses is given by the below equation.

$$-\overline{\rho u_i' u_j'} = \mu_t\left(\frac{\partial u_i}{\partial x_j} + \frac{\partial u_j}{\partial x_i}\right) - \frac{2}{3}\left(\rho k + \mu_t \frac{\partial u_i}{\partial x_i}\right)\delta_{ij} \tag{5}$$

To complete those equations, additional governing equations are proposed. Here, the standard $k-\varepsilon$ turbulence model proposed by Lauder and Spalding [10] is applied. Thus, two additional governing equations are employed; they are the turbulent kinetic energy (k) equation and the turbulent dissipation rate (ε). These equations are defined in the followings.

$$\rho\frac{Dk}{Dt} = \frac{\partial}{\partial x_i}\left[\left(\mu + \frac{\mu_t}{\sigma_k}\right)\frac{\partial k}{\partial x_i}\right] + G_k + G_b - \rho\varepsilon - Y_M \tag{6}$$

$$\rho\frac{D\varepsilon}{Dt} = \frac{\partial}{\partial x_i}\left[\left(\mu + \frac{\mu_t}{\sigma_\varepsilon}\right)\frac{\partial \varepsilon}{\partial x_i}\right] + C_{1\varepsilon}\frac{\varepsilon}{k}\left(G_k + C_{3\varepsilon}G_b\right) - C_{2\varepsilon}\rho\frac{\varepsilon^2}{k} \tag{7}$$

In the above equations, there are two types of viscosities, the actual dynamic viscosity and the turbulent viscosity μ_t. The turbulent viscosity is given by

$$\mu_t = \rho C_\mu \frac{k^2}{\varepsilon} \tag{8}$$

The parameter G_k, G_k and Y_M is defined as generation of turbulent kinetic related to mean velocity gradient, generation of turbulent kinetic energy related to buoyancy and the contribution of fluctuating dilation in compressible turbulence to the overall dissipation rate, respectively. While, the parameter $C_{1\varepsilon}$, $C_{2\varepsilon}$, $C_{3\varepsilon}$ and C_μ are constants. Here, the default number from CFD code are used. The parameter σ_k and σ_ε are the turbulent Prandtl numbers for k and ε equation, respectively.

The method, in glance, explained as follows. All of the governing equations are discretized based on finite volume method. To couple the velocity field and other fields, iterative SIMPLE algorithm is employed. The residual of the continuity and k and ε equations are monitored. The iteration will be terminated when the maximum residual reached a value of less than 10^{-3} or the residual reached a constant value.

3. Results and Discussions

The simulations have been carried out for chord length 30 cm, 45 cm, 60 cm, and 70 cm. Since the diameter of the rotation region is 150 cm, the non-dimensional chord length is 20%, 30%, 40%, and 50% respectively. In all cases the pitch angle was fixed at $\varphi = 8^o$ and the tip speed ratio was also fixed at $\lambda = 2$. The velocity of the wind at the inlet was 5 m/s.

3.1. Numerical Validation

In order to make sure that the developed numerical method is acceptable, a numerical validation has been performed. Lift coefficient (C_L) of an airfoil NACA 4155 has been tested in a wind tunnel at different angle of attack (α) [11]. The results are presented in Figure 2 by blue mark. The present method was used to estimate the Lift coefficient at different angle of attack and the results are also shown in figure 2 by read mark. The figure shows that the present numerical method shows some discrepancies with the experiment at angle of attack higher than 8°. However, for angle of attack less than this value the numerical results do agree well with experiments. Based on this fact, it can be said the results of the present numerical method is acceptable. Thus, it can be used to investigate the effect of chord length.

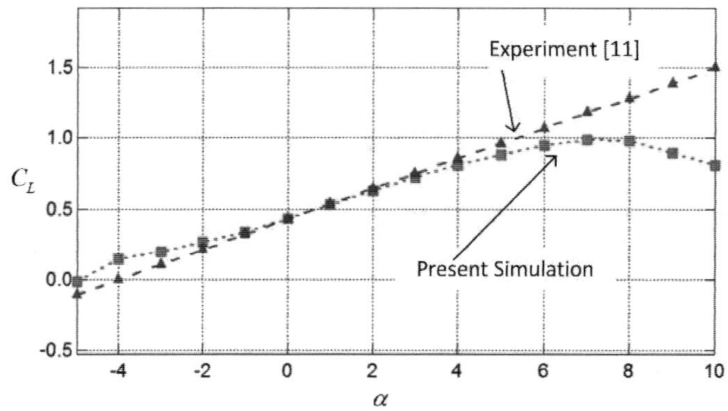

Figure 2. Numerical validation of NACA 4155

3.2. Velocity field

Figure 3 shows the velocity fields resulted by numerical simulation for all cases. Figure 3(a), figure 3(b), figure 3(c), and figure 3(d) shows the contour velocity at non-dimensional chord angle 20%, 30%, 40%, and 50%, respectively. As a note, the red colour indicates high velocity and the blue colour indicates low velocity. It can be seen that the wind velocity decrease in the inner area of rotating regions. In addition, on the inner surface of the air foil the velocity close to zero (dominated by blue colour). On the other hand, the outer surface the velocity increases (dominated by red colour). The difference of velocity from both surfaces results in power of the turbine. The effect of the chord length can be explored by comparing Figure 3(a) the lowest chord length and Figure 3(d) the highest cord length. Figure 3(a) shows that after the rotating region the velocity of the wind is relatively high (shown by green colour). On the other hand, Figure 3(d) shows that after rotating region the velocity of the wind is relatively low (shown by the blue colour).

In order to examine the effect of chord length to the performance, the velocity of the wind leaving the rotating regions are shown in Figure 4. The profile for all chord lengths are shown in the figure. The figure shows that if the chord length increase, the velocity of the wind leaving the turbine decreases. This is because the energy of the wind extracted by the turbine to become mechanical energy in the shaft. However, the velocity profile for chord length 40% and 50% show the almost similar values.

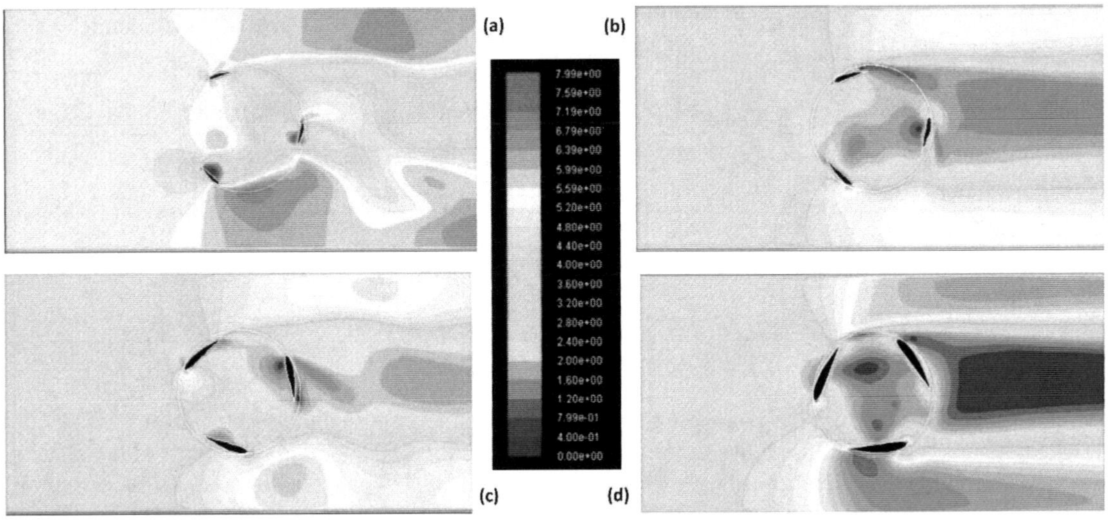

Figure 3. Contour velocity for all chord lengths

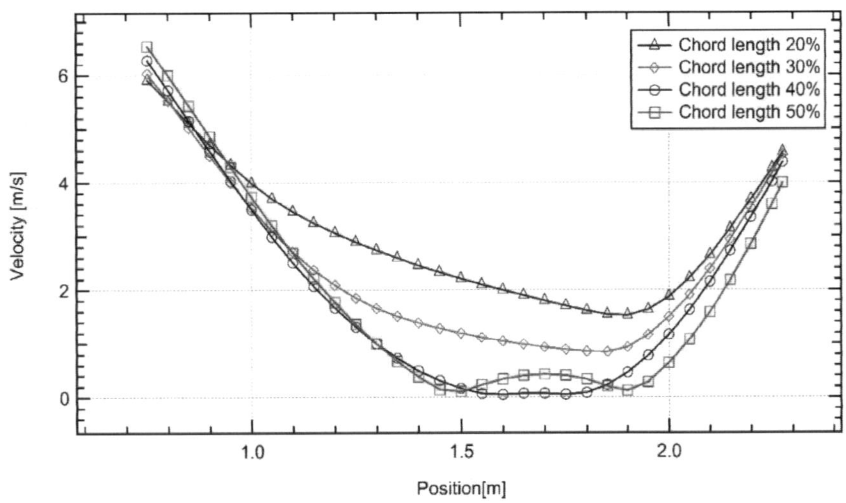

Figure 4. Velocity of the wind leaving the turbine

3.3. *Power coefficient*

The velocity profile of the wind leaving the wind turbine will be used to estimate the power coefficient. The power coefficient (c_p) is estimated using the below equation.

$$c_p = \frac{P}{P_0} = \tfrac{1}{2}\left[1 - \left(\frac{v_2}{v_1}\right)^2\right]\left(1 + \frac{v_2}{v_1}\right) \tag{9}$$

Here, P [Watt] and P_0 [Watt] is the mechanical power extracted from the wind and power in the wind stream, respectively. The velocity v_2 and v_1 is the wind velocity before entering and after leaving the turbine. Table 1 shows the average velocity leaving the turbine. It can be seen that the average velocity of the wind decreases as chord length increase. However, the average velocity leaving the turbine for chord length 0.4 and 0.5 is similar. Power coefficient for all cases are calculated using equation (9) and the results are presented in Table 1. It can be seen that, as expected, Power Coefficient increases as chord length increases. However, the Power Coefficient for chord length 0.4 and 0.5 is the same.

Table 1. The effect of Chord Length on Power coefficient

Non-dimensional Chord Length	Average at inlet v_1 [m/s]	Average velocity leaving the turbine v_2 [m/s]	Power coefficient c_p
0.2	5	3.01	0.51
0.3	5	2.42	0.57
0.4	5	1.99	0.59
0.5	5	1.97	0.59

4. Conclusions

Numerical simulation has been carried out to study the effect of chord length to the performance of the H-Darrieus VAWT. The turbine has diameter of rotating region of 1.5 m and the length of the blade is 1.55 m. The non-dimensional chord length (comparison of chord length to the dimeter of rotating region) was varied from 0.2 to 0.5. The velocity contour and the velocity profile of the wind leaving the turbine are plotted. The conclusion here is that non-dimensional chord length strongly affects the power. Increasing the chord length will increase the coefficient power. However, there exists an optimum chord length it is 0.4. Thus, it is suggested to use the non-dimensional chord length of 0.4.

References

[1] Mollerstrom E, Gipe P, Beurskens J and Ottermo F 2019 *Renewable and Sustainable Energy Reviews* **105** 1-13.
[2] Hashem I and Mohamed MH 2018 *Energy* **142** 531-545
[3] Ali A, Lee S-M and Jang C-M 2018 *Energy Conversion and Management* **171** 1322-1338
[4] Abdalrahman G, Melek W, and Lien F-S 2017 *Renewable Energy* **114** 1353-1362
[5] Bianchini A, Ferrara G and Ferrari L 2015 *Energy Conversion and Management* **89** 690-707
[6] Lanzafame R, Mauro S and Messina M 2014 *Energy Procedia* **45** 131-140
[7] Mohamed MH, Ali AM and Hafiz AA 2015 *Engineering Science and Technology, an International Journal* **18** 1-13
[8] Asr MT, Nezhad EZ, Mustapha F and Wiriadidjaja S 2016 *Energy* **112** 528-537
[9] Ambarita H, Siregar M R and Kawai H 2018 2018 *IOP Conference Series: Materials Science and Engineering* **343** 012025
[10] Launder BE and Spalding DB 1974 *Computer Methods in Applied Mechanics and Engineering* **3(2)** 269 – 289
[11] Abbot IH and Von Doenhoff AE 1959 *Theory of the Wing Sections*, Dover Publication Inc., New York

IOP Conf. Series: Materials Science and Engineering **648** (2019) 012031 doi:10.1088/1757-899X/648/1/012031

Vertical Sterilizer Design of Building Palm Oil Processing Plant to Cost Efficiency

I Rizkya[1]*, K Syahputri[1], Erwin[1]
[1]Departement of Industrial Enginnering, Faculty of Engineering, Universitas Sumatera Utara

*E-mail: indahrizkya@usu.ac.id

Abstract. Research is carried out in the Palm Oil Processing Plant is being built. The establishment of a factory experienced cost problem due to the increasing price of products import. This study describes the application of value engineering to find the highest cost components and alternative designs are able to replace the initial design and the building cost in the factory can be reduced. Vertical sterilizer is known to be the largest cost component followed by the cost of the Valves and Hydraulic System Control Unit. The company is currently designing to replace 4 units vertical sterilizer capacity 10 tons fresh fruit bunches with 3 units vertical sterilizer capacity 15 tons fresh fruit bunches. The change in the quantity and capacity of the vertical sterilizer is followed by a reduction quantity of Control Units c / w Valves and Hydraulic Systems used. As a result of this design change, the initial life cycle cost is IDR 7.531.154.360,- decrease to IDR 6.642.162.754,-. By using value engineering, it is able to make the cost efficiency in a factory about 11,8%.

1. Introduction

Indonesia is the first country in the world's largest palm oil industry. The total land area planted to palm oil in Indonesia has reached 9,7 million hectares [1]. Indonesian country produces around 40-50% of the total world palm oil production. Indonesia is the largest producer and exporter of palm oil and its derivatives. Palm oil is one of the most consumed and produced oils in the world. The palm oil industry as one of Indonesia's leading sectors has a significant contribution to national non-oil and gas exports and every year tends to increase significantly. Indonesia also shows an increasing trend every year with an average increase of 12,97%. This industry contributes US $ 17,6 billion through exports in 2012 [1].

In the long term, the world demand for palm oil shows a tendency to increase in line with the increasing world population. This increase will have an impact on increasing consumption of products with palm oil raw materials. Therefore, companies need to produce value engineering to minimize cost in the development of manufacturing industries [2,3].

Value engineering is one of the techniques applied to analyse the function of an item or process to determine the best value, or the best relationship between worth, resources and cost. The best value is showed by an item or process that consistently performs the required basic function. Value engineering has the lowest life-cycle cost [4]. Value engineering is related to the development of new products or services with an emphasis on the ease and production and delivery costs. While value analysis deals with design modification of existing products or services with the goal of reducing manufacturing costs and product delivery. Value Engineering is currently used in industry particularly in building.

Value Engineering is used to find alternatives or ideas that aim to produce costs are better or lower than previously planned cost with functional limits and work quality [5]. Value engineering can be applied at any time throughout the project duration from the beginning to the completion of the project development until the replacement stage. Value Engineering must begin early in the concept stage and continuously at each stage until the completion of planning [6]. Application of Value Engineering helps in on-going and upcoming construction projects that improve decision-making and leads to optimal expenditure of owner funds. Owner funds directed to meeting required function and quality level [7]. The theory of Value Engineering was introduced into China since 1978. Value engineering has been widely adopted by many companies and has a great effect on economic benefits. 35 years practice experience, the theory and methodology of Value Engineering has been recognized by the many sector especially to academic community, business circles, which has been one of the significant methods to improve product quality and cost reduction [8].

Research on the benefits of value engineering application to costs minimization has been done previously. VE can be used in manufacturing sectors such as bridge construction and various economic activity development processes are aimed at consumer-oriented [9]. Value engineering has been used at drainage project discussed to assess the initial design of the stream viewed from the cost budget plan. Next to find out whether the alternative design is required, find out the ideal alternative design, and get a comparison between the initial design and the alternative design. After comparison between initial design and some alternative design, it is known that alternative designs that can be recommended are the smallest costs alternative while maintaining the main function. [10]. The same research was also carried out by Amruta Chougale, A.K. Gupta and Sneha Patil that presented a cost minimization of residential building is having more cost on existing and the comparing cost from actual and recommendation of the lowest cost of building [11].

Researches about value engineering have been done before, especially in the construction sector. However, there are still few studies discuss of value engineering that applied to manufacturing industries, especially in the Palm Oil Processing Plant. Value engineering carried out in the manufacturing industry is used to obtain alternative design machines and equipment used for the production process. In Palm Oil Processing Plant, plant start-up costs mostly come from bank loans, and the rest of the owners. Increasing interest rates and declining rupiah exchange rate in US dollar have resulted in increased building costs. Value engineering aims to find alternative designs is able to minimize costs without changing the main functions that the company expected. This paper will discuss the changes in the vertical sterilizer design and Sterilizer Valves and Hydraulic System Control Unit as an alternative to reduce the building cost in palm oil processing plant.

2. Methodology

The study was conducted at one of the palm oil processing plants are being built. *Value Engineering* is a solution method to the problem of specific engineering, science, expert teams, creative approach organized. Value engineering aim is to define and eliminate costs that are not needed as costs that do not contribute to the quality, usability, age, and the appearance of the product and the consumer appeal. To use the value engineering method, five stages are generally used, namely the information stage, the creative stage, the evaluation stage, the development stage and the presentation stage. Assessment in value engineering has standard methodology and the tools used to study that can be seen in Figure 1.

The first stage is information which is collecting project information including the structure, functions, and costs of the object being studied. The function analysis phase is done by analyzing the primary and secondary functions of basic design. The creative stage is to develop possible alternatives to fulfill primary and secondary functions. Next step is evaluating the alternatives that have been formed and selects the best alternatives based on costs and changes to functions. The development phase is done by perfecting and adjusting to selected alternatives. The final stage is presentation and recommendations. This stage explains the results of team work to management and answers the questions about which alternative is best, what are the influences of developing alternative ideas, how much they cost, and how they are performed.

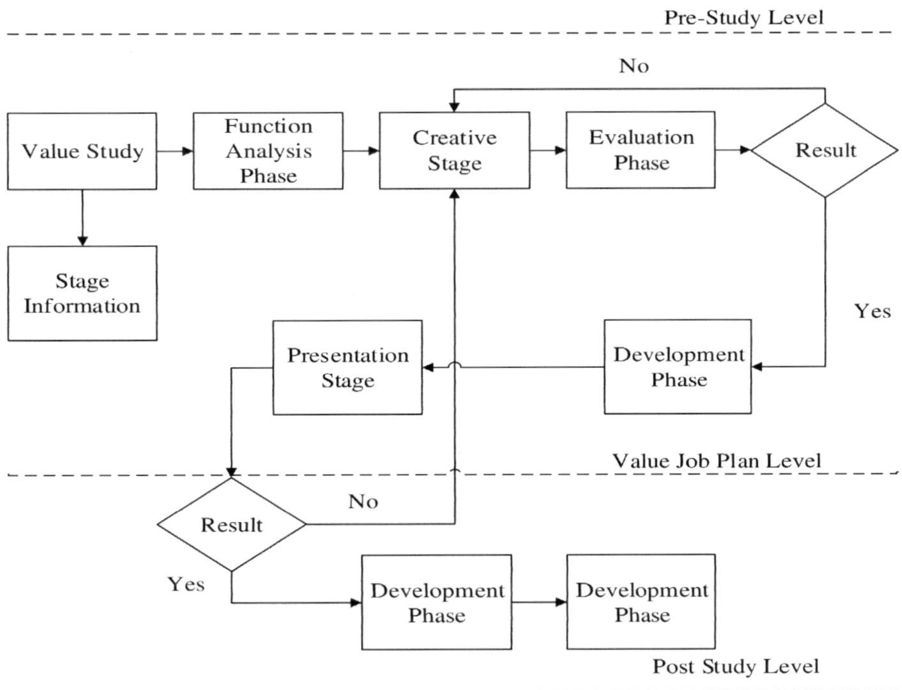

Figure 1. Stage of Value Engineering [12]

3. Result and Discussion

3.1. Building Information

Information collection about project data in general such as, project type, name, location, project scope, and project costs. The research was carried out on the building cost of a Palm Oil Processing Plant into Crude Palm Oil and Palm Kernel, located in Pekan Baru, Riau. The palm oil processing plant is planned to process 40 tons of fresh fruit bunches in one hour. The following is a summary of the building cost in Palm Oil Processing Plant.

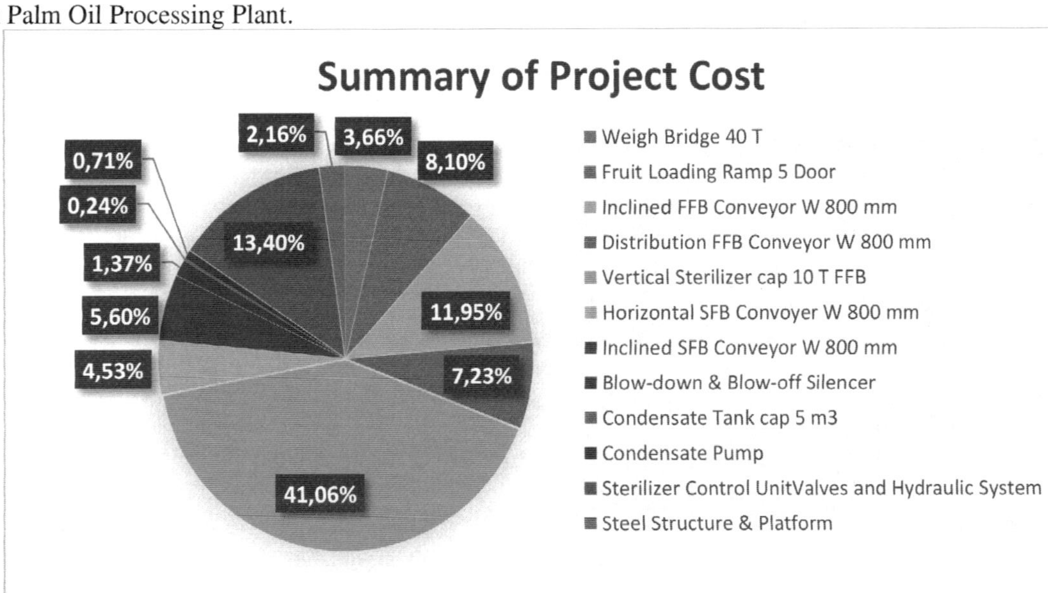

Figure. 2 Summary of Project Cost

Figure 1 shows the costs amount that will be spend to build a Palm Oil Processing Plant for a 40 tons capacity of fresh fruit bunches per hour.

3.2. Pareto Analysis

Pareto analysis is used to find out high-cost items that consume most of the budget plan. The Pareto principle states that 80% of the highest total costs occur in 20% of work activities. Pareto diagram of the work on building cost of a palm oil processing plant can be seen in Figure 3.

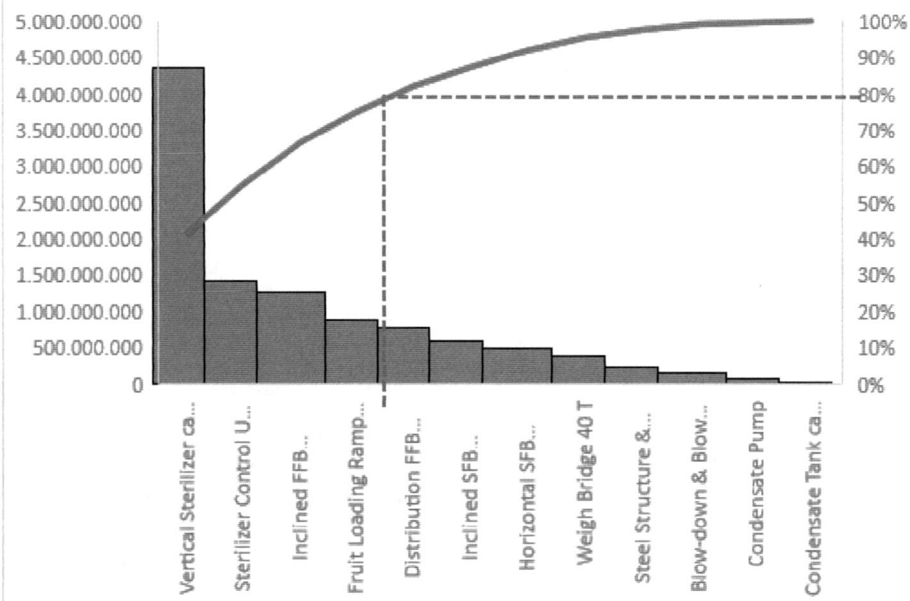

Figure 3. Pareto Diagram of Project Cost

Based on the results of high costs identification, it shows as many as 4 work items that spend almost 80% of the total cost, namely *Vertical Sterilizer 10 T FFB*, *Sterilizer Control Unit Valves and Hydraulic System, Inclined FFB Conveyor W 800 mm*, and *Fruit Loading Ramp 5 Door.* Based on the high cost analysis above, two items of work that spend the highest cost chosen are *Vertical Sterilizer 10 T FFB* and *Sterilizer Control Unit Valves and Hydraulic System,* to do a function analysis at the investigation stage.

The investigation was carried out by analyzing the functions of the vertical sterilizer and sterilizer control unit. *Vertical Sterilizer* is *Steam pressure Vessel* upright secondary shaped used to boil fresh fruit bunches. *Sterilizer Control System* serves to control the boiling process, which is opening and closing the door to the entry of fresh fruit bunches when boiled and opening-closing the door to the expenditure of fresh fruit bunches. *Sterilizer Control System* consist of controllers that can be programmed in a Programmable logic Control and hydrolic system. This system can be operated in full - automatic and semi - automatic. The vapor pressure in the stew and the boiling cycle time is controlled by this system.

3.3. Alternative Design

At the creative stage brainstorming is carried out with experts and stakeholders. This creative stage aims to reduce the quantitiy of ideas that must be generally identified to a short list of ideas with great potential to reduce the cost of building coat in Palm Oil Processing Plant. At the evaluation stage, an analysis of the advantages and disadvantages is carried out to get the most appropriate alternative. One design alternative that is possible to minimize the building cost in Palm Oil Processing Plant is to replace Vertical Sterilizers with a 10 tons capacity of fresh fruit bunches of 4 units with a Vertical Sterilizer with a 15 tons capacity of 3 units. Changes in the quantity of vertical sterilizers will also have an impact on reducing the quantity of Sterilizer Control Unit Valves and Hydraulic Systems.

3.4. Analysis of Life Cycle Cost

Life Cycle Cost is a technique to analyze economically by calculating all relevant costs during the investment period at the time value of money. LCC analysis is based on the assumption of project economic life (n) 10 years, interest value (i) of 13%, and annual maintenance costs of 6% of the work value. Changes in the design of vertical sterilizers and sterilizer control units do not affect on the other cost items. The cost calculation is only focused on the vertical sterilizer and sterilizer control unit. The following is a comparison of the actual cycle cost of the actual design and using alternative design.

Table 1. Comparison of Basic and Alternative Design Cost

NO	DESCRIPTION	BASIC DESIGN (IDR)	Alternative (IDR)
1.	Construction Cost of Vertical Sterilizer	4.364.160.000	3.886.920.000
2.	Redesign Cost of Vertical Sterilizer		85.000.000
3.	Construction Cost of Sterilizer Control Unit	1.424.040.000	1.068.030.000
4.	Redesign Cost of Sterilizer Control Unit		65.000.000
5.	Total Initial Cost/IC (1+2+3+4)	5.788.200.000	5.104.950.000
6.	Annual Maintenance Cost (6% of Total initial cost)	347.292.000	306.297.000
7.	Present value of maintenance cost by n=10 and i=15% (5,0187)	1.742.954.360	1.537.212.754
8.	Total Cost Present Value (5+7)	7.531.154.360	6.642.162.754

In calculation of the life cycle cost, replacement cost, savage cost and operation cost items are not considered. The absence of a replacement cost is because because of the planned 10 years there has been no design change. Savage cost and operational cost are considered non-existent because the estimated operational costs and residual values at the end of the project are the same for the initial design and alternative design.

3.5. Analysis of Cost per Worth

Value engineering analysis is able to provide input in the lowest cost determination to carry out functions, and provide information about costs that can be reduced and eliminated without affecting performance or constraints. Table 2 shows an analysis of cost per worth to see whether the alternatives offered are feasible to apply.

Table 2. Cost per Worth Analysis

NO	DESCRIPTION	BASIC DESIGN (IDR)	Alternative (IDR)	C/W	Summary
1.	Construction Cost of Vertical Sterilizer	4.364.160.000	3.886.920.000	1,122	Worth
2.	Construction Cost of Sterilizer Control Unit	1.424.040.000	1.068.030.000	1,330	Worth
	Total	5.788.200.000	5.104.950.000	1,134	Worth

The main objective in building this palm oil processing plant is to be able to process 40 tons of fresh fruit bunches per hour. Utility value gives an value indication. Value indication means the lowest cost needed for the implementation of a specific function. Utility value is used as a tutorial to identify areas with potential and important savings to the high value objective [12].

Based on the life cycle cost analysis and cost per worth analysis, it is known that the alternate offered is feasible. Design of actual 4 units Vertical Sterilizer 10 tons FFB and 4 units Sterilizer Control Unit Valves and Hydraulic System requires a cost of IDR 7.531.154.360,-. While the alternative design is it can be replaced with 3 units Vertical Sterilizer 15 tons FFB and 3 units Sterilizer Control Unit Valves and Hydraulic System requires a large cost of IDR 6.642.162.754,-. There is a reduction in costs by value engineering application [7,12]. The recommendation offered is to apply an alternative design 3 units Vertical Sterilizer of 15 tons FFB and 3 units Sterilizer Control Unit Valves and Hydraulic Systems due to able of the efficiency costs up to 11,8%.

4. Conclusion

With value engineering, it is known that the cost of vertical sterilizer and sterilizer control unit is the largest cost item with cost of 41,06% and 13,40%. Therefore, in the design of actual 4 units Vertical Sterilizer of 10 tons FFB and 4 units Sterilizer Control Valves and Hydraulic System Units can be replaced with 3 units Vertical Sterilizer of 15 tons FFB and 3 units Sterilizer Control Unit Valves and Hydraulic Systems. The replacement for replacing the vertical sterilizer and Sterilizer Control Unit is known to save costs by 11,8%.

References

[1] Sequiño A C and Jessica M 2015 *Avenido International Journal of Ecology and Conservation* **13** (1) pp.152-164
[2] Ewaldo E 2015 *Jurnal Perdagangan Industri dan Moneter* **3** (1) pp. 10-15.
[3] Tryfono 2006 *Potensi dan Prospek Industri Kelapa Sawit* Jakarta : Gramedia.
[4] Atabay S and Galipogullari N, 2013 *Journal of Traffic and Transportation Engineering* **1** (1) pp. 39-48.
[5] Kembuan A S, 2016 *Journal Sipil Statik*, **4** (2) pp. 95-103.
[6] Khamim M and Harsanti W 2015 Journal Teknik Sipil, **9** (2) 126-135.
[7] Yadav N B, Kacha R, Sharma N D and Rathod H A, *International Journal of Civil, Structural, Environmetal and Infrastructure Engineering Research and Development* **3** (5) 1-8.
[8] Rifin A 2017 *Jurnal Manajemen & Agribisnis*, **20** (2) pp. 53-64.
[9] Bertolini V, Wishnumurti and Zacoeb A, 2016 *Jurnal Iptek*, **14** (2) 103-108.
[10] Ridwan M A, Adi H P and Adhy D S, 2017 *Proceeding of International Conference: Problem Solution and Development of Coastal and Delta Areas Semarang Indonesia* pp.712-717.
[11] Chougale A, Gupta A K and Patil S, 2014 *International Journal of Innovative Research in Advanced Engineering*, **1** (12) pp.115-118.
[12] Harini M W, Erizal and Widyarti M, 2018 *Asian Journal of Applied Sciences* **06** (02) pp. 71-77

Acknowledgement

We are also grateful to Universitas Sumatera Utara for financial supported to publish this paper. We are thankful to our colleagues in Department of Industrial Engineering for providing expertise that is very helpful in this research.

IOP Conf. Series: Materials Science and Engineering **648** (2019) 012032 doi:10.1088/1757-899X/648/1/012032

Data Envelopment Analysis for Productivity on Small and Medium Enterprise: A Measurement

Rahmi M. Sari[1], Erwin[1], Khalida Syahputri[1], Ghassan M. Fattah[1]
[1]Department of Industrial Engineering, Faculty of Engineering, Universitas Sumatera Utara

Email : rahmi.m.sari@usu.ac.id

Abstract. Every operator wants SME that can be work optimally, Thus the SME's productivity projection to have increasement. As a case study, this research conduct in the SMEs that worked in general printing. This research also pursues to measure the productivity of operators. Based on the results of the initial research obtained by the station that most requires physical work is the pod station. The work system in pod station has not designed yet. Therefore, a work system will be designed for the station. First, the operator productivity measurements are carried out. The method used to measure operator productivity is Data Envelopment Analysis. The analysis started by making cycle times, work presentations, and the number of products produced as variables to obtain optimal functions. During the observations made at the SMEs, productive operators were obtained who managed 83.69% on the operator I and II carriers on 79.34%. From the calculation of the DEA, the second operator has the value equal to 1 so that it has passed the threshold of expected quality.

Keywords: Productivity measurement, Data Envelopment Analysis, SME

1. Introduction

Each SME wants an operator that is able to work optimally, so that the productivity of the SMEs is expected to increase of overalls. This research has been done in the SMEs that worked in general printing.

There are seven elements of work for SMEs, that are, operator brings paper and ink to the printing site, measure the amount of ink needed and put it into the printing press, print paper with a printing machine, brings printouts to pod station, cuts the mold, perform printing results, the last element is bring the print out to the storage area. The seven elements of work are done at 4 work stations, which are: printing station, pod station, separation station, and office.

Based on the results of preliminary observations it was found that the station that needed the most physical work was the pod station. This SMEs has a specific problem that they do not have the design of work system especially in pod station. For this reason, a work system will be designed for the station. So that, the aim of this research is to design the work system for the pod station. Previous

First, the operator productivity measurements using Data Envelopment Analysis. It calculated by cycle time, presentation of work, and the number of products produced as variables to get the optimum function.

DEA is a mathematical programming technique based on linear programming is used to evaluate the efficiency of a decision-making unit which is responsible for using a number of inputs to obtain an output target. DEA method was created as a performance evaluation tool in a unit activity entities (organizations), hereinafter referred to Decision Making Unit (DMU). Simply put, this measurement is expressed by the ratio: output / input, which is a measurement of the efficiency or productivity [1].

2. Methodology

In the measurement of efficiency using DEA, there are two models frequently used, namely Constant Return to Scale (CRS) and a Variable Return to Scale (VRS) [2]. However, in this study using a CRS. DEA model is oriented towards input based on assumptions constant return to scale that is known as CCR models. In the constant return to scale models of each DMU will be compared with the rest of the DMU in the sample with the assumption that the internal and external conditions are the same DMU. Criticisms of these assumptions is that the assumption of constant returns to scale is only suitable for the conditions in which the entire DMU operating at optimal scale [3]. However, in reality though the DMU operates with a power source (input) the same and produce the same output, but also internal and external conditions may be different so as to result in a DMU is not operating at optimal scale. According Charnes, Cooper, and Rhodes this model can show the overall technical efficiency or value of profit efficiency for each DMU [4].

DEA assumes that each DMU will choose weights that maximize the efficiency ratio. Because each DMU using a combination of different inputs to produce outputs different combinations, each DMU will choose a set of weights that reflect that diversity. Generally DMU will assign a high weight for the input and output use is little that can be produced by many. Weights is not the economic value of the input and output, but rather as a determinant to maximize the efficiency of a DMU. As gambararan, if a DMU is a company oriented to profit (profit-maximizing firm) and each input and output has a cost per unit and the selling price per unit,

DEA for a DMU may be formulated as a linear program fractional solution can be obtained if the model is transformed into a linear program with weights of input and output such as a decision variable DMU (decision variables) [5] simplex method can be used to solve the model has been transformed into a linear program. DEA requires completion of a linear program for each DMU. The result is a set of weights for a DMU and figures relative efficiencies. DEA has several managerial values[6]. First, DEA efficiency for each DMU, DMU relative to others in the sample. This efficiency rate allows someone DMU analysis to identify the most in need of attention and remedial action plan for the DMU walkin / less efficient.

Second, if a less efficient DMU (efficiency <100%) showed a number of DMU DEA has perfect efficiency (efficient reference sets, efficiency = 100%) and a set multiplier that can be used by managers to develop strategies for improvement. Such information allows one analysis made DMU hypothetical use fewer inputs and produce outputs at least equal to or more than the DMU walkin efficient, so the DMU hypothetical would have perfect efficiency when using weights input and weighting the outputs of DMU inefficient. The approach members strategic direction for managers to improve the efficiency of a DMU walkin efficient through the introduction of too much used input and output production is too low.

Third, DEA provide cross efficiency matrix. A cross-DMU efficiency of the DMU B weighted ratio of output divided by the weighted input which is calculated using the input level and outut DMU A and the weight of input and outp DMU B. Analysis of cross can help one manager to identify an efficient DMU but using a combination of inputs and produce a very different output combinations with other DMU [7]

The approach used is input-oriented DEA assuming variable returns to scale (VRS). Selection of these assumptions are based approach that has not been operating at optimal scale due to the many government policies that affect and limited availability of natural resources. Efficiency scores for each firm to-i has a value between 0 - 1. The score of 1 indicates a point on the frontier where the company has technically efficient. Assuming VRS DEA will produce efficiency score greater than or equal to the value obtained when using the assumption of constant returns to scale (CRS) [8],

Productivity measurement is done with the approach Malmquist Productivity Index. Some of the advantages of this method, among others, can measure the change increase or decrease performance over several time periods. In addition, this method can decompose the changes in productivity into technical efficiency change and technological change. Malmquist Productivity Index between t and t + 1 is expressed as follows [9]:

$$m_i(q_t, q_{t+1}, x_t, x_{t+1}) = \frac{d_i^t(q_t, x_t)}{d_i^{t+1}(q_{t+1}, x_{t+1})} \left[\frac{d_i^{t+1}(q_{t+1}, x_{t+1})}{d_i^t(q_{t+1}, x_{t+1})} x \frac{d_i^{t+1}(q_t, x_t)}{d_i^t(q_t, x_t)} \right]^{1/2} . \quad (1)$$

Where d (x, y) indicates the distance input function
The index value changes in efficiency can be greater than 1 (one) which indicates the level of efficiency increase, equal to 1 (one) means that no changes in efficiency, and less than 1 (one) showing a drop in efficiency between t and t + 1[16], This value indicates how far the position of a company on the production frontier. Just as changes in efficiency, value changes in technology could also be greater, equal to, or less than 1 (one) which indicates whether the frontier moved forward, fixed, or retreat[10], Shifting forward the frontier indicated there are advances in technology and vice versa [11].

In resolving problems or Company level organizational efficiency of each branch or unit, using mathematical problem solving Data Envelopment Analysis (DEA). The data input in the form of input and output factors. Input factor consists of the percentage of Work (X1,%), Cycle (X2, CNN), Output factor used is the amount of product (Y1, Unit). Data processing steps are, first, calculated variable expenses for each DMU to be assessed levels of efficiency. Second, determine a mathematical model that will be used as a means of solving the problem, in this case using a mathematical model of Data Envelopment Analysis (DEA). Third, calculated targets for DMU relatively less efficient for the input obtained from the difference between the actual input value to the value of the input slack.

3. Results and Discussion

Operator start working from 8:00 to 11:59 AM, break from 12:00 to 12:59 PM, followed at 13:00 to 18:30 PM. Having obtained the data from observations on SMEs will be a recapitulation of data collection to facilitate data processing. Recapitulation data operator activity can be seen in table 1.

Table 1. Operator Activity

operator activity	*Work*	*idle*	total
1	77	15	92
2	73	19	92

In Table 2 shows the percentage of operator activity. This table consists of the percentage of operator's wprk and idle. Recapitulation percentage of the activity of each operator can be seen in the following table.

Table 2. Percentage of Operator Activity

operator activity	*Work*	*idle*	total
1	83.69%	16.31%	100%
2	79.34%	20.66%	100%

Target of printed sheet in 8-hours working day are 3000 sheets. When multiplied by the target of the print sheet with 8-hour working day, the first mold then takes 2852.27 seconds can be completed in 1 day 10 Molds.

DMU determination can be obtained by assuming the response to the characteristics of smaller the better as input and set a unit (one) as the output. Table 3 below is data DMU that is already in the set:

Table 3. Determination of DMU Input and Output

input		output
X1	X2	Y1
0.83	1314.45	1
0.79	1537.82	1

The data in Table 3 were determined both responses ie, percentage of work (X1) has the same characteristics, namely bigger the better, and Cycle Time (X2) has the characteristics. Both smaller the better the response included in the DMU input. Unit (one) set as well as the DMU output.

$$\text{Max } \theta = \sum_{r=1}^{s} u_r y_{ro} \quad (2)$$

with constraints,

$$\sum_{i=1}^{m} v_i x_{io} = 1 \sum_{r=1}^{s} u_r y_{rj} \leq \sum_{i=1}^{m} v_i x_{ij} \quad j = 1, ..., n; u1, u2, ..., us \geq 0; \text{ and } v1, v2, ..., vm \geq 0 \quad (3)$$

where the value 1 to the data there are 2 DMU

The calculation using the software lingo and get the following results that shows in Table 4,

Table 4. Calculation Model DEA CCR

DMUj	input		output	Result
	X1	X2	Y1	
DMU1	0.83	1314.45	1	1,000
DMU2	0.79	1537.82	1	1,000

Due to the result value equal to 1, we conclude that both the DMU has passed the threshold Bayas expected quality.

4. Conclusion

In observations carried out in this following SME, obtained percentage of carriers for productive work is 83.69% on the carrier I and 79.34% in the second operator. From the calculation of DEA found that both operators have result value equal to 1. It means that both of them have passed the threshold Bayas expected quality.

References

[1] Filardo A, Negoro NP, Kunaifi A 2017 Application of Data Envelopment Analysis In Product Retailers Efficiency Measurement Vehicle Brands Toyota *Journal of Science and Arts ITS* **6** pp 74-75
[2] Koondhar M A, Li H, Qiu L, Joyo M A, Chandio A A, Liu W, He G 2016 Analysis of the Algicultural Land Use Efficiency Based on DEA Model: A Case Study of Asian agriclutural Developing Countries *Transylvanian Review* **14**(11) pp. 1072-1074
[3] Cooper WW, Seiford LM, Tone K 2009 *Data Envelopment Analysis*, Springer, pp. 1-5
[4] Cook WD, Tone K, Zhu J 2013 Data envelopment analysis: Prior to choosing a model of *Omega* **44** 1-4
[5] Khezrimotlagh D, Zhu J, Cook WD, Toloo M 2019 Data and Big Data Envelopment Analysis *European Journal of Operational Research* **200** pp. 5-7
[6] Liu JS, Lu LYY, Lu WM. 2018 Research Fronts in Data Envelopment Analysis Taiwan: National Taiwan University pp. 3-10
[7] Asosheh A, Nalchigar S, Jamporazmey M 2010 Information technology project evaluation: An integrated Data envelopment analysis and balanced scorecard approach *Expert Systems with Applications* **37** pp. 1-3
[8] Boothroyd G 2011 Product Design for Manufacture and Assembly, CRC Press
[9] Mousavi SHA, Rafiee S, Jafari A, Mohammadi A 2011 Optimization of energy consumption for production of soybean using Data Envelopment Analysis (DEA) approach *Applied Energy* **88** 5-7
[10] Sueyoshi T, Goto M 2012 Efficiency rank-based assessment for electric power industry: A combined use of Data Envelopment Analysis (DEA) and DEA-Discriminant Analysis (DA) *Energy Economics* 634-644
[11] Umri N, Hidayat R, Utami ID 2011 Performance Cost Reduction Methods Data Envelopment

Analysis (DEA) *Scientific Journal of Industrial Engineering* 216-223

Acknowledgement

The author would like to thanks to the Universitas Sumatera Utara who has provided financial support to publish this paper. The authors express their gratitude to the owner of the SMEs who are willing to be the object of research. We appreciate the efforts of all those who have cooperated in conducting this study.

IOP Conf. Series: Materials Science and Engineering **648** (2019) 012033 doi:10.1088/1757-899X/648/1/012033

Relationship Management Accounting and development of information technology

Keulana Erwin[1]
[1]Department of Accounting, Universitas Sumatera Utara, Medan-Indonesia

*Email :keulana@usu.ac.id

Abstract The development of information technology has brought a change significant in business. The problems that occur at this time are still large barriers to management in decision making process. This is because there is still a lack of information that can be obtained quickly with the advent of information technology, there are several types of information systems that we can see, among others Decision Support System, Executive Information System, Accounting Information System, Accounting Information System, Expert System, Electronic Data Processing System, Management Information System, and Electronic Data Processing Systems. The research method used in this study is a qualitative method. The development of information technology too influential on the field of management accounting as a field information producer in the planning framework, controlling and decision making of management. The influence can be of a nature profitable or detrimental to the company.

1. Introduction

At present, most people increasingly feel information as one of the basic needs besides the need for clothing, food and board. Along with that, information has changed into a form tradable commodity. This situation is proven by increasingly the development of information service businesses, such as television stations, newspapers, radio and internet that have entered the joints of human life. Change This rapid, dynamic and extensive environment is supported by technological advancements information in all fields. This has encouraged the transformation of society traditional information society.

The development of information technology has had an impact on life society. Since the discovery of computers in 1955, world civilization has entered the information age. Information technology with computers as motors the driver has changed everything. Computer-based information processing began to be known to people and until now there has been a lot of software that can people use as a data processing tool to produce information. In the field of accounting, a computer-based accounting information processing system many are offered with the aim to provide convenience for the people accountant to produce information that is reliable, relevant, timely, complete, understandable, and tested.

In the era of global business, the influence of information technology advances cannot avoid, such as the use of telephones, facsimiles, computers and satellites inside various activities that communicate with the company. Information Technology allows humans to obtain information from faraway places in a short time and at a low cost.

Organizational management must be responsive to this environmental change if desired the organization can still survive and increase its performance. Management organizations must also be sensitive to the effects of technological developments includes information, technical equipment and processes in converting inputs into output. In addition, management must be able to understand the role of the system well information in organizations [1].

2. Literature Review

Information technology can be defined as a combination of technology computer and telecommunications with other technologies such as software, hardware, network technology, database and telecommunications equipment others. Furthermore, information technology is used in organizational information systems to provide information for users in order to retrieve decision.

Several types of information systems exist, including Decision of Support Systems, Information System of Executive, Information system of accounting, System of Expert, System of Electronic Data Processing, and Information system of Management [2]. Communication channel can be used to communicate is standard telephone lines, coaxial cable, fiber optics, microwave systems, communications satellites, cellular radio and telephone. While network configuration that can be used to communicate is a Wide Area Network (WAN), Local Area Network (LAN), and Client / Server Configurations [3].

One important technology of information is usage Interchange of Electronic Data (EDI). EDI is communication between computers with the purpose of increasing effectiveness and reducing work of clerical. Hansen and Hill [4] define EDI as a movement of business documents in a structured format between various business partners in a organization. With EDI, received of documents can be ordered by the computer automatic. Integrated EDI provides opportunities for managers to fully concentrate on strategic decision making and improving ability to control several activities.

Technology will continue to grow. Strong information technology will become competitive edge for companies and at the same time become an entry barrier [5]. For organizations that want to progress and develop, there is no reason not to using technology as long as it can make it easier for the company adjust to their environment [6][7].

3. Result and Discussion

Information systems of Management Accounting have three main objectives, namely (1) to provide information used in calculating service of costs, products and other desired objectives of management, (2) to provide information used in planning, controlling, evaluating, and continuous improvement, and (3) to provide useful information in decision making [8]. These three goals reveal that managers and other users need management accounting information and need to know how to use it. Management accounting information can help management identify a problem, solve problems, and evaluate performance. Management accounting information is needed and used in all scope of management, including planning, control and decision making. Furthermore, the need for information is not limited to manufacturing organizations, but also includes trade and service organizations.

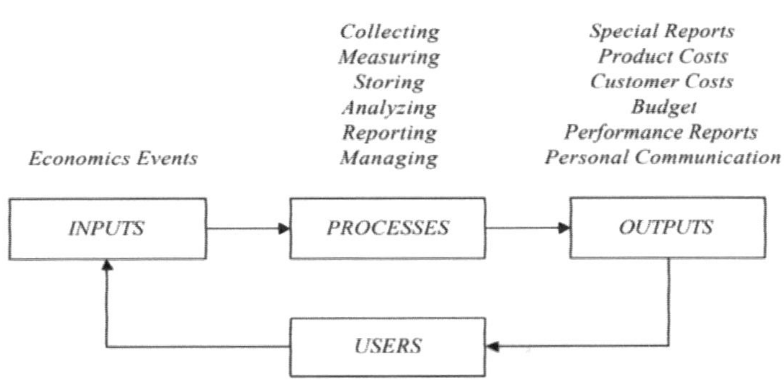

Figure 1. Operational Model : Information System of Managemen Accounting

Information technology progress also has a big impact on developments in the paradigm and in manufacturing technology. Some fundamental management accounting factors experience changes as a result use of information technology. These changes include processes planning, controlling routine activities, organizational structure and work situations. In situations where the environment changes, the organization's plan must also be changed to stay afloat, and the organization remains stable. Organization which faced with environmental changes must be responsive if you don't want to decreased activity that cannot be avoided. This condition requires management to always make innovative improvements continuously in all aspects so that the company can stay within competition is very tight, it can even allow companies to get become a leading company.

Management accounting should pay attention to the application of concepts appropriate management and accounting, and techniques for presenting information finance that is useful for management in planning, supervision, decision making, and the most allocation of resources economical. By considering the capabilities possessed, application information technology is a challenge for deep management accounting face existing technology. It is the management accountant who determines decision to adjust information technology capabilities with ability or capacity of the company. The management accountant is responsible creating a positive climate for making changes in a organization. Clearly, there is an advantage for management to apply information Technology. Information technology can also provide opportunities and support companies to be able to compete more in the era of globalization. The definition of a more competitive ability refers to the provision of facilities to be able to compete through improving service quality.

The following are some examples of the implementation of information technology in management accounting. The EDI application in Just In Time (JIT) offers inventory control, directing orientation to quality and energy efficiency work. EDI also provides opportunities for deep management accountants improve quality related to production, scheduling, sales forecasting, speed up internal response time, connect more closely with customers, and assist management in improving control business activities. In Activity Based Costing (ABC), information technology can apply to process cost information so that it can provide clarity regarding sources or causes of cost posts organized and quickly. Whereas in Total Quality Control (TQC), information technology can be applied to make continuous improvements so as to produce zero-defect products and product quality are the responsibility from all parts. The development of information technology in management accounting can also provide information about the correlation between costs and time with fast and relevant. This makes managers able to respond to changing conditions market quickly and precisely. In addition, the application of manufacturing cells can speed up the time spent on production and reduce costs production. Other implementation of IT related technology is by using RFID Technology to get effective workload for employee in order to count effective production hours in man-machine related application [9]

Management accountants must also consider the rapid development information Technology. They must be able to adjust the level of development company with the availability of existing hardware and software. Information technology implementation must consider not only costs investment, but also maintenance and operating costs, including costs experts and network usage on third parties.

The development of information technology on the one hand benefits management accounting. But on the other hand, it can cause some problems. Even information technology is one of the causes of business pressure on the organization. To overcome various problems that arise due to the development of information technology, several actions are sought. Resistance to change problems must be eliminated because this can lead to decreased productivity, increase absenteeism, and reduce motivation or work strikes [10]. To prevent undesirable conditions, Gordon suggested that members of the organization or workers be involved in carrying out certain tasks and create an environment that supports the quality of members of the organization. In addition, it is necessary to provide awareness to employees that the use of information technology can provide benefits in the long term and show weaknesses of the old system. Furthermore, Gordon proposed several things that must be done to reduce resistance to change towards changes in the implementation of information technology, including communication, educational programs, evolution change, employee involvement, new policies and procedures, staff change, temporary structure and steering committees.

To be able to have the expertise and capabilities of information technology, members of the organization need to get additional education and training and the provision of relevant skills. In addition, before the organization management implements new information technology, they must consider the amount of costs needed and the benefits to be obtained (cost-benefit analysis). Information technology will be applied if the benefits obtained using information technology outweigh the costs of implementing information technology. This is also called the value of information technology. Therefore, management accountants need to consider it well before making a decision. Security must always be improved, to avoid misuse of information technology. For example, by storing computers in a safe place, they should only be used by certain people who are interested, using passwords, and making access control matrices.

4. Conclusion

To maintain the survival of an organization, it must be realized that the business environment will always change, including information technology also experiencing development. With the development of information technology resulting in changes in management accounting. Management accounting will adjust to the development of activities company.

The presence of information technology provides many benefits for companies, like being able to lighten complex business activities and produce information that is reliable, relevant, timely, complete, understandable, and tested in the framework of planning, control and decision-making management. In addition, the efficiency of the company's operations and company performance also can be improved. As a result, companies can survive in the era information and being able to face global market competition. In addition to generating benefits, the development of information technology can also cause some negative impacts on the company, such as the closure employment opportunities, the emergence of resistance to change and the emergence of information technology crimes that can harm the company.

References

[1] Elliot R K 1992. The third wave break on the shore of accounting. *Accounting horizon*. pp. 61
[2] Bodnar, George H. And William S Hopwood , 1998. *Accounting Information System*. 7th edition. Upper Saddle River-New Jersey; Prentice-Hall International, Inc
[3] Romney. Marshall. B. And Paul John Steinbart. 2000. *Accounting Information System*. 8th edition. Upper saddle River-New Jersey; Prentice-Hall International, Inc
[4] Hansens J V and Hill N C 1989 Control and Audit of Electric Data Interchange *MIS Quarterly* pp. 402-403
[5] Regina F 1994 Right Way to go Global; an Interview with Whirpol CEO, Davit Whitman. *Harvard Business Review*, pp. 135-145
[6] Hanscombe, Richard and Philiph Norman. 1989. Strategic Leadership: The Missing Link. International edition, Singapore: Mc Grawhill Book Co.
[7] Gunawan D, Amalia A, Rahmat R F, Muchtar M A, Siregar I 2018 Identifying strengths and weakness of Quality Management Unit University of Sumatera Utara software using SCAMPI C *IOP Conference Series:Materials Science and Engineering, 308(1), 012001*
[8] Hansen, Don R. and Mowen M M 2000 Management Accounting. 5th edition, Cincinnati-Ohio: South-Western College Publishing.
[9] Rahmat RF, Herly E T, Siregar B, Syahputra M F and Sitompul O S 2018 RFID presence monitoring system as an input to measure the workload of employee *International Conference on Computer Applications and Information Processing,* pp 1-6.
[10] Gordon, Judit R. 1993. Organizational Behavior. 4th edition. Needham Height. Allyn and Bacon.

IOP Conf. Series: Materials Science and Engineering **648** (2019) 012033 doi:10.1088/1757-899X/648/1/012033

Acknowledgments
Author is grateful to God, the ruler of the universe who always brings his grace together with guidance, so the writer can complete the research.

Interaction 3D Human Anatomy of Ear, Nose, Throat Using Mix Reality

M F Syahputra[1], R F Rahmat[1], Devira Zahara[2], M Arfiza Putra Saragih[2], M Anggia Lubis[1], M A Muchtar[1], U Andayani[1]

[1]Department of Information Technology, Faculty of Computer Science and Information Technology, Universitas Sumatera Utara, Indonesia
[2]Department of General Medicine, Faculty of Medicine, Universitas Sumatera Utara, Indonesia

Correspondence author: nca.fadly@usu.ac.id

Abstract. The development of information technology in the future will continue to grow and full of innovations, all shortcomings in an application or equipment will continue to be minimized, all ways to simplify and efficiently a technology are continuously looked for and developed by experts in their respective fields. The frequent use of 3D objects and animation in medical learning media is felt to be incomplete without the direct interaction of users with modern computers so that users are more interested in learning for Ear, Nose and Throat. Therefore, in this study, an interaction system was made on the Ear Nose Throat medical learning application using Microsoft HoloLens. In this study used interactions that use hand gesture and voice input. The results of this study, users will be able to interact with the Ear, Nose and Throat using Microsoft HoloLens. With this extensive application, it will facilitate users in Ear, Nose and Throat medical learning more interactively with the unique features of applications that use voice input and hand gesture.

1. Introduction

The development of information technology in the future will continue growing, meet all needs will be increasingly minimized, all ways to facilitate and efficient technology continue to be sought and developed by experts in their respective fields. Ear, nose, throat disease is a disease that is common in Indonesia. Most people do not recognize the types of diseases that promptly attack the ear nose throat (ENT) [6].

Ear, nose, throat are important organs typically found in the human body because they are naturally associated with the auditory and respiratory systems. In examining the ears, nose, throat (ENT) into a unit interconnected with each other, if one part of the organ is disturbed then the other two organs will be affected because it is connected through the "Eustachian tube" channel. Therefore, if the nose has an infection, it can spread to the throat or vice versa [2].

Several studies that have properly used mixed reality in the field of education have been done before, such as Jayaputra et al [3] such as Application of Mixed Reality as a Human Sensory Learning Facility using Hologram Technology, this study makes learning applications about humane visual surveillance using 3D holographic visualization which later from the application is expected to be used as a modern learning tool. In 2018, Lauren Siff conducted academic research in the medical field by displaying 3D models of specific organs and their scenarios. Users can naturally learn about the organs displayed [3]. Based on the background above, the authors propose research to develop an application that will visualize more interactive ENT learning by taking the title "Interaction Development in Ear Nose Throat Medical Learning Applications Using Microsoft HoloLens."

The use of 3D objects is a medical learning technique in properly recognizing the functional anatomy of ENT (Ear, Nose and Throat). But the lack of interaction makes users unable to come into direct contact with these 3D objects. To make virtual objects more interactive, another technique is needed to display virtual objects that can enter direct contact with users so that users can recognize the ear, nose and throat anatomy.

The previous research is conducted by Ferrari et al in 2009 was to develop a 3D visualization device using mixed reality that can be used by doctors to conduct preoperative radiological examinations. This automated system allows users to properly view augmented stereoscopic images, which are made by mixing 3-D anatomical models obtained by preoperative volumetric radiological image processing (computed tomography or MRI) with current images of real patients, captured by the camera [1].

In 2016, Apri Santoso and Elki Noviandi conducted research. The specific purpose of this research is to analyse and design the innovative design of an Augmented Reality Body Organ Learning Application. The methodology used in developing this application is the Microsoft Solution Framework (MSF) methodology. The analysis carried out includes conducting research on the applications that will be built and collecting data to gain the information needed in developing applications. Making augmented reality on this computer uses the ARtoolkit coding application. Three-dimensional objects created using Autodesk 3ds Max. The results of the comprehensive analysis and design of this application are expected to provide necessary convenience, improve effectiveness and efficiency for users to study human organs based on Augmented Reality [7].

In 2017, Kurniawan et al conducted a study entitled Application of Augmented Reality Respiratory System. This modern application can be effective for high school students (high school), who obtain learning materials for the human digestive system. In this study, UNITY Game Engine was intentionally used to build Android-based applications and Vuforia SDK so that the applications developed can represent an Augmented Reality application. Accompanied by 1 marker image contained in the biology textbook which when directed to the application can display the visualization of 3D objects like children's teeth, adult teeth, molars, tongue, animation digestion of the mouth, oesophagus, stomach, small intestine and large intestine [4].

In 2017, Mustaqim and Kurniawan conducted a study entitled Development of Augmented Reality Based Learning Media. This study aims to undoubtedly create a more effective learning media by properly utilizing innovative Augmented Reality technology [5].

In 2018, Samuel Ezzay conducted an academic study entitled 3D ENT (Ear, Nose and Throat) Modeling in Science Learning Media Using 3D Blender. This study aims to show every organ displayed in a 3-dimensional object according to other objects, with the existence of a 3D model that adopts digital technology, it will make it easier for science students to surgeons to carry out maximum organ learning. [8].

2. Methodology

The methodology in this study is explained in the general architecture that can be seen in Figure 1.

IOP Conf. Series: Materials Science and Engineering **648** (2019) 012034 doi:10.1088/1757-899X/648/1/012034

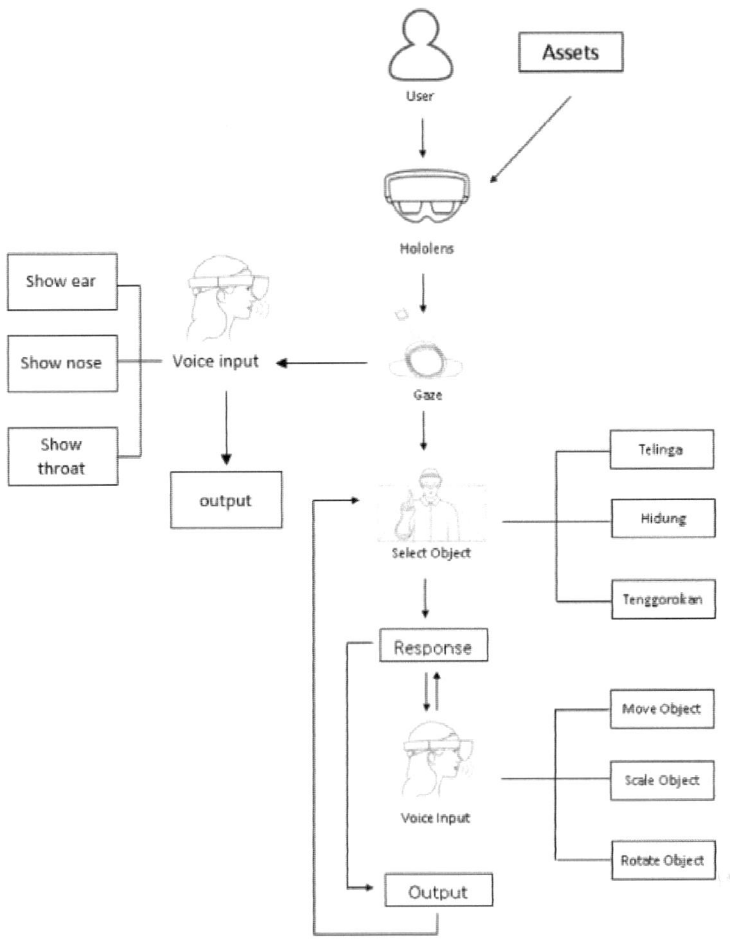

Figure 1. General Architecture

2.1 *Gaze Object*

This stage function to display 3D objects so that the user can instantly see the display using Microsoft Hololens. This process properly uses the hololens camera feature which serves to capture the appearance of 3D objects so that they can be displayed on Microsoft Hololens.

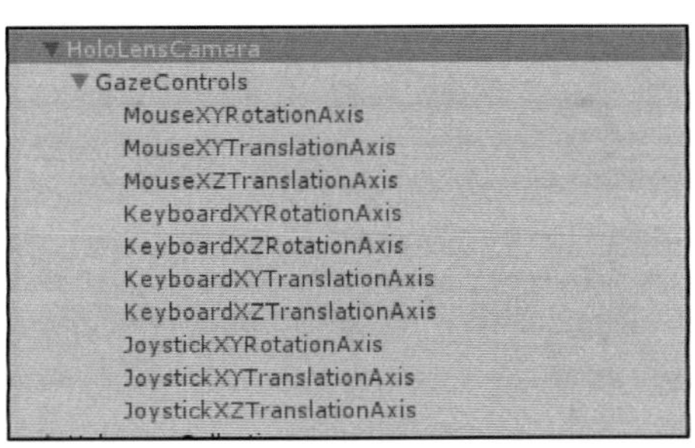

Figure 2. Hierarchy hololens camera in unity

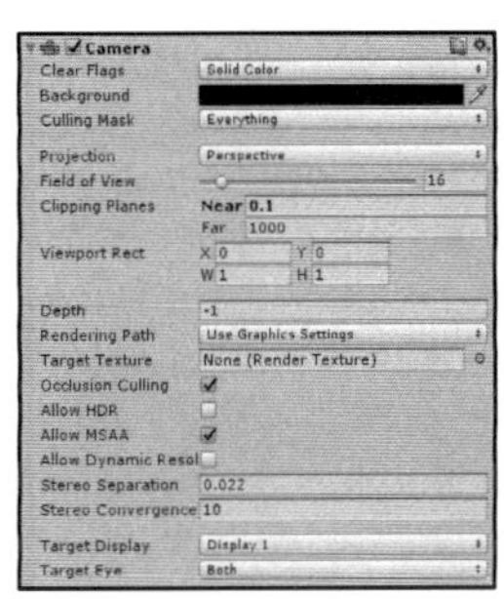

Figure 3. Hololens camera settings in inspector unity

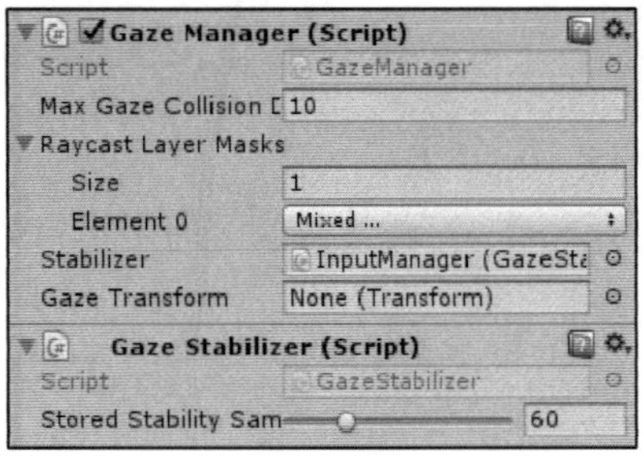

Figure 4. Gaze manager and Gaze stabilizer

Figure 5. Camera preview in scene unity

Figure 2 represent the appearance of the hierarchy hololens specialized camera on unity which serves to regulate the direction of the direct object to be projected by the camera hololens. To set the camera hololens can be set in the inspector unity. After adjusting the direction of the object to be projected and properly adjusting the hololens camera, then the display settings on the object will be performed by entering the 'gaze manager' and gaze stabilizer 'scripts which function to carefully adjust the stability of the object to be captured on the hololens camera. Here is a picture 3.4 that shows the gaze manager and gaze stabilizer settings. After that, in figure 5. you can see the camera preview on the unity scene. The preview camera serves to see objects projected by the hololens camera.

2.2 Selected Object

This stage functions so that the user can carefully choose what object to choose to run the interaction. In hierarchy unity some objects will be inputted by the script 'taptoselect' so that each object can function correctly that can select the object.

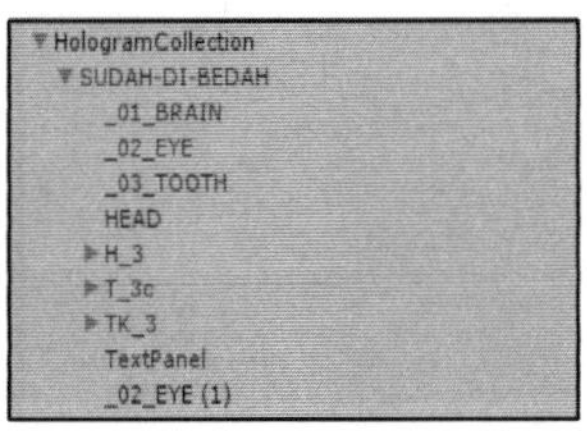

Figure 6. Hierarchy unity

Figure 7. Input script process in inspector unity

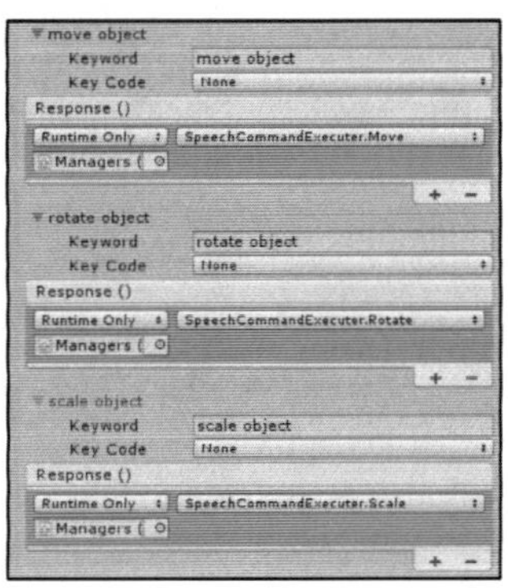

Figure 8. Voice input process

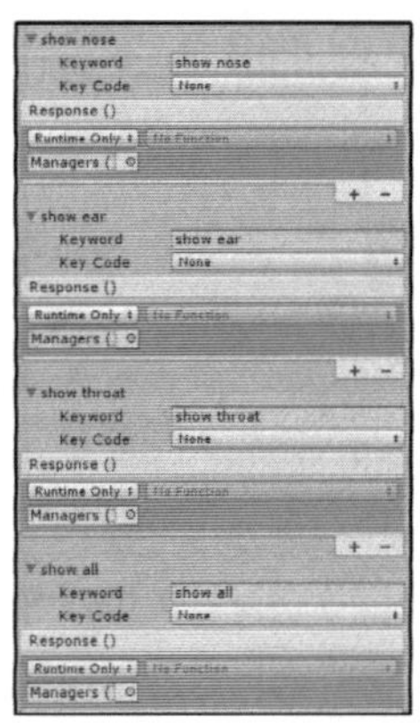

Figure 9. Voice input process

Figure 10. Build Setting

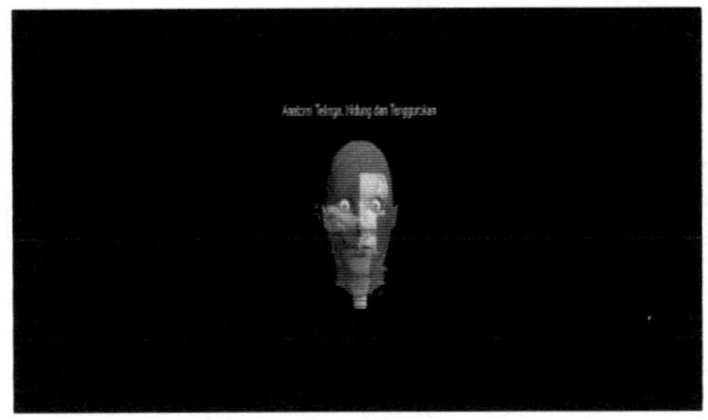

Figure 11. Application Output that will be shown on the game engine in unity

Figure 6 shows the hierarchy in unity which contains objects that will be given a specific function so that they can be selected to carry out direct interactions. In figure 7 shows the inspector for unity, in this process, enter the 'TapToSelect' script on the Ear, Nose and Throat objects that will be given a function to be able to select the object.

2.3 Voice Input

At this stage, the inputting process is properly completed to be used as an interaction on the ENT object which the user will later say. By including the essential Speech Input Source component to the unity application. In figure 8 demonstrates the process of setting voice input. In the picture shown there are 3 input sounds that will be done, namely move object, rotate object and scale object. In this process, a script will also be entered that will perform correctly the voice input function. In figure 9 demonstrates the process of setting voice input. In the picture shown there are 4 input sounds to be performed, namely the show nose, show ear and show throat. In this process, a developed script will also be entered that will perform the voice input function.

2.4 Output

After these processes, a build project will be carried out to conduct the output. In figure 11 shows the application output displayed on the game engine display in unity. The build process of this project will develop projects in the form of mixed reality which will be implemented into Microsoft Hololens. In this process, some settings need to be done precisely such as:

- add open scenes, make sure the target device is hololens
- choose the visual studio version to use visual studio and carefully check the C # project unity
- then in the player settings, make sure the virtual reality SDK is Windows Mixed Reality.

3. Results and Discussion

3.1 Application Display

In the initial display of the application, the user can see the 3D object Ear, Nose and Throat in the 3D object head, this activity can be justly called the gaze object. In this view, the user can move his vision assisted by pointers or blue dots contained in useful applications as markers of objects that we see. The initial appearance of the application can be marked in figure 12.

In this view, the user can manage an interaction that serves to move objects. The methods are:

1. Select the object you want to select by doing water-tap on the desired object
2. After that, the system will respond in the form of sound which means the object is detected by hand movements, and the water-tap has been successfully carried out
3. Next, enter voice input by saying carefully 'move object'
4. If the system detects voice input, the system will respond in the form of sound

5. Finally, the object can be moved around in the desired position by doing a hand gesture that is pulling the object to the desired position.

Figure 12. Initial display of the application / *Gaze Object*

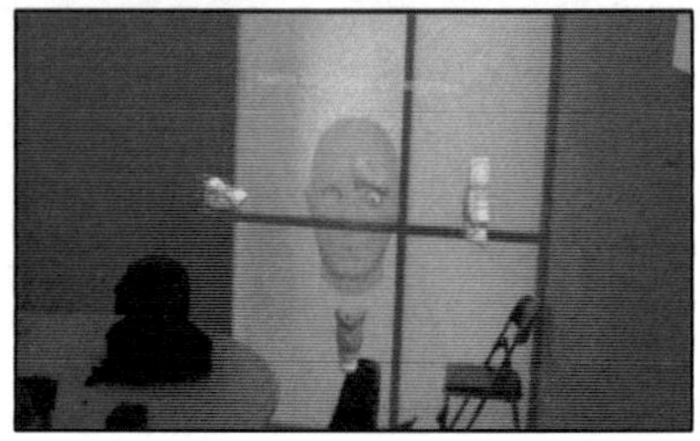

Figure 13. Interaction display *'move object'*

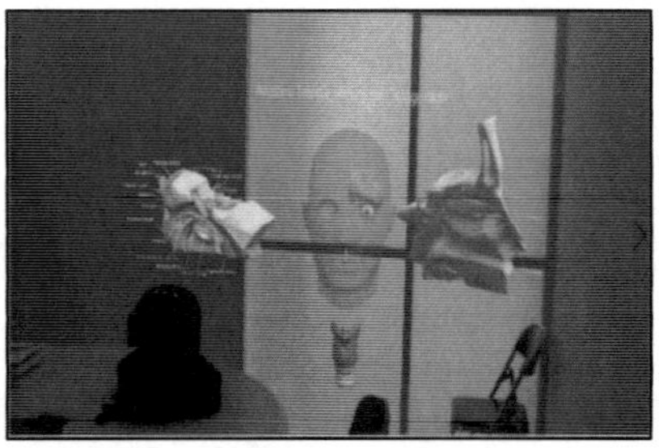

Figure 14. Interaction display *'scale object'*

Figure 15. Interaction of *'rotate object'* that is applied in Nose *3D object*

Figure 16. Interaction of *'rotate object'* that is applied in Ear *3D object*

Figure 17. Voice input interaction *'show ear'*

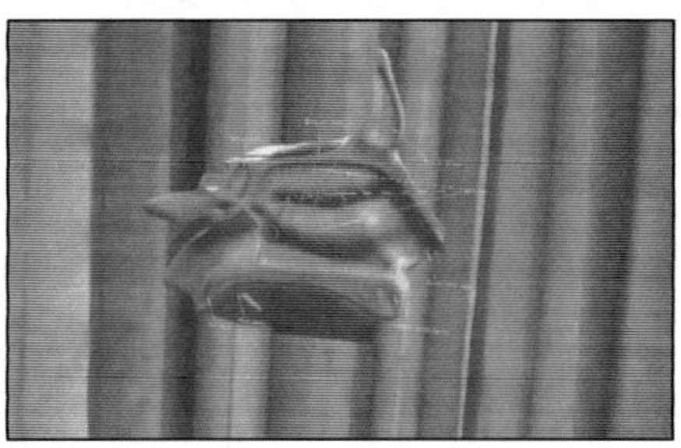

Figure 18. Voice input interaction 'show nose'

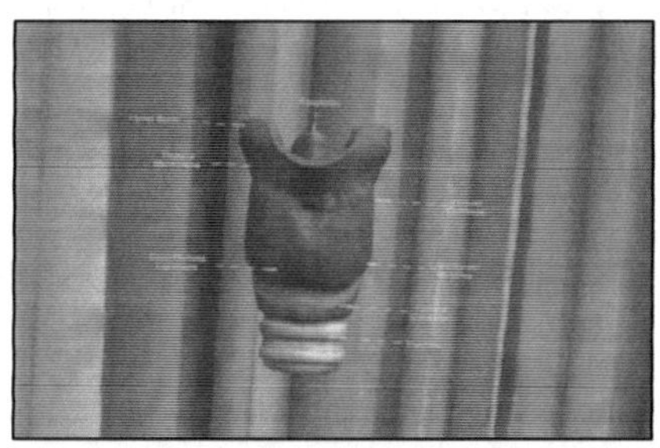

Figure 19. Voice input interaction *'show throat'*

Figure 17 is an application display when performing voice input interaction 'show ear' which serves to prominently display 3D Ear objects. Figure 18 is an application display when performing voice interaction on the 'show nose' input which serves to display 3D Nose objects. Figure 19 is an application display when performing a voice input 'show throat' interaction which serves promptly to display a 3D object throat.

3.2 System Performance Testing

System performance testing is done to test the components in the system that have been designed and implemented and ensure that each component can function properly.
In this testing system, the author applies augmented reality evaluation using the user report method using a questionnaire [8]. From the results of the questionnaire, the authors grouped the results of the answers into 3 groups, namely clarity of medical information, features in the application and assessment of user experience.

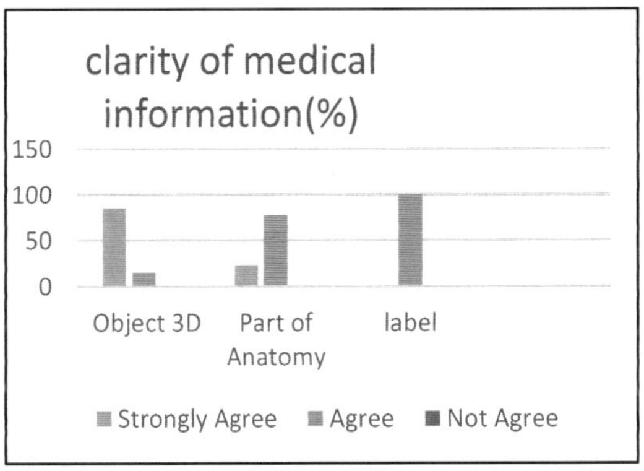

Figure 20. Graph of clarity of medical information

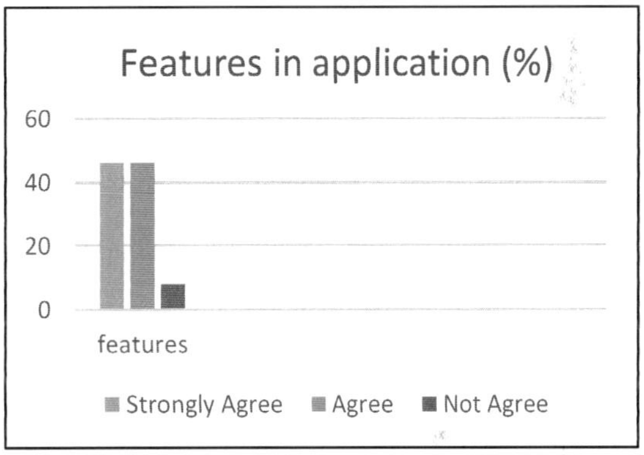

Figure 21. Application feature graph

Based on figure 20 shows 85% of selected respondents stated strongly agree that 3D Ear, Nose and Throat objects in the application are clearly displayed. 15% of respondents agreed that 3D Ear, Nose and Throat objects in the application are clearly displayed and 0% of respondents said they did not agree that the 3D Ear, Nose and Throat objects in the application are clearly displayed. Based on the direct results of the questionnaire, there were typically no problems experienced by respondents for the clarity of medical information in the form of 3D Ear, Nose and Throat objects.

For the specific section of each anatomy, 23% of selected respondents stated strongly agree the anatomical parts were in accordance with the original, 77% of respondents agreed the anatomical parts indicated in the application were in accordance with the original and 0% of respondents stated no agree the anatomical parts shown in the application are in accordance with the original. Based on the results of the questionnaire, there were precisely no problems experienced by respondents in the ear, nose and throat anatomical parts displayed by the application.

For the accuracy of the label position in the anatomy section, 100% of the respondents agreed that the label on the 3D object had clearly explained the anatomical parts. Based on the questionnaire, respondents did not have a problem with the exact position of the labels displayed correctly by the application.

Based on Figure 21 shows 46% of respondents stated strongly agree that features in the application such as hand gesture and voice input run smoothly. 46% of respondents agree that the features of the application run smoothly and 8% of respondents do not agree that the features of the application run smoothly. Based on the observations of researchers, respondents who disagree that the features of the application run smoothly due to the hand gesture performed by the user is not running properly, this is due to the user's hand is not right in front of the Microsoft HoloLens sensor so the sensor cannot detect hand gesture. Some of hand gesture problem has been solved in other research conducted by Rahmat with the use of skin color segmentation [10]. It can be included as future research later. Another thing experienced by respondents who disagree that the features of the application run smoothly is the voice input feature. Voice input made by the user is difficult to respond to by Microsoft HoloLens because of several factors, namely the pronunciation of voice input that is unappropriated, therefore, the user who has a problem must repeatedly repeat voice input until the exact pronunciation and Microsoft HoloLens can detect the voice input. Another obstacle experienced by the user is the sound sensitivity level, Microsoft HoloLens is difficult to detect voice input in the state of the room that produces noise. Therefore, the user who encounters this problem must repeat voice input several times until Microsoft HoloLens detects the voice input that is done.

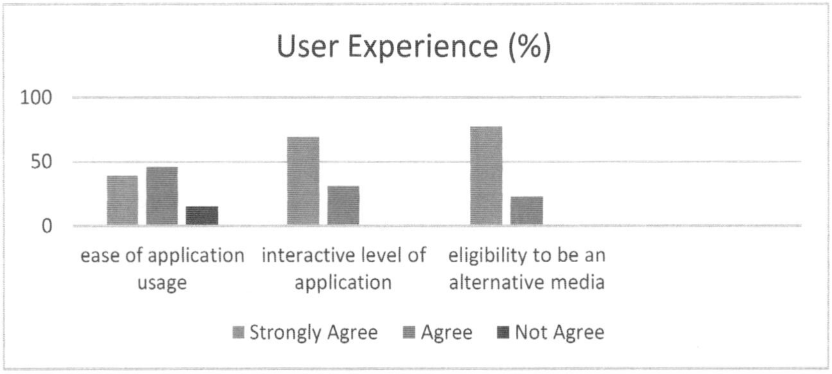

Figure 22. User experience graph

Based on Figure 22, 39% of respondents strongly agree that the application is easy to use, and 46% of respondents agree that the application is easy to use and 15% of respondents do not agree that the application is easy to use. Based on observations of researchers, 85% of respondents did not have constraints in the possible use of applications, while 15% of respondents experienced difficulties in managing the application, the practical difficulties experienced were those when using the hand gesture and voice input features. The difficulty when using the hand gesture feature is that the user is not right to carefully put his hand in front of the HoloLens sensor so that the sensor has difficulty detecting the gesture that will be applied. Another difficulty experienced by users is the effective use of voice input features, the difficulty when using the voice input feature is that the user does not say voice input properly so the HoloLens system does not detect sound properly and the user must repeat voice input until proper pronunciation and HoloLens system can detect voice the input. Another possible obstacle experienced when using voice input is the level of sound sensitivity, Microsoft HoloLens is difficult to detect voice input in a room that produces noise. Therefore, the user who encounters this problem must repeat precisely voice input several times until Microsoft HoloLens detects the voice input that is done.

For application performance, 69% of selected respondents stated strongly agree that the application runs interactively, 31% of respondents agree that the application runs interactively and 0% of respondents do not agree that the application runs interactively.

Based on the independent assessment conducted by respondents, 77% stated strongly agree the application could be an alternative media to be a learning method, 23% agreed the application could be an alternative medium to become a learning method and 0% of respondents did not agree the application could be an alternative medium to become a method learning.

4. Conclusion

Based on the results of the performing phase and system testing, some conclusions were acquired in this study, namely:

- The application of medical learning Ear, nose and throat using Microsoft Hololens can make it easier for users to interact and identify specific organs in the Ear, Nose and Throat.
- This application can run properly when it instantly detects hand gesture when selecting objects
- Noise levels are sensitive when doing voice input, users must be in a place where the noise level is low so the system can respond clearly to voice commands. If the user is in a noisy place, the system will be difficult to respond to voice commands.
  Based on the questionnaire conducted, it can be reasonably concluded that this application can be managed as an alternative media for learning efficient methods.

Acknowledgments

The Universitas Sumatera Utara supported this research. All lecturers and staff members of the Information Technology Department.

References

[1] Ferrari, V., Megali, G., Troia, E., Pietrabissa, A. & Mosca, F. 2009. A 3-D Mixed Reality System for Stereoscopic Visualization of Medical Dataset.
[2] Hakim, F, A., Hidayat, N. & Dewi, R,K. 2018. Sistem Pakar Diagnosis Penyakit Telinga Hidung Tenggorokan (THT) Menggunakan Metode Naive Bayes Berbasis Android. *Jurnal Pengembangan Teknologi Informasi dan Ilmu Komputer*, Vol. 2,No. 4, April2018, hlm. 1492-1500.
[3] Jayaputra, A., Tolle, H. & Wardhono, W.S. 2017. Penerapan Mixed Reality Sebagai Sarana Pembelajaran Indera Penglihatan Manusia Menggunakan Teknologi Hologram. *Jurnal Pengembangan Teknologi Informasi dan Ilmu Komputer*, Vol. 1, No. 9, Juni 2017, hlm. 715-722.
[4] Kurniawan, M.J., Anra, H. & Pratama, E.E. 2017. Aplikasi Augmented Reality Sistem Pernafasan.
[5] Mustaqim, I. & Kurniawan, N. 2017. Pengembangan Media Pembelajaran Berbasis Augmented Reality.
[6] Nurwahida, Khairina, *D.M. & Astuti, I.F. 2015.* Sistem Informasi Penyakit Telinga Hidung Tenggorokan (THT) Berbasis Web, Vol 10, No 2 (2015).
[7] Santoso, A. & Noviandi, E. 2016. Rancang Bangun Aplikasi Pembelajaran Organ Tubuh Berbasis Augmented Reality.
[8] Schaffer, S.A. 2014. Usability Evaluation for Augmented Reality.
[9] Tarigan, S.E.B.R. 2018. Pemodelan 3D THT (Telinga, Hidung, dan Tenggorokan) pada Media Pembelajaran Sains menggunakan Blender 3D.
[10] Rahamt R F, Chairunnisa T, Gunawan D, Pasha M F, Budiarto R 2019 Hand gesture recognition with improved skin color segmentation in human-computer interaction applications *Journal of Theoretical and Applied Information Technology* **97**(3) pp. 727-739

IOP Conf. Series: Materials Science and Engineering **648** (2019) 012035 doi:10.1088/1757-899X/648/1/012035

3D Modelling Intestine Anatomy with Augmented Reality for Interactive Medical Learning

U Andayani[1], M F Syahputra[1], M A Muchtar[1], M Sattar[1], Santi Prayudani[2], F Fahmi[3]

[1]Department of Information Technology, Universitas Sumatera Utara, Indonesia
[2]Department of Computer and Informatics, Politeknik Negeri Medan, Indonesia
[3]Department of Electrical Engineering, Universitas Sumatera Utara, Indonesia

Correspondence author: ulfi.andayani@usu.ac.id

Abstract. Humans, one of the heterotrophic living things must meet energy needs by consuming food. The food is then described in the digestive system to become an energy source. The digestive system is composed of the digestive tract and digestive gland glands. In the excerpt of the book "Prometheus" human anatomy, the digestive tract acts as a tool for receiving, chewing, delivering, storing, digesting, absorbing (absorbing) and eliminating food. The small intestine, large intestine, and rectum are organs included in the digestive process. This organ is one of the materials taught to science students to the level of students, especially in medicine. Submission of material about the digestive system in humans themselves is still through conventional media such as blackboards, and images contained in books. Several innovations have been carried out, but still, the development of learning innovations must continue to be developed, so that less interesting subject matter makes it easier for students to know a subject matter with the help of technology and multimedia. The conventional method can only be told without being visualized. If using Augmented Reality (AR) has advantages, namely virtual storage (memory), can be learned at any time, can only be installed on a smartphone, each of them can be studied without thematic tools, and can be visualized by scenarios or stories about problems. The marker was used from the Prometheus book [10] then the AR as the visualization of the organs. The results by developing a 3D model using blender and unity showed that the optimal models could be created by combining both their applications and the testing result of marker detection stable from 15-45 cm and tilt marker detection between 45^0 - 150^0.

1. Introduction

Augmented Reality is a technology that displays virtual objects precisely in a real object in real time; these virtual objects are projected in real time through a marker or marker directed to the camera. By using this technology, students can see 3D objects in the digestive organs that will be applied to mobile devices to create more interactive learning. The emergence of various devices that support 3D graphics at an affordable price has prompted developers to replace the physical model into a virtual model. The medical field was not spared from this development. Various types of digitization have developed rapidly in recent years. The 3D Virtual Model of medical organs is trusted by researchers as an alternative that can help students, especially in the field of science, to be more familiar with more tangible organs in the process of honing their abilities. One of the developments in multimedia technology related to learning in medical science is learning with an anatomical model from the results of 3D print, which is one of the solutions in helping the team of doctors perform patient intervention planning [1]. One method of learning application about the human senses using 3D holographic visualization [7]. Whereas another method that can be used to search the location is Augmented Reality, a technology that combines two-dimensional virtual object into a real environment and projecting a three-dimensional virtual objects in real-time [9] .Not only for medical science, but also a survey of Augmented Reality use motion tracking [13] that describes the cultural preservation [12], manufacturing, visualization, path planning, entertainment, marketing [6], and military applications have been explored [2]. 3D human model is created using open source software (Make Human) [11].

Some previous studies using Augmented Reality, one of the organs was also research from M Abdul Gani [3] Augmented Reality Anatomy of the Respiratory System Using Leap Motion Controller as a Medical Learning Media. The purpose of the research to be carried out is the use of Leap Motion Controller as a tool for the recognition system of human anatomy, especially 3D lungs. Other researchers use Augmented Reality, one of which is research from Rahmat Heriza [6] Implementation of Augmented Reality in the Enclosure of the Aceh Museum Collection as a Multimedia Learning Facility Based on Android. The purpose of this study is to provide education and information to the public to learn about the collections of the Aceh Museum by utilizing Augmented Reality technology.

Another study by Abdul Aziz et al., (2016) Implementation of markerless Augmented Reality Using Android Sensors for Identification of Buildings in Eleven March University. The aim is to identify buildings that are located at Universitas Sebelas Maret so that the general public from outside the city or for the first time entering does not find it difficult to go to the desired building.

Whereas in this study an application will be developed that displays or visualizes the anatomy of the human digestive system organs especially in intestines and their parts in detail, accompanied by labeling them. Then in the study, several additional or complementary objects will be used, so that visualization of the intestine's details was more exciting and easier to understand as the interactive learning media based on books description and better than only used teaching aids.

2. Methodology

2.1. Data Collection

The data used as the labeling and description of intestine details was from Prometheus book edition 3 (three) human anatomy [6], and the marker of the digital image obtained from that book also used AR marker. In taking a marker, images must be considered the distance between the camera, position, and lighting of camera devices.

2.2. System Design

Android mobile-based programming and used marker were developed for Augmented Reality by displaying the labeling of anatomy intestine in details which divided into 3 (three) parts such as large intestine, small intestine, and rectum. The AR method divided into 3 parts production and final reconstruction that can be seen in Figure 1 and Figure 2.

The production description parts as follows:

- Ideas and concepts: discussing and develop ideas from various references. The conclusions from the discussion are poured into a concept, where the concept in question is in the form of a digestive organ (small intestine, large intestine, and rectum).

- Collecting Materials: conducting research related to the expected results. Collection of material in the form of collecting 2D sketches from the organs that will be the object to be built, data collection of organs that can be modelled in context does not reduce the objective of the model, collection of free supporting assets from internet networks that can be used to complete and perfect the model, to the learning module that can enhance the making of 3D models.

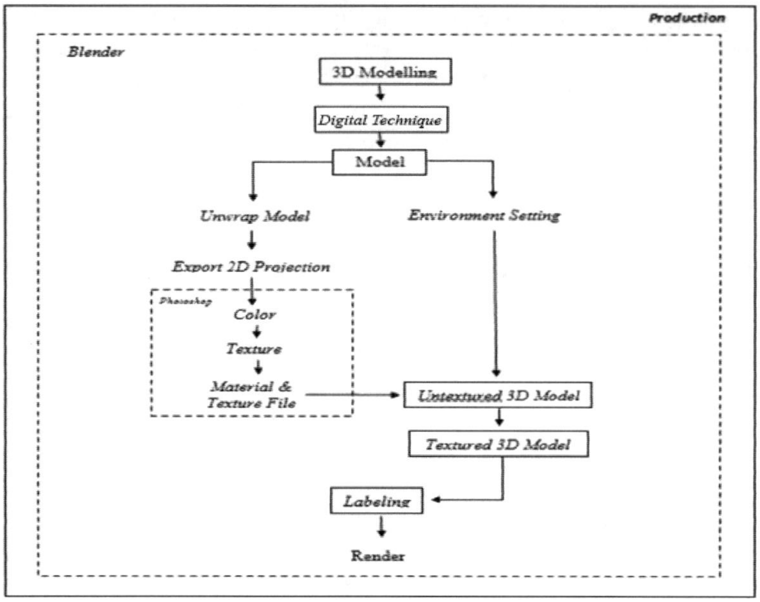

Figure 1. Production

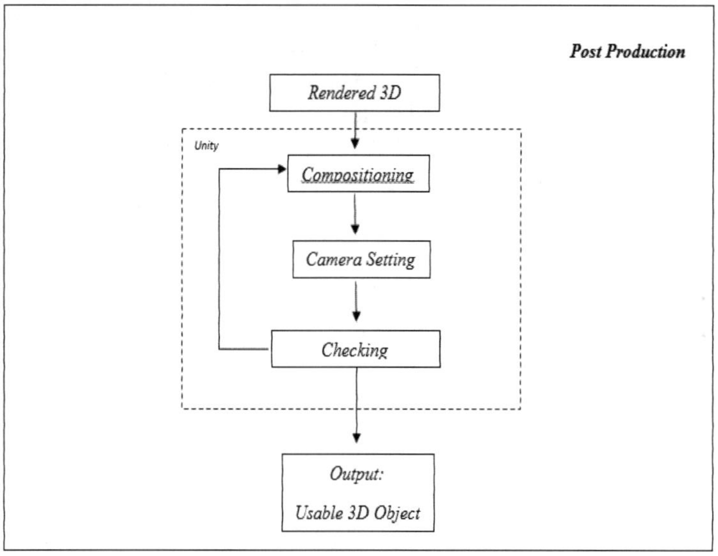

Figure 2. Post Production

- Data Filtering: conducting the discharge of any data that has been collected, including filtering the assets that have been obtained

2.2.1. Production

- 3D Modelling: Sketch 2 dimensions are collected and designed, then the 3D modeling process is done using the method of digital Sculpting.
- Material / Texturing: The stage that is carried out after modeling, the texturing object is done on the object to make it more like the real one. In the process of texturing, the modeled object must undergo an unwrapping process. This unwrap process is the stage where 3D objects are automatically

converted using the blender application to form a 2D layer. Furthermore, the coloring of the 3D artist uses the Photoshop application, so that the unwrapped model resembles the original.
- Labeling: The labeling stage is the process of giving the names of each component that is built on the table of relations between the name and code pinned on each object.

2.2.2. Post-Production

Model improvement, where lighting, camera position, and recheck are performed on the model.

Finally, the reconstruction needed after the production steps were done.

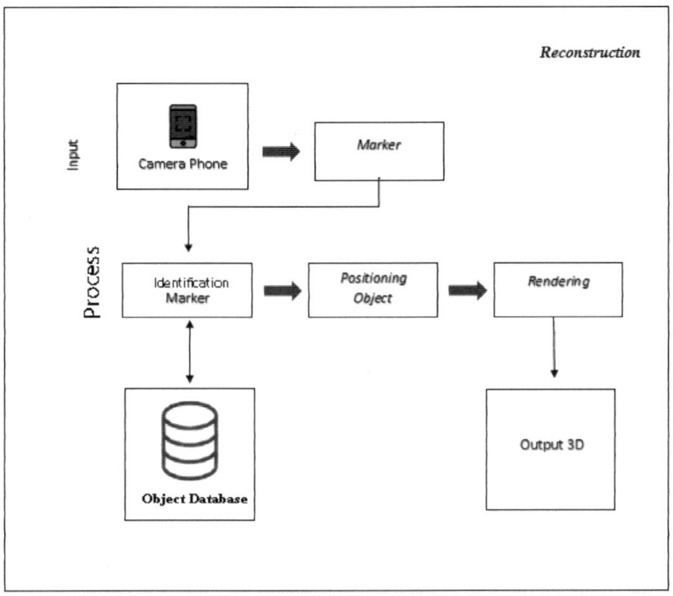

Figure 3. Reconstruction.

After production, then connect the 3D modeling that has been made. The camera will capture the marker; then the marker will be identified to match the specified object to be displayed above the marker. After identification, a database of markers and objects will be checked which will be displayed later. Then the location of the 3D object against the marker will be determined. After that, a rendering process will be performed to display the output, which is a 3D object that will be displayed above the marker.
By using the steps of modelling, rigging, texturing, simulation, rendering, composition, and game creator in engine game using both windows, Linux and Mac OS X (windows) were the process of developing mathematical representations of each surface of a three-dimensional object (either dead or alive) through specialized software, the process of embedding human skeletons into human models that have been built using third-party software. The joints of the human skeleton must be placed in a suitable position in the human model to make it easier and to make the human model built more real [14]. The process of making and giving color and material to the object that has been modeled in order to show a real impression. Giving material to 3D objects will define the appearance and type of material from 3D objects [4]. The simulation can make objects interact with other objects like the real world, collisions, weight, elasticity, and strength can be arranged in such a way. In addition, this feature can also create animated water, fire and smoke. The process of forming an image from a model formed by animation software, the model contains the geometry data, points of view, texture, and light needed to make a complete picture. 3D rendering is an essential process and has been used for a

IOP Conf. Series: Materials Science and Engineering **648** (2019) 012035 doi:10.1088/1757-899X/648/1/012035

variety of users, such as computer game programs, special effects on films, and simulation programs [8]. Combining and using across platform system operation.

3. Results and Discussion

3.1. Scene 3D

The process of making the 3D can be seen in Table 1. There will find Blender column show the creating the models, unity column shows the coloring and texturing model, and 3D view shows the final of the model before being labeled of their anatomy in details.

Table 1. Creating models based on resource.

No.	Textbook Resource	Blender	Unity	3d View
1.	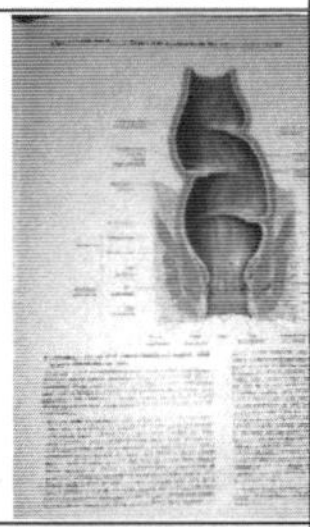	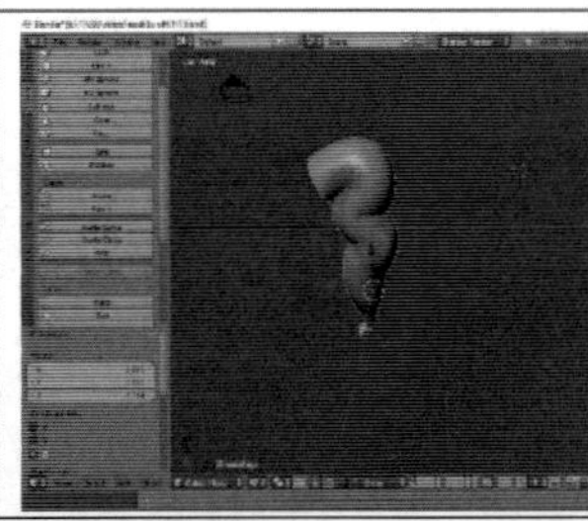	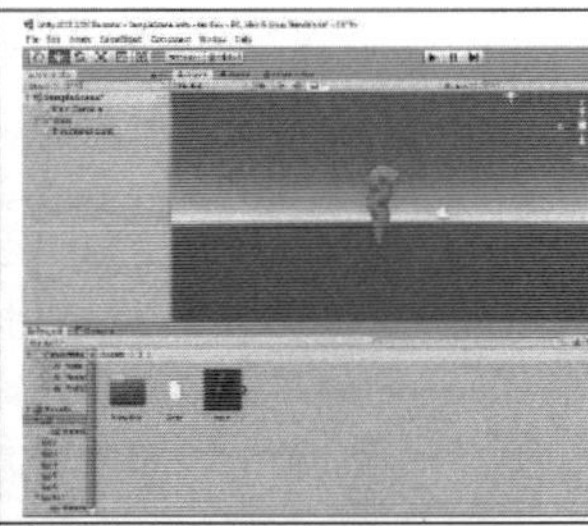	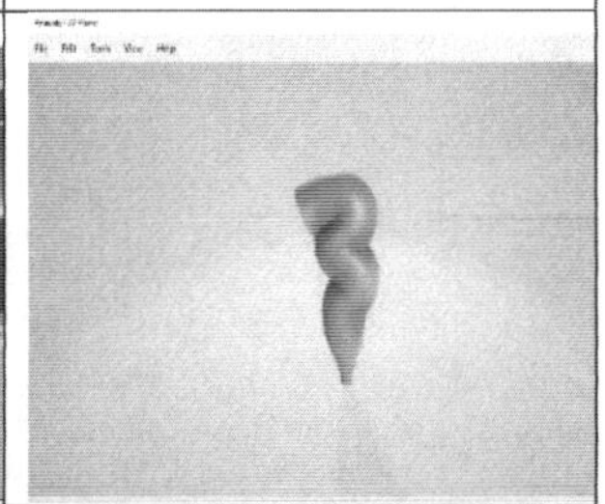
2.	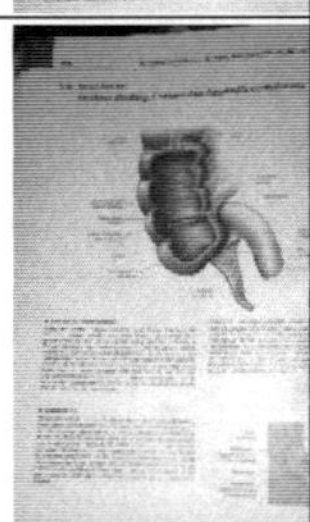	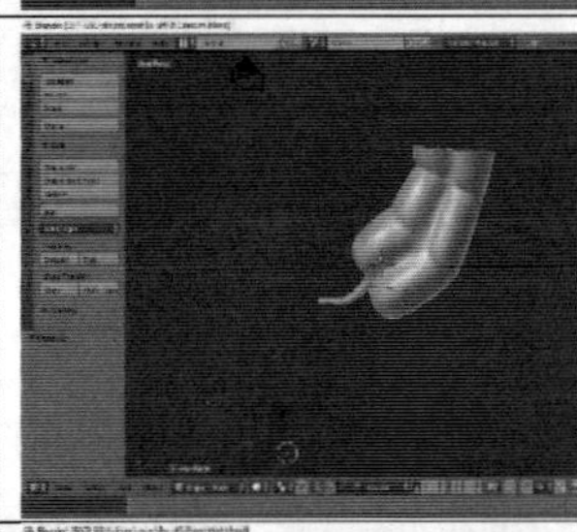	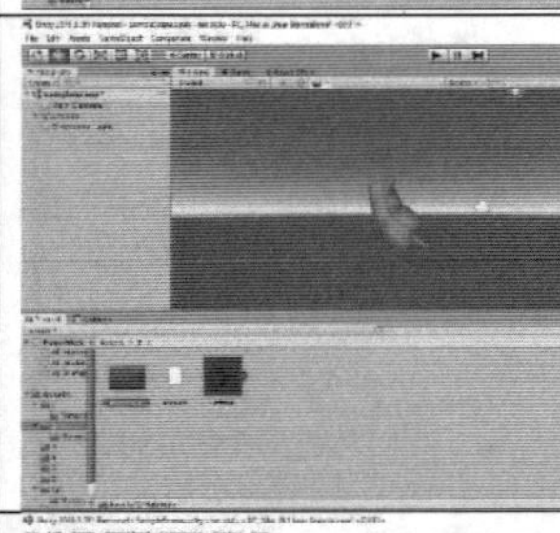	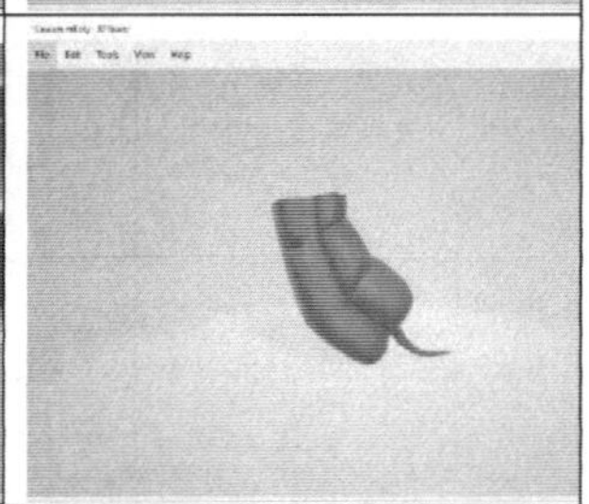
3.	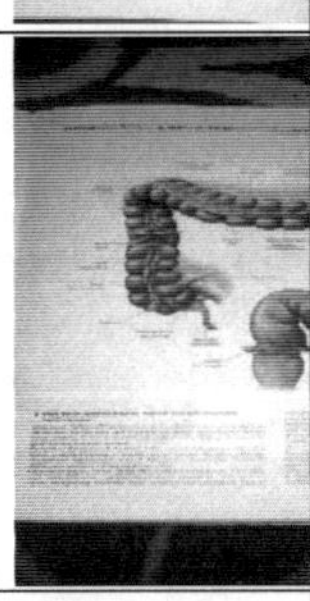	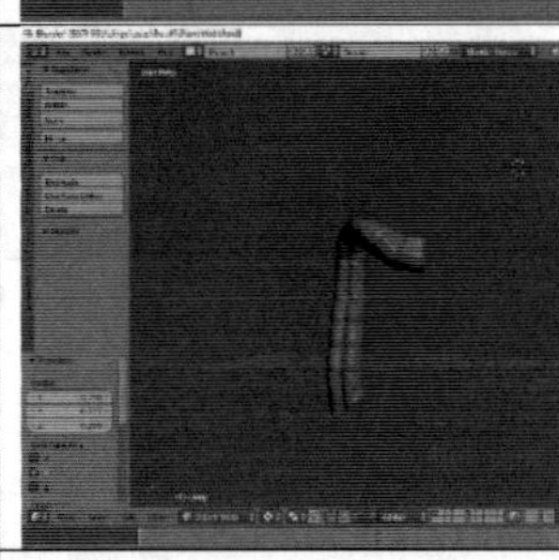		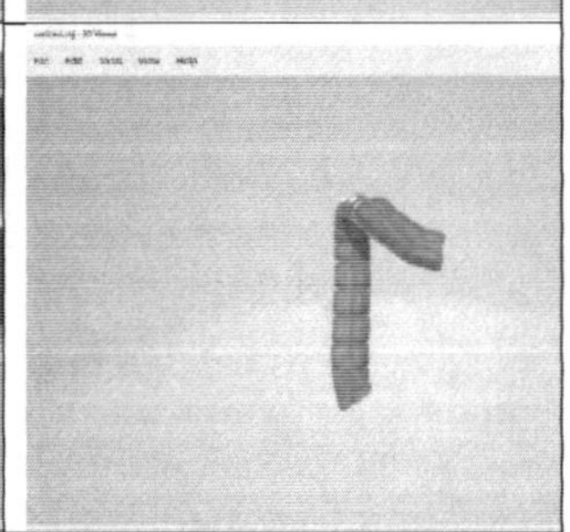
4.	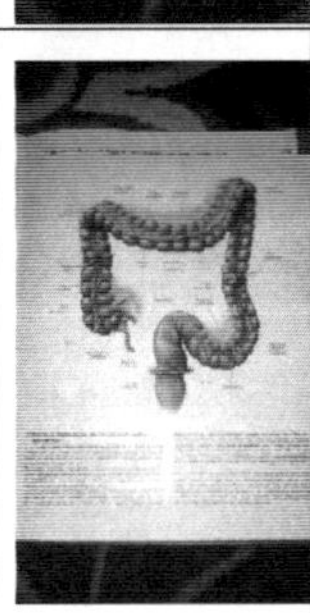	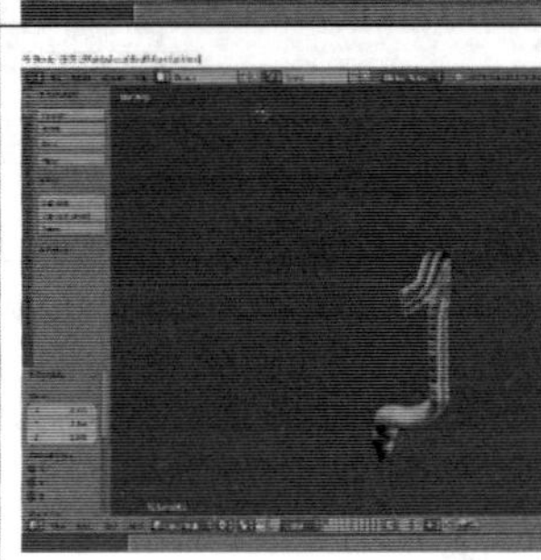	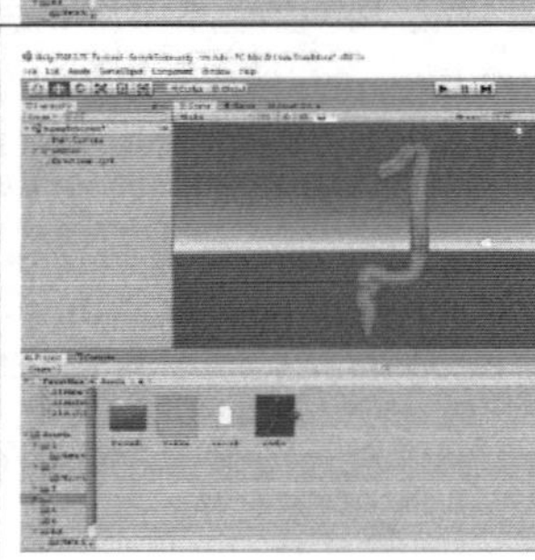	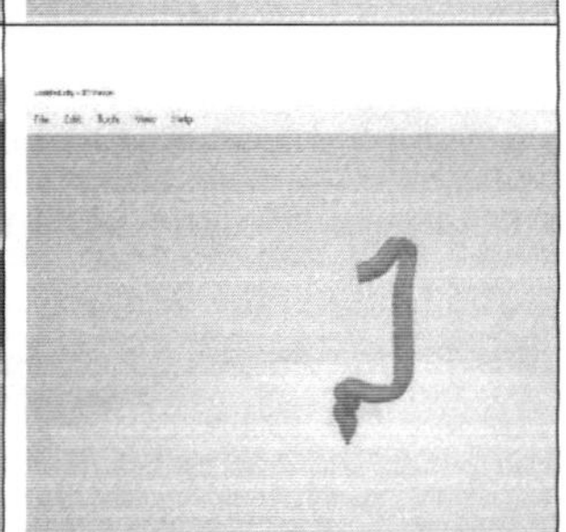

IOP Conf. Series: Materials Science and Engineering **648** (2019) 012035 doi:10.1088/1757-899X/648/1/012035

5.	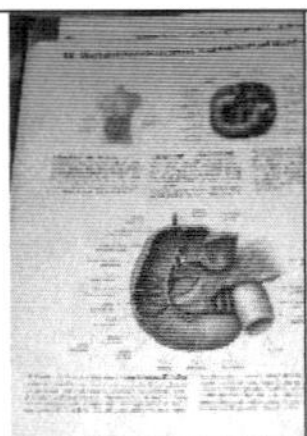	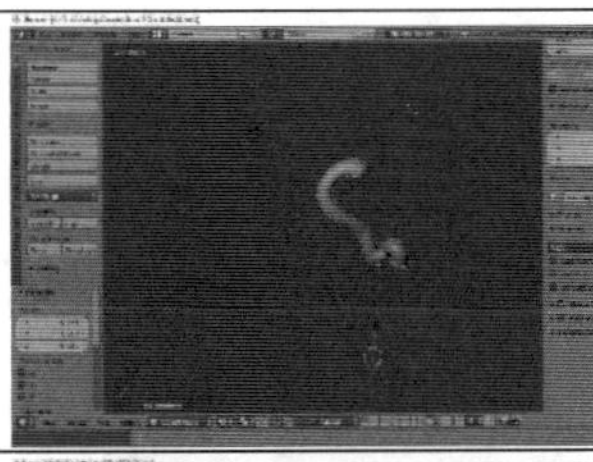	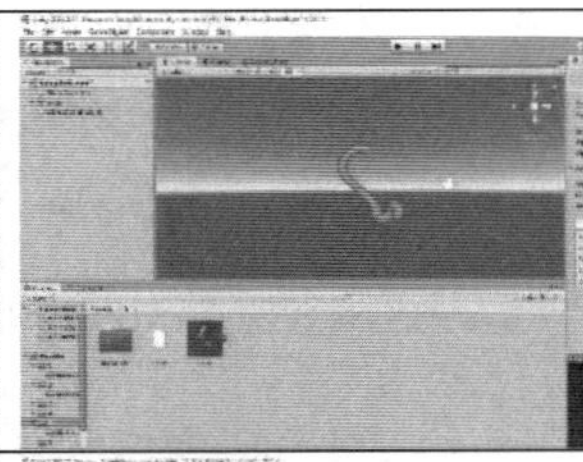	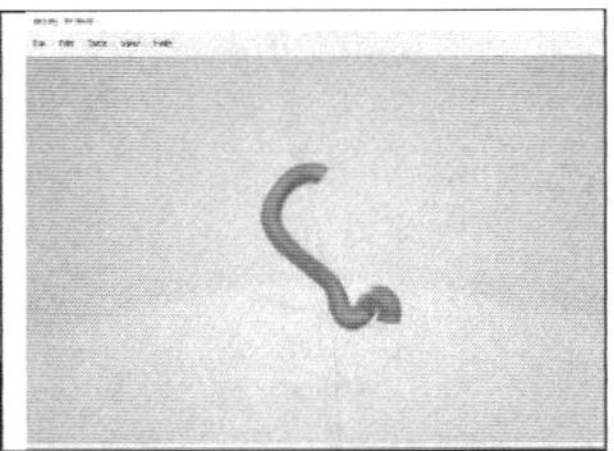
6.	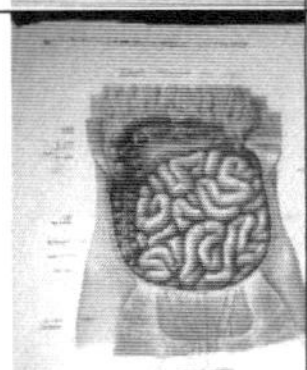	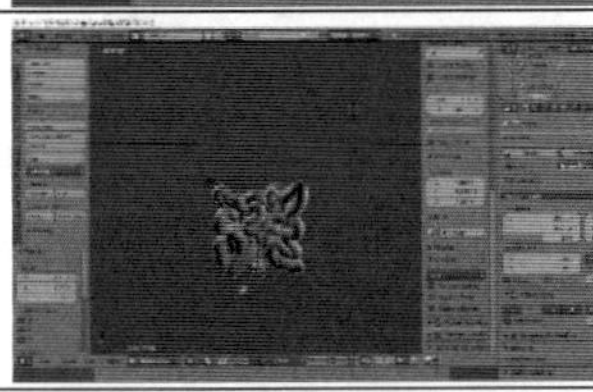		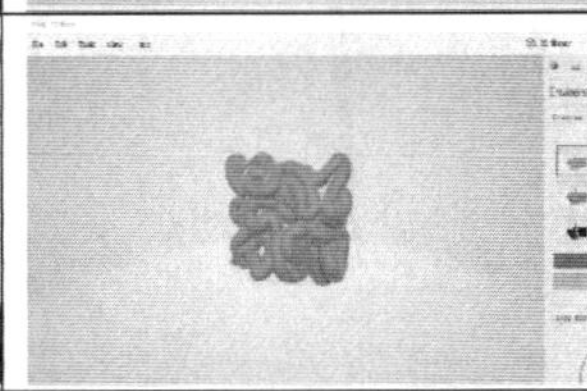
7.			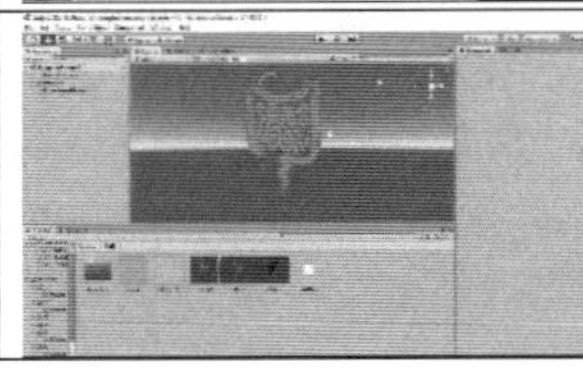	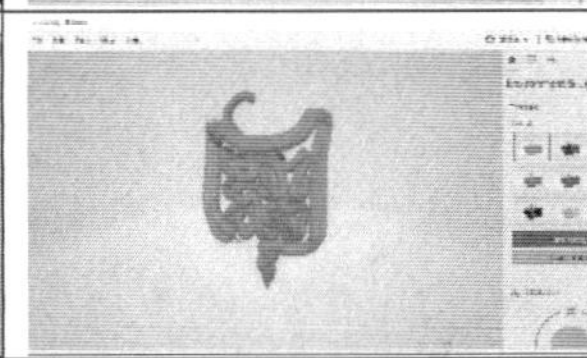

The 3D scene used based on a marker figure 4. After scanning marker, the 3D model will show the main menu to be chosen in figure 5 and figure 6. There are 2 option in main menu, such as "Play AR" and "How to Play". In "Play AR" menu, scene 3D can be shown whereas the other menu is an instruction how to play the application. The label details of intestine such as figure 7, 8, 9, 10, 11, 12 can be seen below:

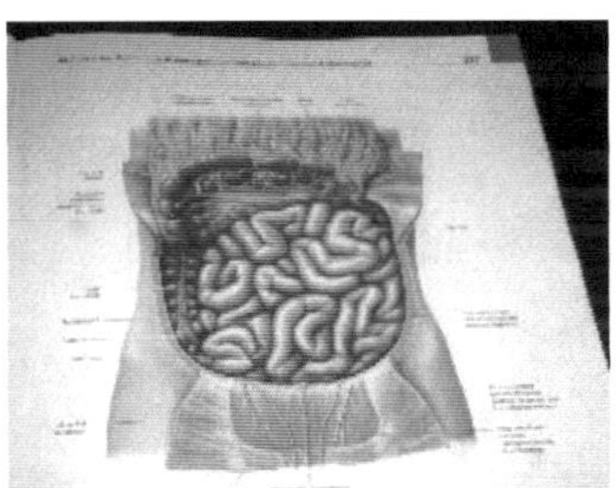

Figure 4. Marker.

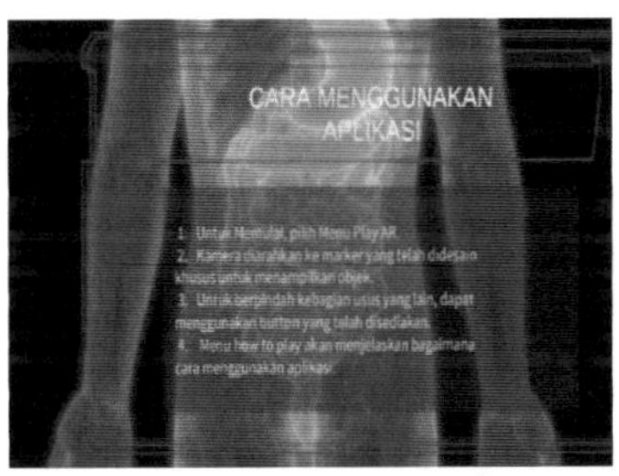

Figure 5. Play AR Menu.

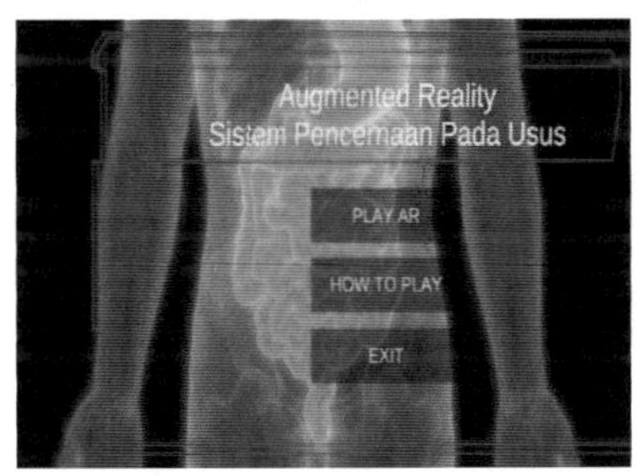

Figure 6. How to Play menu.

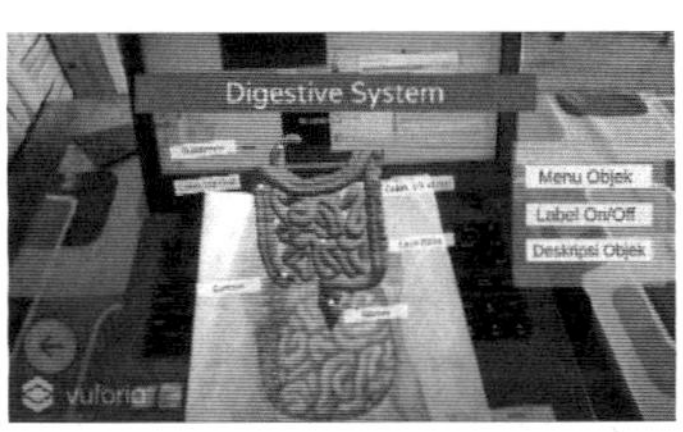

Figure 7. Digestive System

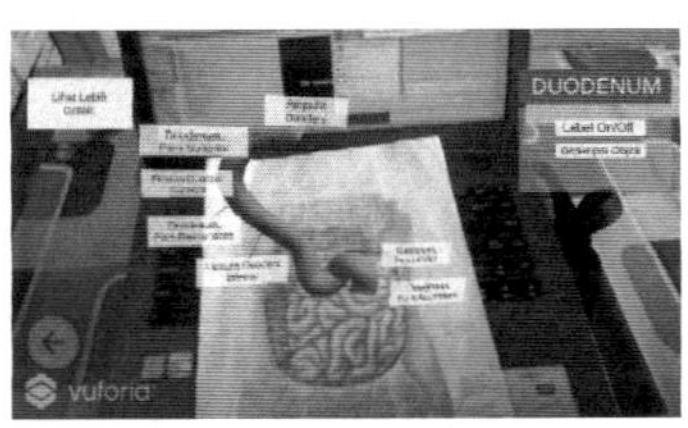

Figure 8. Duodenum and labels details.

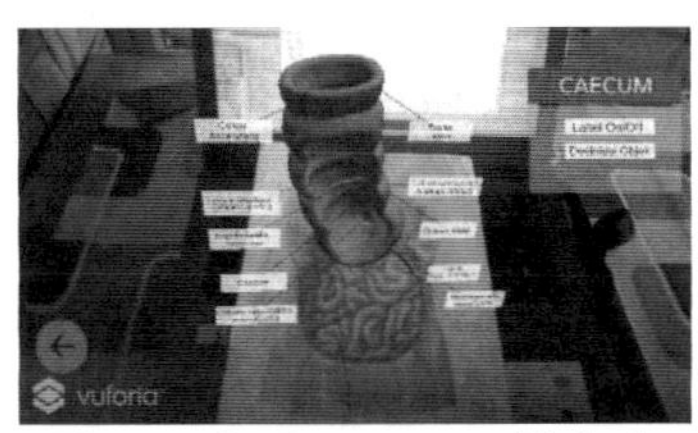

Figure 9. Caecum and labels detail

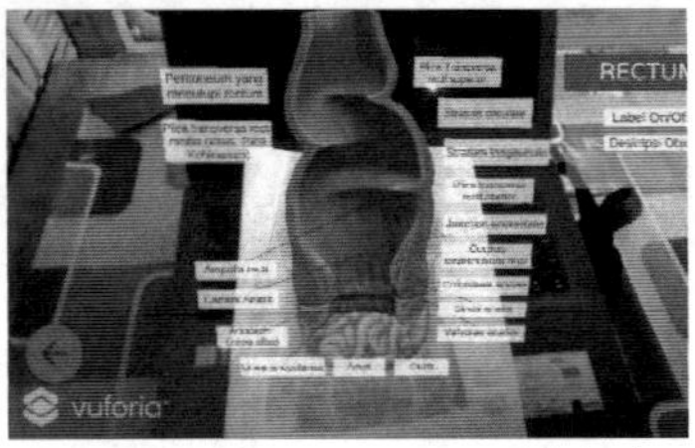

Figure 10. Rectum and labels detail.

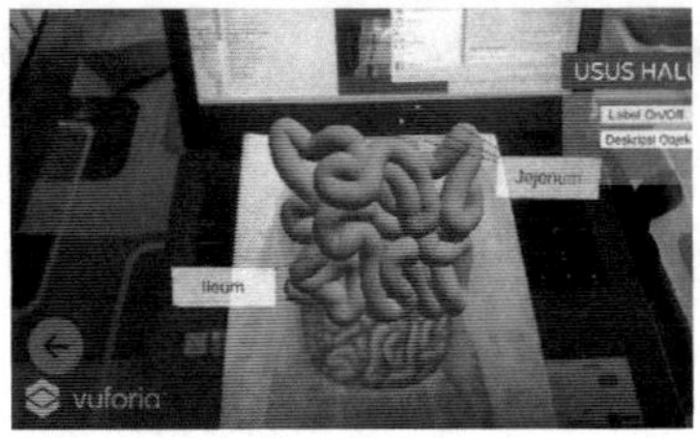

Figure 11. Small Intestine and labels detail

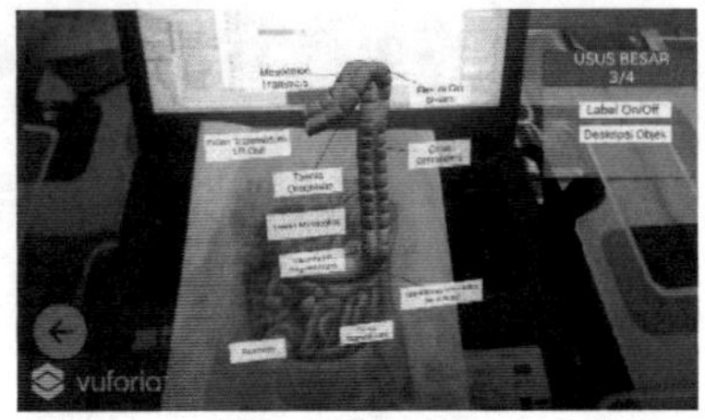

Figure 12. Colon and labels detail

3.2 Augmented Reality Marker Testing

Marker testing was conducted to determine the level detect marker image. The marker size is picture from a book of *Prometheus LernAtlas der Anatomie:Innere Organe, 3 Auflage* [10]. By using mobile phone camera for detecting the marker image and display the object. Marker used already registered in Vuforia markers and inserted into the database of Unity software.

3.2.1 Distance Detection

The distance detection testing is conducted to detect the level of the marker image with a certain distance. The aim of the testing is to know how far the camera can detect the marker image in centimetres.

Table 2. Distance marker detection test.

NO.	DISTANCE (CENTIMETRES)	NO. OF TEST	CORRECT	FALSE	TEST RESULT
1	15	15	15	0	Detected and Stable
2	20	15	15	0	Detected and Stable
3	25	15	15	0	Detected and Stable
4	30	15	15	0	Detected and Stable
5	35	15	15	0	Detected and Stable
6	45	15	15	0	Detected and Stable
7	50	15	9	6	Unstable

From the experimental of distance detection, in 15 cm the camera phone can stable detecting the object. The first step is by scanning the marker used the camera phone showed that the 3D model can appear stably. So, it means the stable distance of the marker detection is from 15-45 cm.

3.2.2 Slope Distance Detection

Tests conducted to determine the distance detects the marker level with a certain distance. To detect the marker by testing how much tilt the camera done.

Table 3. Tilt marker detection test.

No.	Slope Distance (^{0})	No. of Test	Correct	False	Test Result
1	0	15	0	15	Not detected

2	25	15	0	15	Not detected
3	45	15	10	0	Detected and Stable
4	75	15	15	0	Detected and Stable
5	110	15	15	0	Detected and Stable
6	135	15	15	0	Detected and Stable
7	150	15	15	0	Detected and Stable
8	155	15	7	8	Detected and unstable
9	180	15	0	15	Not detected

From the point of the camera to the marker experimental, in The camera can detect but not stable in 155^0 and the camera can detect and stable when the distance between 45^0 - 150^0.

4. Conclusions

- The file size used 3D blender was small
- Model Interface was user-friendly and neatly organized using 3D Blender.
- The tool for creating complete 3D objects includes modelling, UV mapping.
- Cross Platform, with uniform GUI and supports all platforms. Blender 3D.
- The quality of 3D architecture is of high quality and can be done more quickly and efficiently.
- The testing result of marker detection stable from 15-45 cm and tilt marker detection between 45^0 - 150^0.

Future works

In further research, improving by adding some animation, converted into mix reality so that the application is more interesting.

Acknowledgments

Universitas Sumatera Utara supported this research. All lecturers and staff members of the Information Technology Department.

References

[1] Amin, N 2017 Replikasi Anatomi Jantung pada Pasien Penderita *Ventrice Septum Defect (VSD)* dengan 3D Painting. Universitas Gadjah Mada. 374-380

[2] Azuma, R T 1997 "A Survey of Augmented Reality". Presence: Teleoperators and Virtual Environments.

[3] Gani A 2018 *Augmented Reality* Anotomi Sistem Pernapasan Menggunakan *Leap Motion Controller* sebagai Media Pembelajaran. Universitas Sumatera Utara.

[4] Handayani T K 2011 Pembuatan Animasi 3D Organ Reproduksi Manusia Untuk Meningkatkan Pemahaman Remaja Dalam Penyuluhan Kesehatan Reproduksi Remaja. UIN Syarif Hidayatullah Jakarta.

[5] Heriza R 2016 Implementasi *Augmented Reality* pada Pengenalan Koleksi Museum Aceh Sebagai Sarana Pembelajaran Multimedia Berbasis Android. Universitas Sumatera Utara.

[6] Hermawan L 2015 Pemanfaatan *Augmented Reality* Sebagai Media Informasi Kampus Menggunakan Brosur. Yogyakarta.

[7] Jayaputra A 2017 Penerapan *Mixed Reality* Sebagai Sarana Pembelajaran Indera Penglihatan Manusia Menggunakan Teknologi Hologram. Universitas Brawijaya.
[8] Jostonchoniv and Windarto Y 2013 Pembuatan Film Animasi 3D Cerita Rakyat "The Legend of Toba Lake". Universitas Kristen Krida Jakarta.
[9] Nugroho R K 2016 *Implementation Markerless Augmented Reality Using Android Sensors for Identification of Buildings in* Sebelas Maret *University.* Universitas Sebelas Maret.
[10] Schunke M 2011*Prometheus LernAtlas der Anatomie:Innere Organe, 3 Auflage.*
[11] Shingade A and Archana G 2014 Animation of 3D Human Model Using Markerless Motion Capture Applied to Sports. *International Journal of Computer Graphics & Animation (IJCGA)* Vol.**4**, No.1, Januari
[12] Syahputra M F, Lumbantobing N P, Siregar B, Rahmat R F, Andayani U 2018 Implementation of Augmented Reality to Models Sultan Deli *Journal of Physics: Conference Series* **978**(1) 012108
[13] Syahputra M F, Fatimah S, Rahmat R F 2018 Interaction on Augmented Reality with Finger Detection and Hand Movement Recognition. *Lecture Notes in Computer Science LNCS* 10851, pp 702-712

Designing the quality of coffee bean detection application using Hue Saturation Intensity

T H Nasution[1*], M Rumansa[1], Lukman Adlin Harahap[2]

[1]Department of Electrical Engineering, Universitas Sumatera Utara, Medan, Indonesia
[2]Department of Mathematics and Computer Science, Faculty of Science, Chulalongkorn University, Bangkok, Thailand

*Email: tigor.nasution@usu.ac.id

Abstract. This study discusses the design of applications that can detect the quality of raw coffee beans both good and bad based on the value of HSI (Hue Saturation Intensity) in the coffee beans using digital image processing and Backpropagation artificial neural networks. The process of identifying coffee beans is based on the intensity of the HSI value that is owned by coffee beans. The HSI value is converted from RGB values, then the training process is performed on backpropagation artificial neural networks to recognize good quality seeds and poor-quality seeds. The testing phase is carried out using an interface designed in the MATLAB R2013a software. Based on the results of testing, it was found that this application was able to detect samples of coffee beans that were not properly trained.

1. Introduction

Coffee is one of the superior agricultural commodities[1–3]. The economic value of coffee beans is strongly influenced by its quality. However, not all coffee beans produced by coffee farmers are of good quality. Coffee beans can be damaged due to several things such as being attacked by pests or because of the post-harvest process, for example the process of stripping fruit skin is less than perfect and the process of washing and drying is not evenly distributed. This causes some parts or all parts of the coffee beans to become damaged. Therefore, it is necessary to select quality coffee beans before making a sale to increase the selling value of the coffee beans.

Generally, the selection of quality coffee beans is still done manually by visually looking at coffee beans[4]. In this way, the selection of good coffee beans can only be done by people who are knowledgeable about the characteristics of good coffee beans. Based on these reasons, in this study we designed a software to recognize good quality or damaged coffee beans. With this software ordinary people will be able to distinguish the quality of coffee beans from the visual without needing to know about the characteristics of coffee beans that are visually good.

This research is based on several previous studies about the quality of coffee beans. Previous studies have been carried out to evaluate the quality of coffee beans using infrared spectroscopy[5]. In addition, there is also the introduction of good coffee beans using image processing based on RGB color components[6]. We have also conducted research using image processing to determine the type of coffee roasted[7]. In contrast to some previous studies we will design software to recognize the quality of coffee beans by using image processing and artificial neural networks based on the intensity of HSI (Hue Saturation Intensity) values.

Coffee beans can be damaged due to several things such as being attacked by pests or because of the post-harvest process, for example the process of stripping fruit skin is less than perfect and the process of washing and drying is not evenly distributed. This causes some parts or all parts of the coffee beans to become damaged.

Damage to coffee beans can be detected, one of them by looking at the spread of colors that are less uniform in coffee beans. Visually poor-quality coffee beans generally have black patches and tend to have dull and dry skin on the surface. This is very different from good quality coffee beans, where the surface of the skin looks fresher and has a more evenly spread color on its surface as shown in Figure 1.

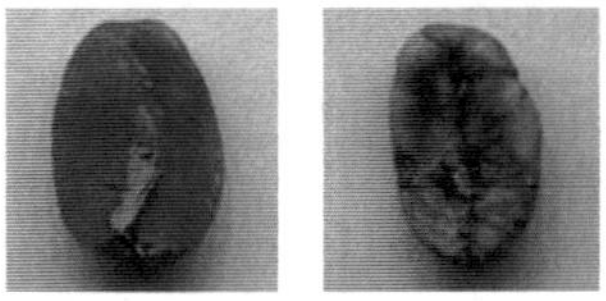

Figure 1. Good quality and bad quality coffee beans

In this research, good and bad quality coffee beans will be detected based on the color distribution. Based on SNI 01-2907-2008[8], regarding the determination of the value of defective coffee beans, several parameters will be used which can determine a poor-quality coffee bean based on the distribution of colors such as: full and partial black seeds, full brown beans and some as shown in Figure 2.

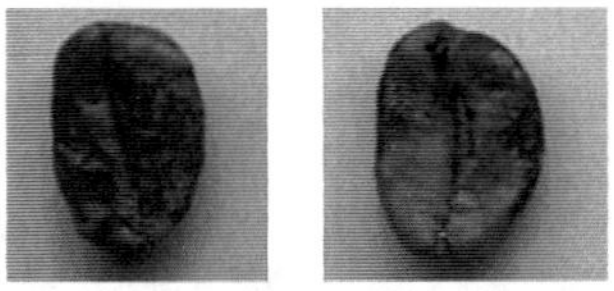

Figure 2. Black and brown coffee beans

2. Method

This research stage begins with the acquisition of the image of the beans, and then process the data pre-processing resulting features/characteristics that can distinguish the quality of the coffee beans. Then the training phase is done using backpropagation artificial neural networks. At the last stage, the creation of an interface makes it easier in the testing process. The stages of this research flow can be seen in Figure 3.

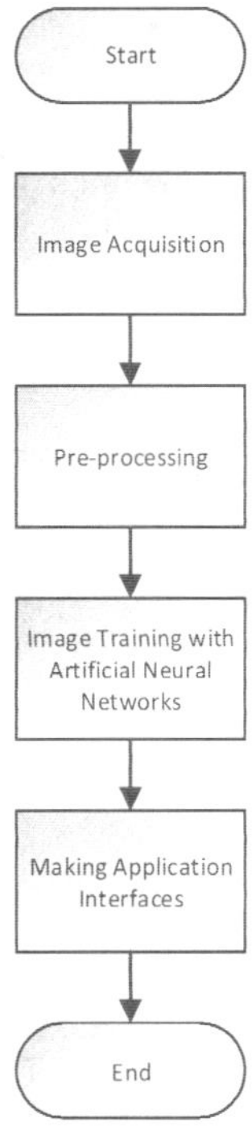

Figure 3. Research flow chart

2.1. Image acquisition

In the image acquisition stage, the image of coffee beans is taken as training material. Image capture is done using the Xiaomi Redmi Note 5 smartphone camera with 12MP f1/9 resolution with lighting using LED lights. Examples of image acquisition results can be seen in Figure 4.

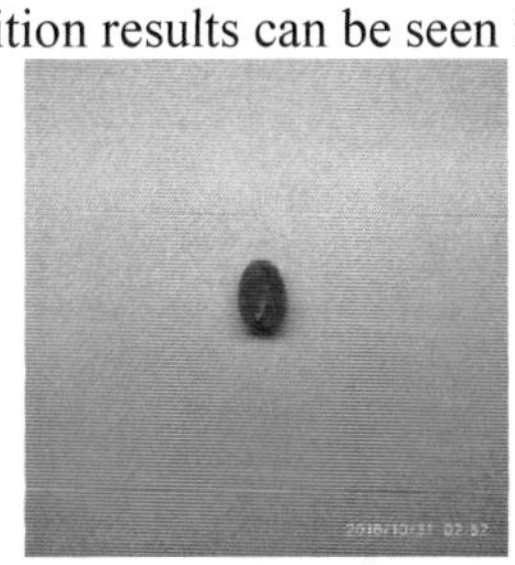

Figure 4. Example image

2.2. Pre-processing

Pre-processing starts with cropping and normalizing the size to uniform image size. This is also done because the resolution of the image produced by the camera is large enough to slow down the subsequent processing. In the cropping process the image size is uniformed to 150 x 150 pixels as shown in Figure 5.

Figure 5. Results of cropping the image

The next step is to get the HSI value from the image RGB value. Previously the RGB value of the image was normalized first from 0-255 to 0-1. This is because the range of saturation and intensity values in the HSI ranges from 0-1[9–12]. The following is the equation used to convert the values of R, G, B to H, S, I.

$$\theta = cos^{-1}\left\{\frac{\frac{1}{2}[(R-G)+(R-B)]}{\left[(R-G)^2+(R-B)(G-B)^{\frac{1}{2}}\right]}\right\} \quad (1)$$

$$H(Hue) = \begin{cases} \theta & if\ B \leq G \\ 360-\theta & if\ B > G \end{cases} \quad (2)$$

$$S(Saturation) = 1 - \frac{3}{R+G+B}[min(R,G,B)] \quad (3)$$

$$I(Intensity) = \frac{1}{3}(R+G+B) \quad (4)$$

The Hue value that ranges from 00 - 3600 is also normalized so that the value becomes 0-1 with the equation $H = \frac{H}{360}$. The saturation value that has been obtained is used to separate objects (foreground) in the background using the thresholding technique.

The next process is the thresholding process. Thresholding is used to separate objects (foreground) from the background[13,14]. This process is done by setting a threshold value, where the RGB value that is equal to or more than the threshold value will be changed to white (1), while the RGB value that is less than the threshold value will be changed to black (0). In this case the threshold value is based on the saturation value that was obtained earlier. Figure 6 shows (a) the thresholding image and (b) RGB on the thresholding (c) HSI image in the Thresholding image.

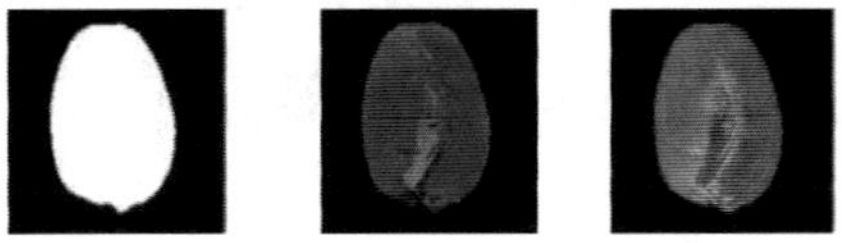

(a) (b) (c)

Figure 6. Image thresholding results

2.3. *Image training with artificial neural networks*

The next stage is image training using artificial neural networks. At this stage the HSI value that has been successfully obtained in the image will be used as training input on artificial neural networks. The network architecture designed is 3 neurons in the input layer, 10 neurons in the hidden layer and 1 neuron in the output layer. The artificial neural network method used in this study is backpropagation with the sigmoid activation function[15–17]. Figure 7 shows the network training process with momentum 0.7 and performance goal 0.001. This network training uses the supervised learning method[18–20], so that from the beginning the network training output value has been determined first. In this network training the output value of good quality coffee beans = 1 and poor-quality coffee beans = 0.

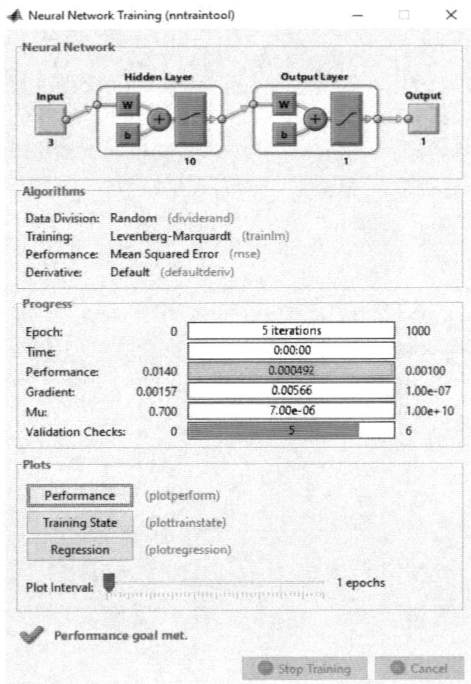

Figure 7. Artificial neural network training

2.4. Interface design

The next step is designing the application interface. At this stage we design the application interface for testing using MATLAB software. In figure 8 we can input the image of the coffee beans by using the 'load' button so that the image of the coffee beans will appear in the input image. Then the 'process' button will be processed in the input image so that the segmented image will be displayed in the box next to it. The results dialog box will display the results of detecting coffee beans whether good or bad.

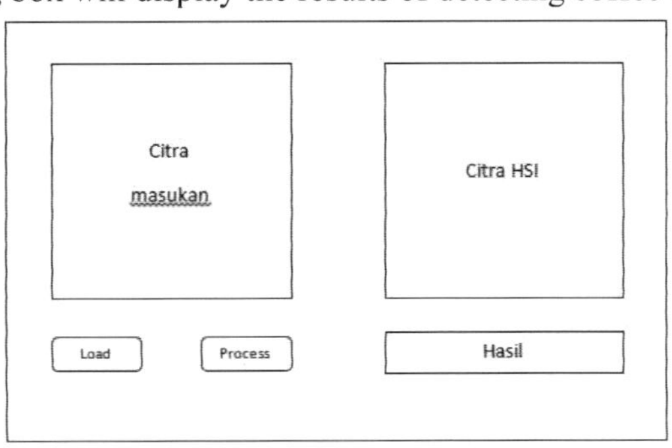

Figure 8. App interface display design

3. Results and discussion

3.1. Training testing

The coffee bean samples used in this study were 100 raw coffee beans with the distribution of 40 good quality coffee beans and 40 poor quality coffee beans which had been separated to be used in the training process and in the testing process 10 good quality non-trained coffee beans were used and 10 poor quality non-trained coffee beans. The testing of the detection of the quality of coffee beans is divided into 2 parts:

1. The trial uses 30 training samples by dividing 15 good quality coffee beans and 15 poor quality coffee beans.
2. The trial uses 20 non-train samples by dividing 10 good-quality coffee beans and 10 poor-quality coffee beans. As shown in Table 1.

Table 1. Test results of 20 non-train samples

Coffee Beans	H	S	I	Output
Biji_Baik1	0.1170	0.3604	0.2376	1
Biji_Baik2	0.1187	0.2973	0.2524	1
Biji_Baik3	0.1166	0.3664	0.2367	1
Biji_Baik4	0.1049	0.3831	0.2279	1
Biji_Baik5	0.1119	0.4024	0.2336	1
Biji_Baik6	0.1200	0.3307	0.2476	1
Biji_Baik7	0.1234	0.3317	0.2316	1
Biji_Baik8	0.1181	0.3762	0.2301	1
Biji_Baik9	0.1195	0.3760	0.3019	0.996
Biji_Baik10	0.1207	0.3025	0.2085	1
Biji_Buruk11	0.0956	0.4726	0.1774	0.027
Biji_Buruk12	0.1174	0.5604	0.2462	0
Biji_Buruk13	0.0835	0.5374	0.1902	0
Biji_Buruk14	0.1102	0.5782	0.2647	0
Biji_Buruk15	0.1145	0.5815	0.2882	0
Biji_Buruk16	0.1154	0.5435	0.2890	0
Biji_Buruk17	0.1040	0.5612	0.2238	0.0612
Biji_Buruk18	0.0995	0.5887	0.1961	0
Biji_Buruk19	0.1055	0.5500	0.2608	0
Biji_Buruk20	0.1026	0.5599	0.2493	0

Based on Table 1, it can be explained that the algorithm that has been designed is able to detect non-training data properly. This might be due to not too much non-training data. With more non-training data being tested, a detection error might be seen.

3.2. Application interface testing

The interface that has been designed successfully runs properly. Figure 9 shows the interface that was designed during the process of detecting the quality of coffee beans. After the image is successfully inputted, the next step is processing the image with the 'PROCESS' button. The results will display a segmented HSI image along with the results of quality detection in the image of the coffee bean.

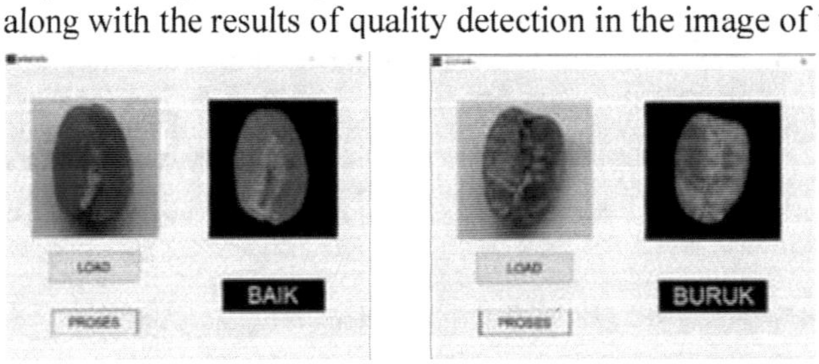

Figure 9. Application interface testing

4. Conclusion

Based on the discussion above, the conclusions are as follows:

1. The quality detection system of coffee beans based on quality can be created based on differences in the intensity of the HSI value (Hue Saturation Intensity) in the image of coffee beans with digital image processing and backpropagation artificial neural networks.
2. Training of artificial neural networks designed in this study using optimal network architecture with Performance goals 0.001, Momentum 0.7.
3. For testing data that is not trained the system is able to detect quite well.

Acknowledgments

The authors are grateful to the Universitas Sumatra Utara for supporting this research. We also thank all the lecturers and staff of the Computer Laboratory and the Arrangement of the Department of Electrical Engineering at the Universitas Sumatra Utara who have helped this research.

References

[1] Wang N and Lim L T 2012 Fourier transform infrared and physicochemical analyses of roasted coffee *J. Agric. Food Chem.* **60** 5446–53
[2] Nasution T H, Putramas A, Soeharwinto, Fahmi and Siregar I 2018 Automatic coffee roaster design using Arduino *AIP Conference Proceedings*
[3] Kelly C and Scott J 2015 Microwave aquametry of roasting coffee beans *Asia-Pacific Microwave Conference Proceedings, APMC* vol 3
[4] Kleinwächter M, Bytof G and Selmar D 2014 Coffee Beans and Processing *Coffee in Health and Disease Prevention*
[5] Santos J R, Sarraguça M C, Rangel A O S S and Lopes J A 2012 Evaluation of green coffee beans quality using near infrared spectroscopy: A quantitative approach *Food Chem.*
[6] Arboleda E R, Fajardo A C and Medina R P 2018 An image processing technique for coffee black beans identification *2018 IEEE International Conference on Innovative Research and Development, ICIRD 2018*
[7] Nasution T H and Andayani U 2017 Recognition of roasted coffee bean levels using image processing and neural network *IOP Conference Series: Materials Science and Engineering*
[8] Indonesia S N 2008 *Biji Kopi*
[9] Rao R A A, Mashor M Y, Ahmad R B, Noor S S M and KOsman M 2010 Comparison of colour thresholding method using RGB and HSI information for Ziehl-Neelsen sputum slide images *10th International Conference on Information Sciences, Signal Processing and their Applications, ISSPA 2010*
[10] Chien C L and Tseng D C 2011 Color image enhancement with exact HSI color model *Int. J. Innov. Comput. Inf. Control*
[11] Blotta E, Bouchet A, Ballarin V and Pastore J 2011 Enhancement of medical images in HSI color space *Journal of Physics: Conference Series*
[12] Rahmat R F, Chairunnisa T, Gunawan D, Pasha M F, Budiarto R 2019 Hand gestures recognition with improved skin color segmentation in human-computer interaction applications *Journal of Theoretical and Applied Information Technology* **97**(3), pp. 727-739
[13] Mitchell H B and Mitchell H B 2010 Image Thresholding *Image Fusion*
[14] Russ J and Neal F 2015 Segmentation and Thresholding *The Image Processing Handbook, Seventh Edition*
[15] Rojas R and Rojas R 2011 The Backpropagation Algorithm *Neural Networks*
[16] Han J and Moraga C 2012 The influence of the sigmoid function parameters on the speed of backpropagation learning
[17] Andayani U, Nababan E B, Siregar B, Muchtar M A, Nasution T H and Siregar I 2017 Optimization backpropagation algorithm based on Nguyen-Widrom adaptive weight and adaptive learning rate *2017 4th International Conference on Industrial Engineering and Applications, ICIEA 2017*
[18] Baştanlar Y and Özuysal M 2014 Introduction to machine learning *Methods Mol. Biol.*
[19] Chelly S M and Denis C 2016 Applying supervised learning *MathWorks, Inc.*
[20] Anon 2010 Supervised learning *Springer Tracts in Advanced Robotics*

Designing an IoT-based air quality monitoring system

T H Nasution[1*], M A Muchtar[2] and A Simon[1]

[1]Department of Electrical Engineering, Universitas Sumatera Utara, Medan, Indonesia
[2]Department of Information Technology, Faculty of Computer Science and Information Technology, Universitas Sumatera Utara, Indonesia

*Email: tigor.nasution@usu.ac.id

Abstract. The air quality in an area very influential on the state of the population in an area because of the quality can affect both the health of humans, animals or plants. Therefore, it is necessary to periodically monitor air quality conditions in an area. In this study an IoT-based air quality monitoring system was designed to determine the air quality conditions in an area. The system will monitor using sensors to see the levels of several substances in the air including O3, SO2, CO and particulates. Reading sensor data using an Arduino microcontroller. Then the data sent to the ThingSpeak cloud system uses a WIFI module on Arduino by accessing the API provided by the ThingSpeak cloud service. The monitoring results will be visible through a web page provided by the ThingSpeak cloud service.

1. Introduction

The air quality in an area very influential on the state of the population in an area because of the quality can affect both the health of humans, animals or plants[1–3]. Poor air quality can impact on health is not good for all living organisms. To prevent something that can worsen health conditions, it is necessary to know the condition of air quality in an area. Therefore, it is necessary to periodically monitor air quality conditions in an area.

Air quality monitoring can be done by measuring several parameters including temperature, humidity and air-contained compounds including O3 (Ozone), SO2 (Sulfur Dioxide), CO (Carbon Monoxide) and particulates[4–6]. At this time, advances in sensor technology can help us to measure the proficiency level parameters to determine the condition of the air quality[7,8]. In addition, the development of IoT (Internet of things) also helps with remote monitoring technology[9,10]. In this study we design an IoT-based monitoring system to monitor air quality in an area on a regular basis. Monitoring is done by collecting data from several sensors that can be accessed remotely.

This research was conducted based on several previous existing studies. Previously there were studies that conducted control and monitoring of air quality in the room[11,12]. This research is also based on our research in designing remote communication for monitoring air quality[13] . In contrast to previous studies we designed an outdoor air quality monitoring system. The system will measure several compounds in the air including O3, SO2, CO and particles. Air quality will be monitored remotely from web pages.

2. Method

2.1. System overview

General overview designed system can be seen in the block diagram in Figure 1. Figure 1 shows a system designed to collect air quality data from various sensors that will measure the levels of O3, SO2,

CO and particles in the air. The measurement results of these sensors will be read by the Arduino Uno controller. Arduino Uno is an ATmega328 based microcontroller board which is widely used for automatic control and monitoring[14,15,16]. Arduino Uno has 14 digital input/output pins (of which 6 can be used as PWM output), 6 analog inputs, 16 MHz ceramic resonator, USB connection, power jack, ICSP header, and reset button[17,18].

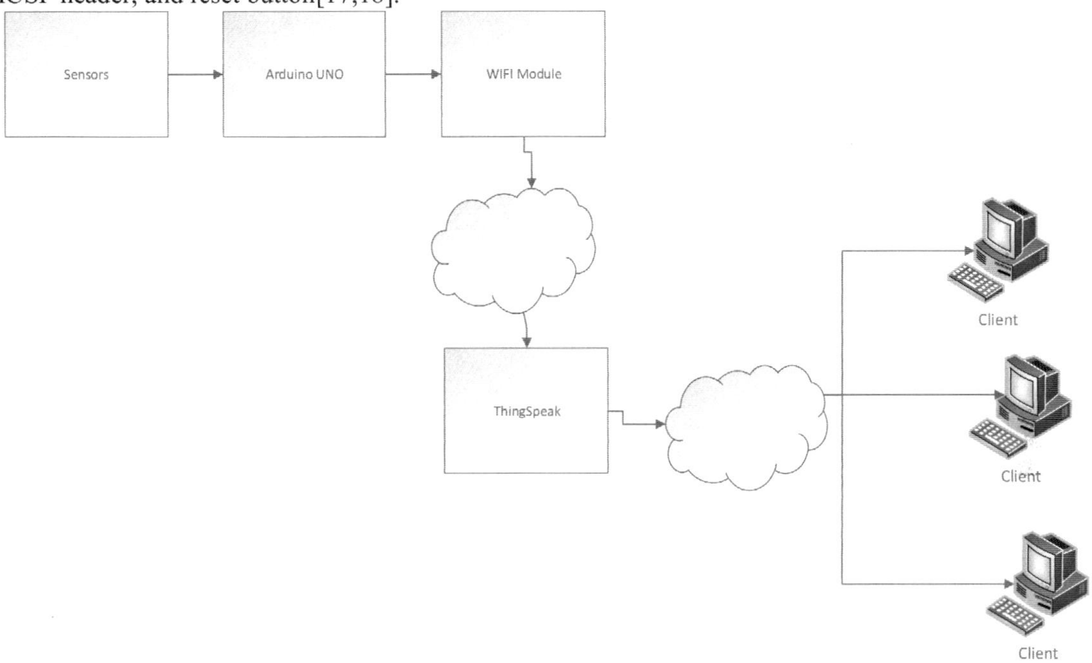

Figure 1. Block diagram of the system

Furthermore, Arduino Uno will send data to the cloud system using WIFI module. We use ThingSpeak for cloud systems. ThingSpeak is an open source IoT application platform and API for storing and retrieving data from things using the HTTP protocol over the Internet or through a Local Area Network[19]. The monitoring results can be seen on the web page provided by ThingSpeak in graphical form.

2.2. The design of the system

The first thing to do in system design is to make a schematic range as shown in Figure 2. We use 4 types of sensors to measure several types that are in the air, namely MQ-7, MQ1-131, MQ-135 and Pm10. Then do the coding on Arduino Uno using Arduino IDE. Arduino IDE is an editor used to write programs to Arduino[20]. Furthermore, create a channel in ThingSpeak to collect measurement data sent by the Arduino. Data transmission from Arduino is done by accessing the ThingSpeak API URL obtained after we create a channel to get data. Arduino accesses the ThingSpeak API using the WIFI ESP8266 module[21].

The MQ-7 sensor is used to read the levels of CO compounds in the air. MQ-7 is an analog sensor. The features of the MQ7 gas sensor are having high sensitivity to CO, stable, and long life[22]. This sensor uses heater power supply: 5V AC / DC and uses circuit power supply: 5VDC, measuring distance: 20 - 2000 ppm to be able to measure carbon monoxide gas. In Figure 2 MQ-7 is connected to the A2 analog pin on Arduino.

To measure Ozone levels, we use MQ-131 which is an analog sensor to measure Ozone levels in the air[23]. MQ-131 works with a power supply of 5V (VCC) which is connected to a VCC pin on the microcontroller. If the sensor detects O3 gas in the air, the output voltage on the sensor will increase, so that the gas concentration will decrease and a deoxidation process will occur. O3 gas concentration value is obtained by calculating the ratio between the sensor resistance value and sensor resistance when the air is clean. In Figure 2 the MQ-131 is connected to the A0 analog pin on Arduino.

SO2 levels in the air are measured using the MQ-135 sensor which is connected to an analog A1 pin on Arduino. MQ-135 sensor is a gas sensor whose working principle is to detect the quality of air in the room so that the characteristics of the sensor as a resistor will turn into a semiconductor or voltage conductor[24]. Besides SO2 MQ-135 can also be used to measure ammonia, aromatics, benzene vapor, and other gases.

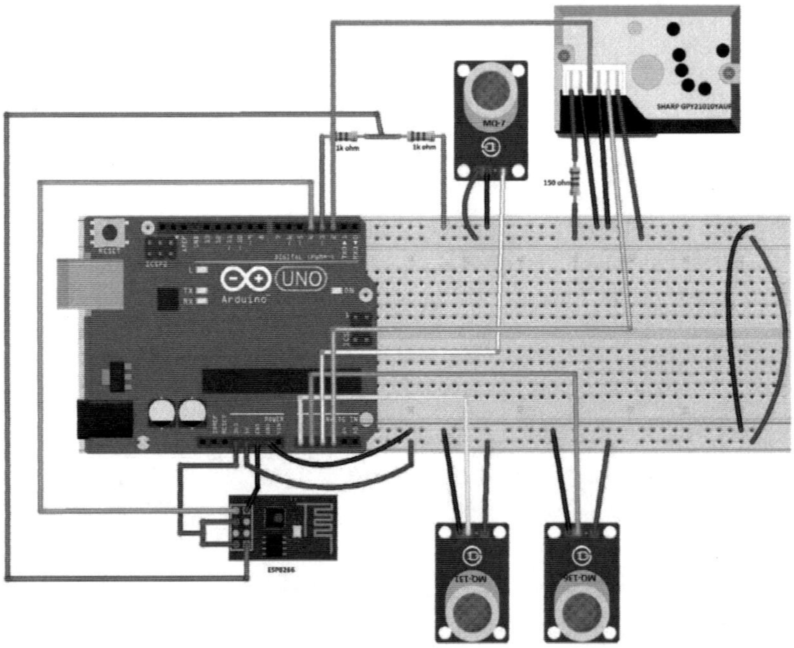

Figure 2. Schematic circuit

In addition to measuring the levels of some gases, we also measure particles in the air using PM10. PM10 sensor is a sensor that can measure airborne particles smaller than 10 microns (micrometres) [25]. In the schematic of the circuit in Figure 2 the Pm10 sensor is connected to an A3 analog pin on Arduino.

3. Results and discussion

From the results of the design that has been done, a system trial has been carried out that has been designed. The trial was carried out in the campus environment to monitor air quality conditions on the campus. From the results of the trial monitored data of compounds in the air that were measured were data on levels of Ozone, SO2, CO and PM10 particles. Data is received by ThingSpeak every one minute. Data received by ThingSpeak can be monitored on the channel pages provided. On the web page data is presented in graphical form. Web pages can be accessed from the internet network by accessing the monitoring channel web pages as shown in Figure 3. The monitoring results can also be exported to table forms in the excel, xml or Json file format. Examples of export data in tables can be seen in Table 1.

Table 1. Data from monitoring

created_at	entry_id	field1 (O3)	field2 (SO2)	field3 (CO)	field4 (Partikulat)
2019-05-26 15:37:54 UTC	1	0.2	1.03	2.03	3.03
2019-05-26 15:38:13 UTC	2	0.1	1.02	2.02	3.02
2019-05-26 15:39:05 UTC	3	0.0	1.01	2.01	3.01
2019-05-27 16:42:56 UTC	4	0.1	1.00	2.00	3.00
2019-05-27 16:43:13 UTC	5	0.1	1.00	2.00	3.00
2019-05-27 16:43:29 UTC	6	0.1	1.00	2.00	3.00
2019-05-27 16:43:46 UTC	7	0.1	1.00	2.00	3.00

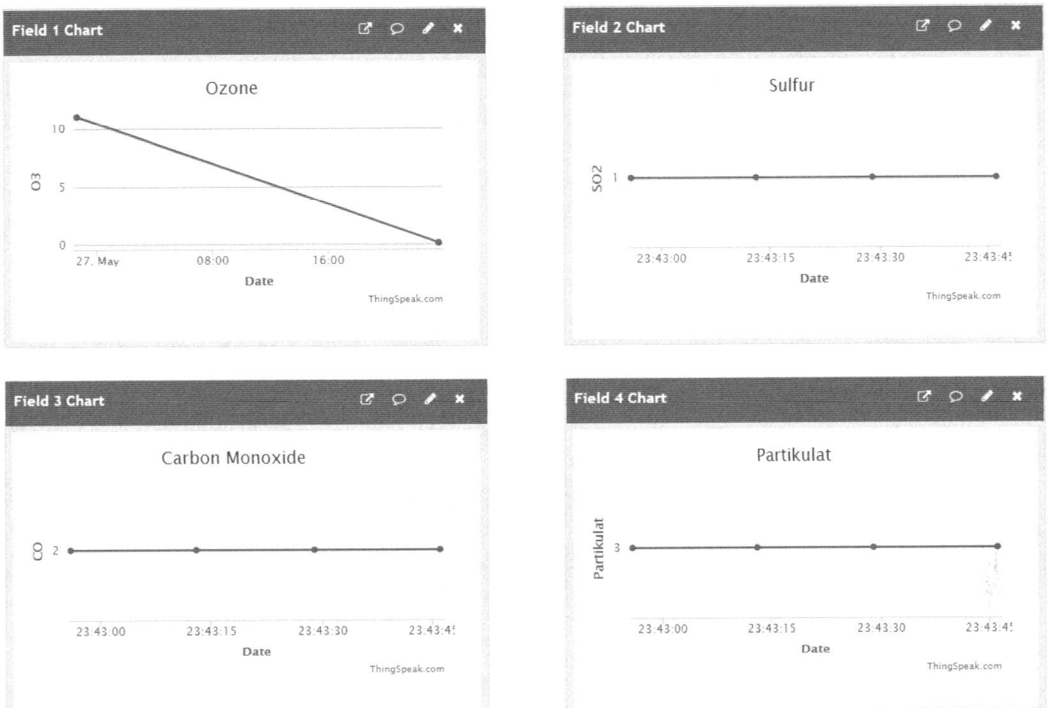

Figure 3. Graph data from monitoring

4. Conclusion

From the trials conducted, the results of the design can work well. It seems that the data obtained from the sensor can be seen from the webpage provided. Data changes can be monitored every once a minute. Data can also be exported into Excel data format, XML or in Json.

From the results obtained, further development can be done by making a mobile-based client application. This is supported because ThingSpeak can also provide API access to get data that has been collected on the server.

Acknowledgement

This research was carried out with the help of the TALENTA Young Lecturer Research Grant Universitas Sumatra Utara. The authors are grateful to the Universitas Sumatra Utara for supporting this research. We also thank all the lecturers and staff of the Computer Laboratory and the Arrangement of the Department of Electrical Engineering at the Universitas Sumatra Utara who have helped this research.

References

[1] de Gennaro G, Dambruoso P R, Loiotile A D, Di Gilio A, Giungato P, Tutino M, Marzocca A, Mazzone A, Palmisani J and Porcelli F 2014 Indoor air quality in schools *Environ. Chem. Lett.*

[2] European Environmental Agency (EEA) 2015 *Air quality in Europe — 2015 report*

[3] WHO | Ambient (outdoor) air quality and health 2019 WHO | Ambient (outdoor) air quality and health *Who*

[4] Mabahwi N A B, Leh O L H and Omar D 2014 Human Health and Wellbeing: Human Health Effect of Air Pollution *Procedia - Soc. Behav. Sci.*

[5] Fiore A M, Naik V and Leibensperger E M 2015 Air quality and climate connections *J. Air Waste Manag. Assoc.*

[6] Wolkoff P 2018 Indoor air humidity, air quality, and health – An overview *Int. J. Hyg. Environ. Health*

[7] EPA 2013 Air Sensor Guidebook *J. Chem. Inf. Model.*

[8] Schneider P, Castell N, Vogt M, Dauge F R, Lahoz W A and Bartonova A 2017 Mapping urban air quality in near real-time using observations from low-cost sensors and model information *Environ. Int.*

[9] Gubbi J, Buyya R, Marusic S and Palaniswami M 2013 Internet of Things (IoT): A vision, architectural elements, and future directions *Futur. Gener. Comput. Syst.* **29** 1645–60

[10] Nasution T H, Siagian E C, Tanjung K and Soeharwinto 2018 Design of river height and speed monitoring system by using Arduino *IOP Conference Series: Materials Science and Engineering* vol 308

[11] Zhi S, Wei Y, Cao Z and Hou C 2018 Intelligent controlling of indoor air quality based on remote monitoring platform by considering building environment *2017 4th International Conference on Systems and Informatics, ICSAI 2017*

[12] Jia F 2018 Home Network Monitoring System Based on Internet of Things *IOP Conf. Ser. Mater. Sci. Eng.* **452** 042118

[13] Nasution T H, Siregar I and Yasir M 2017 UAV telemetry communications using ZigBee protocol *J. Phys. Conf. Ser.* **914**

[14] Chaudhari A, Rodrigues B and More S 2016 Automated IOT based system for home automation and prediction of electricity usage and comparative analysis of various electricity providers: SmartPlug *Proceedings of the 2016 2nd International Conference on Contemporary Computing and Informatics, IC3I 2016* pp 390–2

[15] Nasution T H, Putramas A, Soeharwinto, Fahmi and Siregar I 2018 Automatic coffee roaster design using Arduino *AIP Conference Proceedings*

[16] Rahmat R F, Athmanathan, Syahputra M F, and Lydia M S 2017 Real time monitoring system for water pollution in Lake Toba *International Conference on Informatics and Computing ICIC 2016* pp. 383-388

[17] Drymonitis A 2015 *Introduction to Arduino*

[18] Nasution T H, Muchtar M A, Siregar I, Andayani U, Christian E and Sinulingga E P 2017 Electrical appliances control prototype by using GSM module and Arduino *2017 4th International Conference on Industrial Engineering and Applications, ICIEA 2017*

[19] MathWorks ThingSpeak Documentation

[20] Arduino 2017 Introduction *www.arduino.cc*

[21] 8 K 2015 ESP8266 WiFi Module Quick Start Guide *ESP8266 WiFi Modul. Quick Start Guid.*

[22] Borges M A, Melo G F, De Massaki C, Igarashi O, Lopes P B and Silva L A 2017 An architecture for the internet of things and the use of big data techniques in the analysis of carbon monoxide *Proceedings - 2017 IEEE International Conference on Information Reuse and Integration, IRI 2017*

[23] Zhao Y, Guo Z-H, Meng F-L and Liu J-H 2007 Dynamic detection and recognition system based on the segmental average differentiation *Chinese J. Sensors Actuators*

[24] Vandana K, Baweja C, Simmarpreet and Chopra S 2016 Influence of Temperature and Humidity on the Output Resistance Ratio of the MQ-135 Sensor *Int. J. Adv. Res. Comput. Sci. Softw. Eng.*

[25] Cavaliere A, Carotenuto F, Di Gennaro F, Gioli B, Gualtieri G, Martelli F, Matese A, Toscano P, Vagnoli C and Zaldei A 2018 Development of low-cost air quality stations for next generation monitoring networks: Calibration and validation of PM 2.5 and PM 10 sensors *Sensors (Switzerland)*

Comparative analysis of technical efficiency between organic and non-organic rice farming in North Sumatera Indonesia

Rulianda P Wibowo[1], Ar Raihan[1], Sumono[2], Dani Gunawan[3]

[1]Department of Agribusiness, Faculty of Agriculture, Universitas Sumatera Utara, Medan, Indonesia
[2]Department of Agricultural Engineering, Faculty of Agriculture, Universitas Sumatera Utara, Medan, Indonesia
[3]Department of Information Technology, Faculty of Agriculture, Universitas Sumatera Utara, Medan, Indonesia

Correspondace Email : rulianda_wibowo@usu.ac.id

Abstract. Non-organic farming will deteriorate the land and decrease the total factor productivity, which may result to higher production cost and low crop yield in the long run. However, Indonesian government initiative to become less dependent on non-organic rice farming by encouraging farmer to plant organic rice seems unsuccessful. Nevertheless, the most rice consumed in Indonesia is from non-organic farming. This study uses data envelopment analysis (DEA) based on non-parametric approach to measure technical, allocative and economic efficiency. This study found that pesticide and irrigation is underutilized, but fertilizer and labor are being overutilized. Even though most farmer are technically efficient, better efficient input allocation can improve the efficiency. Organic rice farming is less efficient than non-organic farming due to its low crop yield. Farmer with longer education background tend to be less efficient and more inclination toward planting organic rice.

1. Introduction

The majority of rice farmer in Indonesia are not using organic farming system. The organic farming system is considered impractical because of the low yield result compared to non-organic farming system. Arguably, most of the rice consumed in Indonesia is from non-organic farming system, agricultural practice with the use of chemical fertilizers and chemical pesticides [1]. However, the continuous use of chemical fertilizers and chemical pesticides has worsened the structure and composition of nutrients and soil fertility, which indirectly affects the land productivity [2]. The use of chemical fertilizer causes the soil depends on those substances to sustain the land productivity. In addition, [2] mention that the use of pesticides is also harmful for the environment because of its residual effects.

Non-organic rice farming system had depleted the land productivity in Indonesia, which is shown by a decline in factor productivity for non-organic wetland paddy [1]. It is a possibility that non-organic farming system might have a higher production cost and lower yield when farmers had applied it for every now and then on their land. The land deterioration causes farmer to use chemical substances in larger proportion to maintain high crop yield. Indonesian government

IOP Conf. Series: Materials Science and Engineering **648** (2019) 012038 doi:10.1088/1757-899X/648/1/012038

through "Go Organic 2010" had promoted organic agriculture with the intention of being a major marketer in organic crop market [1]. The use of organic farming system is expected to make farmer less dependent on chemical substances and maintain the land productivity in the long run.

Serdang Berdagai is one of the main districts in North Sumatera Province that plant paddy as a main crop. [3] mention that most of the rice farming are non-organic farming system, which lead to a question whether organic rice farming is less profitable and less efficient than non-organic rice farming. The farmer in Serdang Berdagai had applied the non-organic rice farming for sometimes, which should result in lower land productivity and high production cost.

The objectives of this study are; 1) to observe the efficiency on rice farming, 2) to analyze the the comparison of efficiency between organic and non-organic rice farming, 3) to observe the impact of farm characteristics and input use to efficiency.

2. Data and Methods

2.1 Data

Farm level data were collected from rice farmer in Serdang Berdagai District in 2017. The are 51 farm data comprises 20 organic rice farm and 31 non-organic rice farms. Five inputs were defined: seed, fertilizer, pesticide, labor and irrigation. As suggested by [4], this study assumes law of one price, because of the planting, harvest and selling agricultural product in the same district are in the same period. Farm characteristic data are collected to seek the correlation between operator and characteristic of the farm with overall level of efficiency. Farm characteristics are defined as: age, gender, farming type, experience, family member and education.

2.2 Methods

The data envelopment analysis is used to analyze the input-output efficiency comparison between organic and non-organic farming system. The comparison between those two-farming system will examine which production method is more efficient and provide explanation of cost minimization behavior.

Farrell explained that the efficiency of a decision-making unit (DMU) can be compared with another decision-making unit within a given group [5]. He proposed the measurement of efficiency by assuming a particular production function. There are three types of efficiency measures comprises technical efficiency (TE), allocative efficiency (AE) and Economic Efficiency (EE).

The relative efficiency can be measured not only by assuming a given production function, named as parametric approach, but also can be measured using data envelopment analysis (DEA) based non-parametric approach [6]. This study uses data envelopment analysis (DEA) based on non-parametric approach proposed by [7]. This study follows [8] and [9] for non-parametric analysis that will measure economic, allocative and technical efficiency with the assumption of cost minimization behavior of DMU. [8] stated that farmer behavior tends more to exhibit cost minimization behavior.

Technical efficiency (TE) measures the ability to produce a given level amount of output with the minimum input [5]. Consider the DEA for estimating technical efficiency using linear programming for DMUs at group J as follows:

$$D^N\left(x_j^t, y_j^t\right) = \min \theta_j \tag{1}$$

$$s.t \ \ \sum_{j=1}^{N} \lambda_j^t x_{m,j}^t - \theta_j^t x_{m,j}^t \leq \qquad , for\ m = 1, \ldots, M \tag{2}$$

$$\sum_{j=1}^{N} \lambda_j^t y_{l,j}^t - y_{l,j}^t \geq \quad , \quad for\ l = 1, \dots, L \tag{3}$$

$$\sum_{j=1}^{J} \lambda_j = 1 \tag{4}$$

$$\lambda_j^t \geq 0 \quad , \ for\ j = 1, \dots N \tag{5}$$

Economic efficiency (EE) shows the minimum cost $C_j(w, y, T_c)$ of producing a given quantity of output (y), given input prices (w), and input (x) under constant return to scale technology. Consider the cost minimization DEA for measure economic efficiency by obtain cost minimization for DMUs at group J using linear programming as follows:

$$\min_{\lambda_j^t x_{m,j}^{t*}} C_j(w, y, T_c) = \sum_{m=1}^{M} w_m^t x_{m,j}^{t*} \tag{6}$$

$$s.t \ \sum_{j=1}^{N} \lambda_j^t x_{m,j}^t - x_{m,j}^{t*} \leq 0 \quad , \ for\ m = 1, \dots, M \tag{7}$$

$$\sum_{j=1}^{N} \lambda_j^t y_{l,j}^t - y_{l,j}^t \geq 0 \quad , \quad for\ l = 1, \dots, L \tag{8}$$

$$\sum_{j=1}^{J} \lambda_j = 1 \tag{9}$$

$$\lambda_j^t \geq 0 \quad , \ for\ j = 1, \dots N \tag{10}$$

The economic efficiency is calculated as $EE_j = \frac{C_j(w,y,T_c)}{\sum_{m=1}^{M} w_m^t x_{m,j}^t} = \frac{\sum_{m=1}^{M} w_m^t x_{m,j}^{t*}}{\sum_{m=1}^{M} w_m^t x_{m,j}^t}$

Allocative efficiency (AE) shows the minimum cost of producing a given level of output given input prices under variable return to scale technology [5]. Allocative efficiency (AE) is calculated using the technical efficiency and economic efficiency as follow:

$$AE_j = \frac{EE_j}{TE_j} \tag{11}$$

2.3 Tobit Model

This study will analyze the relationship between efficiency measures with farm characteristic using Tobit model. The latent dependent variables of the Tobit model are overall, technical and allocative, which are measured using non-parametric analysis. The Tobit model [10] was estimated as follows:

$$O_j = \sum_{j=1}^{T} C_i G_i + \varepsilon_i \quad if \quad \sum_{i=1}^{T} C_i G_i + \varepsilon_i < 1, \ = 1 \ otherwise \tag{12}$$

O_i is the measure of efficiency (Pure Technical, Allocative and Economic Efficiency) for each farm , C_i is an estimated parameter, and the explanatory variable (G_i)

The Tobit model will identify the factors related to inefficiency for different farming system. The explanatory variables used in Tobit model are the age, gender, farming type, experience, family member and the education level of farmer. Tobit model also used to analyze the relationship between efficiency and input use. The explanatory variables for input analysis are fertilizer, seed, pesticide, irrigation, and labor. The Tobit models will identify the importance of each input to efficiency analysis. The models also will show which inputs are being under-utilized or over-utilized. Thus, it will give suggestions to farmers of how to increase the efficiency of input use.

3. Result and Discussion

The average output of non-organic farming is more than a double of organic farming (see table 1). The average output for non-organic farming is Rp. 17,707,669 per hectare, in comparison with lower organic farming average output of Rp. 7,849,650. The price of organic rice is not significantly different than non-organic rice, at the other hand the yield is much higher for non-organic rice [3]. Unexpectedly, the cost of fertilizer is higher for organic farming because the use of costly natural fertilizer, fertilizer that made from the plant and animal.

The irrigation and pesticide cost are higher for non-organic farming. Non-organic farmer tends to apply more irrigation and pesticide to reduce risk in crop production. Risk averse farmer will significantly increase total water use for irrigation [11]. However, the most risk efficient irrigation for rice farm is the application of 1 cm height irrigation [12].

Farmer with organic farming have older age and longer period of farming experience (see table 1). On average, organic farmer are 7 years older and having 4 more years' experience compared to non-organic farmer. Arguably, organic farmer has longer run objective in terms of land productivity compared to non-organic farmer that are more focused on short term output. In addition, farmer who apply organic farming is more educated than non-organic farmer.

Table 1. Summary statistic of output, input and farm characteristic use in DEA Analysis

	Organic Farming, n=20				Non-organic farming, n=31			
Variable	Mean	Std. Dev.	Min	Max	Mean	Std. Dev.	Min	Max
Output/Input								
Output	7,849,650	7,975,035	1,720,000	33,600,000	17,707,669	15,607,486	4,050,000	86,775,000
Seed	79,625	76,706	12,000	300,000	274,806	273,296	13,500	1,210,000
Fertilizer	929,700	719,902	120,000	3,310,000	578,674	616,869	75,000	2,802,400
Pesticide	147,000	208,125	0	875,000	561,516	1,501,244	75,000	8,609,000
Irrigation	100,800	95,504	24,000	396,000	166,452	148,377	36,000	804,000
Labor	1,952,713	1,797,614	427,500	7,455,000	3,269,516	3,337,165	645,000	18,085,000
Farm Characteristic								
Age	55.4	5.55	45.00	65.00	48.39	14.32	21.00	80.00
Experience	20.95	8.73	10.00	40.00	16.87	8.18	7.00	50.00
Family Member	2.8	0.60	2.00	4.00	2.32	0.93	0.00	4.00
Education	10.7	3.51	6.00	16.00	8.74	2.65	6.00	16.00

The DEA estimates of Technical, allocative and economic efficiency are summarized in Table 2. The technical efficiency measures show that most farmer are efficient. The technical efficiency score has an average of 88.4%, which implies it can be improved with the reduction of the input usage by 11.6% (i.e. if each farmer produced on the production frontier). Technical efficiency scores were higher than allocative efficiency; there are 10% of the farms has allocative efficiency measures smaller than 50%, while all technical efficiency scores are greater than 50%. The mean of allocative score is 79.5%, which suggests the rice farmer allocate inadequate input resources with a given input price by 20.5% higher than cost-minimizing level. The allocative efficiency has wide-ranging distribution from 37.3% to 100% that infers most farmers are less efficient. The mean of economic efficiency score is 67%, which means the overall cost of rice production can be scaled down on average by 33% to achieve the same level of crop yield.

Table 2. The summary statistic of technical, allocative and economic efficiency scores

Efficiency	Technical	Allocative	Economic Efficiency
<50%		5	12
51%-75%	17	12	21
76%-100%	34	34	18
Min	0.553	0.373	0.358
Max	1	1	1
Average	0.844	0.795	0.67

The technical efficiency results suggest that it is more beneficial to drive the efficiency improvement through the increase of labor use. Using the efficiency score in table 3, it seems lower usage of pesticide and lower irrigation tend to be associated with the increase in technical efficiency. Furthermore, results in table 3 indicated that pesticide and irrigation is underutilized, but fertilizer and labor are being overutilized. With a given input price, the allocative efficiency can be increased if farmers enlarge the use of pesticide and apply higher rate of irrigation but reduce the usage of other inputs comprises labor and fertilizer. Irrigated farmers should focus on reducing fertilizer to increase economic efficiency.

Table 3. Tobit analysis for efficiency measures and farm inputs

Input	TE	AE	EE
Seed	-1.23E-08	-1.55E-07	-9.58E-08
Fertilizer	-7.89E-08	-2.14E-07***	-1.98E-07***
Pesticide	-2.37E-07**	1.94E-07*	2.96E-08
Irrigation	-0.0000131**	5.15E-06**	-4.49E-07
Labor	7.32E-07**	-2.14E-07*	7.81E-08
_cons	0.9079	0.791	0.680

Result in table 4 suggest that a significant positive performance on allocative and economic efficiency comes from planting non-organic rice. Non-organic rice farmer has significantly higher crop yield which result in more efficient input allocation and producing on production frontier. The estimation results reveal a consistently significant positive relationship between longer experience with technical efficiency measures in estimated models. This could be farmer with longer experience adopting more efficient agricultural practices. The result show positive relationship between allocative efficiency and age of the farmer. Older farmer is more efficient in allocating input resources to achieve cost-minimizing level on producing a given crop yield.

Table 4. Tobit analysis for efficiency measures and farm characteristic

Farm characteristic	TE	AE	EE
Padi_Dummy	-0.023	-0.2416***	-0.196***

Gender	0.091	-0.0298	0.056
Age	0.00087	0.0031*	0.0031
Experience	0.0182**	-0.0029	0.0042
Family Member	-0.0578	-0.0064	-0.0404
Education	-0.0439**	-0.0044	-0.0234**
_cons	1.022	0.8786	0.789

We found a negative correlation between technical and economic efficiency with education. This finding is not surprising, because farmer with longer education background have more interest on planting less efficient organic rice. More educated farmer tends to have more awareness to maintain long term land factor productivity by applying sustainable agricultural practices, which considerably produce lower crop yield.

4. Conclusion

This study uses non-parametric analysis to examine the efficiency of organic and non-organic rice farming). The technical efficiency measures show that most farmer are efficient. Technical efficiency can be improved with the reduction of the input usage. The allocative score suggest that the rice farmer allocate inadequate input resources which result in producing higher than cost-minimizing level. With a given input price, the allocative efficiency can be increased if farmers increase the use of pesticide and apply higher rate of irrigation but reduce the use of labor and fertilizer. Irrigated farmers should focus on reducing fertilizer to increase economic efficiency. Older farmer is more efficient in allocating input resources to achieve cost-minimizing level on producing a given crop yield. Farmer with longer term of education tends to have more inclination toward planting organic rice.

5. References

[1] Hidayat A S, and Lesmana, T 2011 The development of organic rice farming in Indonesia. *Revi Indonesian Econ Business Stud*, **2**(1), 71-87.
[2] Irsal L K S, and Setiyanto A P 2006 Isu dan pengelolaan lingkungan dalam revitalisasi pertanian. *Jurnal Litbang Pertanian*, **25**(3), 107.
[3] Nasution A R S 2018 Analisis Perbandingan Kelayakan Antara Usaha Tani Padi Organik dan Padi Non Organik (Studi Kasus: Desa Lubuk Bayas, Kecamatan Perbaungan, Kabupaten Serdang Bedagai).
[4] Chavas J P and Aliber M 1993 An Analysis Of Economic Efficiency In Agriculture: A Nonparametric Approach *Journal of Agricultural and Resource Economics* **18**(01).
[5] Farrell M J 1957 The measurement of productive efficiency. *Journal of the Royal Statistical Society: Series A (General)*, **120**(3), 253-281.
[6] Coelli T J 1995 Recent Developments in Frontier Modeling and Efficiency Measurement *Australian Journal of Agricultural Economics* **39**:219–215.
[7] Färe R, Grosskopf S, and Lovell C A K 1985 *The Measurement of Efficiency of Production*. Dordrecht: Springer Netherlands. Available at: http://link.springer.com/10.1007/978-94-015-7721-2 [Accessed July 17, 2016].
[8] Featherstone A M, Langemeier M R, and Ismet M 1997 A nonparametric analysis of efficiency for a sample of Kansas beef cow farms *Journal of Agricultural and Applied Economics*,**29**(1),

175-184.
[9] Wibowo R P 2019 Is irrigated farm more efficient than non-irrigated no-tillage farm and non-irrigated conventional farm for a sample of Kansas corn farmer?. *IOP Conference Series: Earth and Environmental Science* **260(**1) 012019.
[10]Maddala, G S 1983 *Limited-Dependent and Qualitative Variables in Econometrics.* Cambridge, UK:Cambridge University Press
[11] Wibowo R P, Hendricks N P, Kisekka I, and Araya, A 2017 Using a crop simulation model to understand the impact of risk aversion on optimal irrigation management *Transactions of the ASABE* **60**(6),pp 2111-2122.
[12] Wibowo R P and Rizaldi, T 2019 Deficit irrigation for rice farming with production risk due to weather variability. *IOP Conference Series: Earth and Environmental Science* **260**(1), p. 012021

IOP Conf. Series: Materials Science and Engineering **648** (2019) 012039 doi:10.1088/1757-899X/648/1/012039

Design of Hardware Module for the Vehicle Condition Monitoring System Based on the Internet of Things

F Fahmi[1*], F Nurmayadi[1], B Siregar[2], M Yazid[3], E Susanto[4]

[1]Department of Electrical Engineering, Faculty of Engineering, Universitas Sumatera Utara, Indonesia
[2]Department of Computer Science, Universitas Sumatera Utara, Indonesia
[3]Department of Electrical Engineering, Institut Teknologi Sepuluh Nopember, Indonesia
[4]Department of Physics, Faculty of Science and Technology, Airlangga University, Surabaya 60115, Indonesia

*Correspondence Email: fahmimn@usu.ac.id

Abstract. One problem in the development stage of driverless cars on public roads is how to monitor the condition of the vehicle that is affected by different weather and environment, including dry, snowy or heavy rain conditions, which are sometimes difficult to predict and require monitoring systems that should be real-time and accurate. For this reason, an internet of things-based car condition monitoring system is needed to be developed. The development covers three essential parts, which include the development of hardware and digital communication modules, web server development and data analysis and mobile application development for monitoring. The purpose of this study is to design a hardware module needed for car vehicle condition monitoring system; a prototype completed with the onboard diagnostic module and embedded system that will send data to the server in real time. The system was successfully designed and developed using ELM327 and ESP32 with SIM808 for communication module. The benefits of this research are as a tool for asset management of car vehicles both for personal and corporate needs. This study also supports the development of driverless cars that require real-time data input between the vehicle and the server.

1. Introduction

In implementing industry 4.0, the main supporting factor is the availability of digital infrastructure; one of them is the internet of things (IoT). IoT refers to a network of physical devices, vehicles, household appliances, and other items planted with electronic devices, software, sensors, actuators, and connectivity that make it possible to connect to the internet network and collect and exchange data.

Over time, cars are an indispensable asset to move safely and comfortably. The intensity of the use of private vehicles, primarily cars, as a means of transportation in everyday life is increasing. Monitoring and maintenance are essential to maintain user performance and comfort. Thus, it is necessary to use technology to support this, one of which is the use of Internet of Things (IoT) -based technology.

On the other hand, automotive manufacturers are working to develop unmanned car technology (driverless car), the vehicles that will no longer require drivers because the function has been taken over by artificial intelligent systems (AI). Now one of the technology producers from the United

States, Alphabet Inc's Waymo under the auspices of Google, will soon launch a shutter vehicle service without humans behind the wheel. One obstacle that arises in the testing phase on public roads is the condition of various terrain affected by different weather and environment, including dry, snowy or heavy rain conditions which are sometimes difficult to predict and require real-time and accurate monitoring system. For this reason, an internet of things-based car condition monitoring system is needed to develop the driverless car.

Previous research shows the possibility of these purposes. Expert researchers have proven successful work in the field of wireless networks and smart systems such as the open garden sensor [1], water billing system [2], person locator tracking system [3], integrated smart plug [4], solar radiation monitoring [5], waste transport monitoring [6], pollution monitoring system [7], power meter monitoring [8], and water pollution monitoring system [9]. Srinivasan and others also proposed an IoT cloud-based real-time automobile monitoring system that utilized all possible IoT solution to monitor the vehicle [10].

The purpose of this study is to design and develop hardware module needed for car vehicle condition monitoring system prototype complete with ion bard diagnostic module that sends data to the server in real time.

2. Material and Methods

The system was designed by using the receiving device and data processor for the reading of machine data from OBD devices by adding RTC modules and GPS to get data accurately in time and place, then the data can be stored in the Micro SD memory card and sent to the database server using the GPRS module. The existing data, both in SD memory and database, is then processed using programs to do data display in real time, as well as in-depth analysis to obtain patterns of changes and predictions of future engine performance.

The results of the analysis obtained from existing data are used for reference to predict when the engine needs to be maintained or predict when the engine components must be replaced so that it can prevent damage to the car while traveling. Data that is already in the database will be displayed on a website page built with a programming language. Because data processing is on the server so that it can be accessed using a variety of smart devices such as PCs or smartphones.
For this design to run correctly and directed, the steps taken to make the system are as follows:

2.1. System Design
At this stage, the system model of diagnostic data reading was made from the machine so that the description of the system requirements was designed, in both hardware and software simultaneously.

2.2. Design of hardware modules and data processing
At this stage, the data reader system design and integration of the hardware used in the design are carried out. The design of this device refers to ESP32 as a microcontroller.

2.3. Server and Website Configuration and monitoring
At this stage, the configuration is done by configuring the web server and database server. After the server has been configured, then the system that has been built is stored on the server so that the system can run online and be accessed by users using the internet.

The following figure shows a picture of a block diagram and a series of systems that want to be developed: consisting of an integrated internet system (IoT System), a system for reading data from ECU and a data communication system.

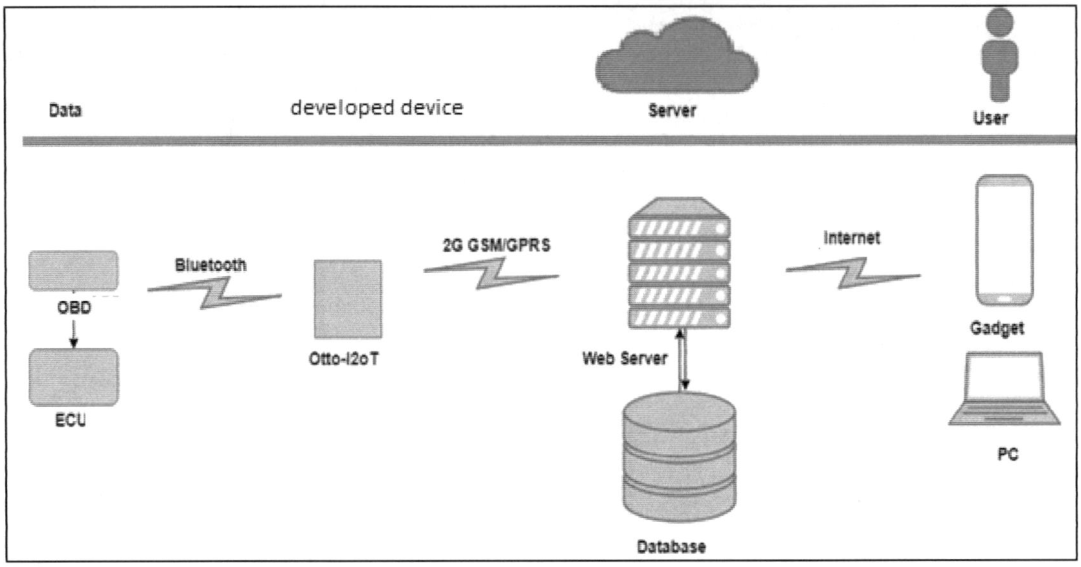

Figure 1. Block Diagram for the vehicle monitoring system

3. Results and Discussion

The hardware module developed consists of an onboard diagnostic unit or known as OBD. OBD is an automotive term that refers to the ability to diagnose a system on a vehicle. The OBD system provides reports on vehicle status to vehicle owners or technicians. Furthermore, a series of low-cost and low-power microcontroller systems will be developed with integrated Wi-Fi and dual Bluetooth modes. Whereas for communication modules, a complete Quad-Band GSM / GPRS module is used, which combines GPS technology for satellite navigation.

3.1 On-Board Diagnostic (OBD)

OBD is an automotive term that refers to the ability to diagnose a system on a vehicle. The OBD system provides reports on vehicle status to vehicle owners or technicians. OBD connected directly to the Electronic Control Unit of the car. The following figure 2 is ELM37 as one type of OBD

3.2 Processor Module

For the processor, ESP32 was used. ESP32 is a series of low-cost and low-power microcontroller systems on chips with integrated Wi-FI and dual Bluetooth mode. The following figure 3 is a type of ESP32

Figure 2. ELM 37

Figure 3. ESP32

3.3 Communication Module

SIM808 is a complete Quad-Band GSM / GPRS module that combines GPS technology for satellite navigation. Together with the external Micro SD for storage and clock timer RTC, it built as a communication module. The following figure 4 is a type of SIM808

Figure 4. SIM 808

3.4 Integrated Design for Hardware Module

The final step of the design stage was to integrate the whole components into a hardware module system as designed. The following figure 5 shows the hardware design and digital communication module that will be developed. And figure 6 shows the prototype design planned for the hardware module.

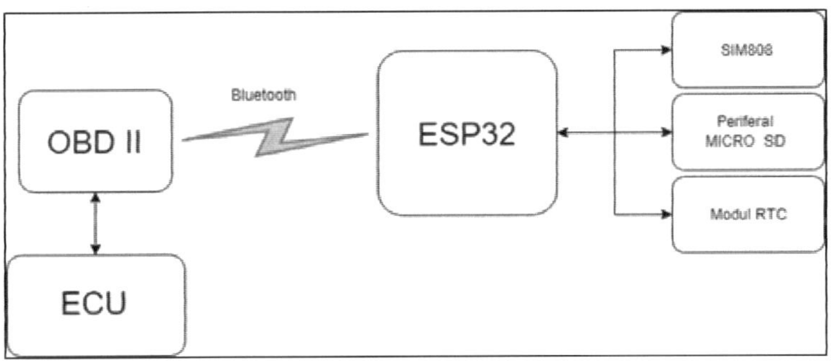

Figure 5. Hardware design with a communication module

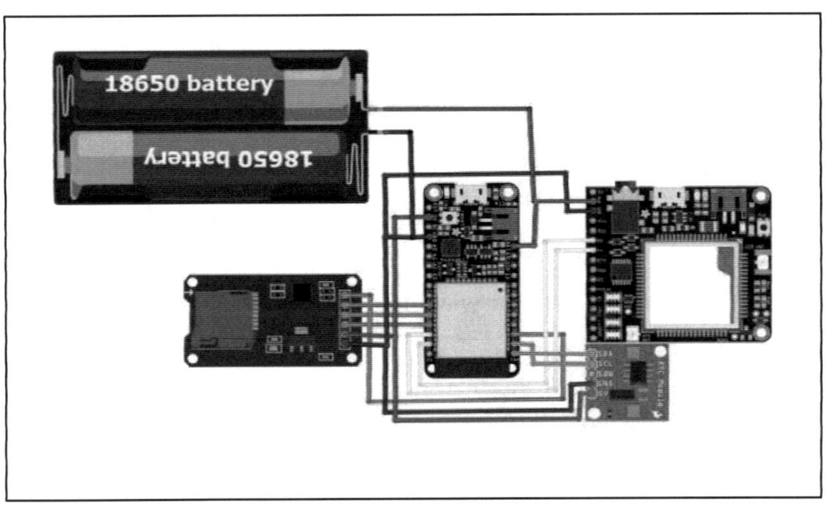

Figure 6. A prototype of the hardware module

4. Conclusions

We have successfully designed a hardware module for vehicle monitoring system based on IoT. The hardware module developed consists of an onboard diagnostic unit or known as OBD. OBD is an automotive term that refers to the ability to diagnose a system on a vehicle. The OBD system provides reports on vehicle status to vehicle owners or technicians. Furthermore, a series of low-cost and low-power microcontroller systems will be developed with integrated Wi-Fi and dual Bluetooth modes. Whereas for communication modules, a complete Quad-Band GSM / GPRS module is used, which combines GPS technology for satellite navigation.

Acknowledgments
Authors wishing to acknowledge the Ministry of Research, Technology and Higher Education (Kemenristekdikti) Government of Indonesia who funded this research through ***2019 Program Penelitian Kolaborasi Indonesia (PPKI)*** scheme.

References

[1] Nasution S Siregar B Kurniawan M Pranoto H Andayani U and Fahmi F 2018 *Application of Open Garden Sensor on Hydroponic Maintenance Management* (Journal of Physics: Conference Series Vol 978) https://doi.org/10.1088/1742-6596/978/1/012091

[2] Fahmi F Hizriadi A Khairani F Andayani U and Siregar B 2018 *Clean water billing monitoring system using flow liquid meter sensor and SMS gateway* (Journal of Physics: Conference Series 978(1) p0–7) https://doi.org/10.1088/1742-6596/978/1/012111

[3] Fahmi F Siregar B Evelyn S Gunawan D and Andayani U 2018 *Person Locator Using GPS Module and GSM Shield Applied for Children Protection* (6th International Conference on Information and Communication Technology (ICoICT) p194–198)

[4] Nasution TH Hanafi M and Tanjung K 2018 *Integrated Smart Plug Design* (ICMSC 2018) https://doi.org/10.1051/matecconf/201822005001

[5] Siregar B Fadli F Andayani U Harahap LA and Fahmi F 2017 *Monitoring of Solar Radiation Intensity using Wireless Sensor Network for Plant Growing* (Journal of Physics: Conference Series 801(1)) https://doi.org/10.1088/1742-6596/801/1/012087

[6] Siregar B Menen K Efendi S Andayani U and Fahmi F 2018 *Monitoring quality standard of waste water using wireless sensor network technology for smart environment* (2017 International Conference on ICT for Smart Society ICISS) https://doi.org/10.1109/ICTSS.2017.8288865

[7] Siregar B Nasution ABA and Fahmi F 2016 *Integrated pollution monitoring system for smart city* (International Conference on ICT for Smart Society ICISS 2016) https://doi.org/10.1109/ICTSS.2016.7792847

[8] Soeharwinto Ariska MFF Siregar B Andyani U and Fahmi F 2018 *Power meter monitoring for home appliances based on mobile data communication* (Proceeding of 2017 International Conference on Smart Cities Automation and Intelligent Computing Systems ICON-SONICS 2017) https://doi.org/10.1109/ICON-SONICS.2017.8267832

[9] Rahmat R F, Athmanathan, Syahputra M F, Lydia M S 2017 Real Time Monitoring system for water pollution in Lake Toba *International Conference on Informatics and Computing, ICIC 2016* pp. 383-388

[10] Srinivasan A 2018 *Iot cloud based real time automobile monitoring system* (3rd IEEE International Conference on Intelligent Transportation Engineering (ICITE) pages 231–235. IEEE) doi: 10.1109/ICITE.2018.8492706M.

The impact of risk and uncertainty on irrigation decision for paddy production in North Sumatera Indonesia

Rulianda P Wibowo[1*], Taufik Rizaldi[1], Sumono[1], Ikhsan Siregar[2]

[1]Faculty of Agriculture, Universitas Sumatera Utara, Medan, Indonesia
[2]Industrial Engineering Department, Faculty of Engineering, Universitas Sumatera Utara, Medan, Indonesia

[*]Email: rulianda_wibowo@usu.ac.id

Abstract. The risk in agricultural profitability is mainly due to risk in crop yield and volatility in crop prices. The risk in agricultural profitability would change the optimal choices of input use, especially water as a main resource for crop growth. Weather variability and limited water resources are main issue in paddy farming. The optimal choices of irrigation intensity are determined by agricultural profitability and farmer's risk preference. Paddy is a water-sensitive crops that require intensive irrigation. The SERF analysis found risk neutral farmer are fond of irrigation intensity at 2.5 cm, but greater risk averse preference would change the optimal choice to irrigation intensity of 1 cm. The increase in risk would affect higher risk averse preference to change the optimal choice from risk neutral behavior.

1. Introduction

The risks in agricultural production are mainly due to weather variability, climate change, and limited irrigation water resources. High volatility on weather creates a big challenge for farmers to succeed the desired crop production. Climate change and high weather variability result in a high risk of failure in crop production. In addition, another risk faced by farmers is the risk of variability in crop prices. The high risk in crop production and crop price variability will affect the profitability of farming. Moreover, Production risk and price risk will govern farmers' decisions on irrigation intensity.

Paddy is the main food crop in Indonesia that requires intensive irrigation. Paddy growth and productivity is sensitive to the availability of irrigation sources during the planting season. One of the characteristics of paddy crop that makes it able to grow in conditions of inundation is because of an aerenchyma channel shaped like a pipe on its root that extends to the tip of the leaf functioning as a provider of oxygen from the dying area [1].

This study focuses on paddy farming and farmer decisions on irrigation intensity influenced by his risk preference. This research was carried out by developing the previous optimality irrigation model, which only focused on production variability, by including price variability factors on farmer irrigation strategy decision [2]. Previous study found the most optimum irrigation intensity is not at the highest rate [2], [3], [4].

The most typical irrigation strategy, that farmers can implement in defiance of weather variability and limited water availability, is to change the intensity of water (intensive irrigation), which is known as deficit irrigation. In the case of limited water resources, farmer also may change irrigation acreage area to compensate the increase on irrigation intensity [3], [5], [6].

This study objectives focuses on risk analysis for optimal irrigation choices affected by profitability and risk preference. Risk analysis is carried out using the Stochastic Efficiency Respect to Function (SERF) method with the assumption that farmers have risk preferences with the characteristics of Constant Absolute Risk Aversion (CARA) criterion. The results of this study are not only be used for the development of risk-based farmers' decision models in Indonesia but can also be used as input for the development of government policies in resource sharing and the establishment of subsidies and price stability programs for agricultural products.

2. Methodology

The optimization on irrigation intensity choices is developed based on utility maximization model. Research takes into account the risk factors due to weather uncertainty and crop price volatility. The paddy farming is presented with a random production function [7]:

$$y = f(w, \mu) \tag{1}$$

where y is production, w is irrigation intensity, and μ is an uncertainty factor caused by weather and price volatility. Bras and Cordova [8] state that the optimal irrigation intensity is influenced by changes in soil moisture that vary randomly affected by weather variability. Farmers choose an optimal irrigation intensity with the aim of maximizing net profit based on random variations in soil moisture [8].

Farmers are considered to be making decisions on resources that are used with the aim of maximizing mean utilities considering his risk preference shown as [7]:

$$maxEU(\pi, \text{p}) = Maxx\ \{EU\ ([\text{h}\ (\mu)\ (f\ (\text{w}, \mu) - \text{x}zi, \text{p})\}, \tag{2}$$

where p is the coefficient that defines the farmer's risk preference , z is the input cost and π is the net profit. h (μ) shows the volatility of paddy price in the market. The risk coefficient (p) does not depend on the initial conditions of net profit (π), presume that farmer has Constant Absolute Risk Aversion (CARA) preferences. Based on the assumption utility mean model corresponds to the exponential negative utility model with SERF analysis as shown below:

$$\text{Max}_{\{w\}} CE(RAC) \tag{3}$$

$$s.t. \tag{4}$$

$$CE(RAC) = \ln(1 - E[u(\pi_i)])^{-1/RAC} \tag{5}$$

$$E[u(\pi_i)] = \int u(\pi_i)\, dF(z, p_i) \tag{6}$$

$$u(\pi_i) = 1 - e^{-r*\pi_i} \tag{7}$$

$$\pi_i = p_i * f_i(\text{w}, \mu) - \text{xzi} \tag{8}$$

SERF analysis is considering the risk and the farmers risk preference. The nature of farmers' preference for risk is represented by Absolute Risk Averse Coefficient (ARAC). The ARAC range is from 0-4, which 0 ARAC states for risk neutral preference and the higher ARAC indicates more risk averse preference. The SERF analysis assumes farmer have a Constant Absolute Risk Aversion (CARA) preference or the risk preferences are not affected by the initial wealth.

The weather uncertainty is exhibited by the difference in soil moisture for land block with different treatment level. The study uses experimental methods with non-factorial Randomized Block Design (RBD) with one treatment into waterlogging consisting of 8 treatment levels and 4 replications. The level of treatment was determined based on the water needs of the paddy crop.

Table 1. Randomized Block Design for Irrigation Treatment

Application	T_1	T_2	T_3	T_4	T_5	T_6	T_7	T_8

Irrigaiton Intensity	–2,5cm =1.16 l/sec/ha	0,0cm =1.18 l/sec/ha	2,5cm =1.20 l/sec/ha	5,0cm =1.22 l/sec/ha	7,5cm =1.24 l/sec/ha	10,0cm =1.26 l/sec/ha	12,5cm =1.28 l/sec/ha	15,0cm =1.30 l/sec/ha

The paddy prices are a producer price that are collected by farmer. The paddy price data is collected from Statistical Bureau Center (BPS) for 2008-2016. Data in table 2 shows the paddy price has an increasing trend between 2008-2016.

Table 2. Paddy Price 2008-2016 (in Rupiah)

Year	Paddy Price
2008	2 811.95
2009	2 987.22
2010	3 547.93
2011	4 046.03
2012	4 463.04
2013	4 592.50
2014	4 766.58
2015	5 303.46
2016	5 455.38

Source: BPS, 2019

3. Result and Discussion

The highest production is found when farmer apply 2.5 cm irrigation intensity (see table 3). Similarly, the highest revenue also found when farmer apply 2.5 cm irrigation. Similar finding on production and revenue because the estimation of revenue uses Statistical Bureau Center (BPS) paddy on-farm estimation standard [9]. The estimation of revenue is assuming the cost of input depends solely on amount of crop production and no economics of scale on production cost.

Table 3. Production (tones), Revenue (Rupiah) and Profit (Rupiah) by Irrigation Intensity

Irigation intensity	Indicator	Min	Max	Mean	Standard deviation
	Production	4.50	9.30	7.32	1.61
0	Revenue	12,653,775	50,735,034	30,885,593	9,539,139
	Profit	(905,225)	37,176,034	17,326,593	9,539,139
	Production	7.70	11.60	9.26	1.49
1	Revenue	21,652,015	63,282,408	39,071,119	10,479,930
	Profit	8,093,015	49,723,408	25,512,119	10,479,930
	Production	7.20	12.30	9.64	1.68
2.5	Revenue	20,246,040	67,101,174	40,674,470	11,243,760
	Profit	6,687,040	53,542,174	27,115,470	11,243,760
	Production	5.30	7.90	6.24	0.94
5	Revenue	14,903,335	43,097,502	26,328,702	6,879,825
	Profit	1,344,335	29,538,502	12,769,702	6,879,825
	Production	4.80	7.70	6.38	0.95
7.5	Revenue	13,497,360	42,006,426	26,919,410	7,007,055
	Profit	(61,640)	28,447,426	13,360,410	7,007,055
	Production	3.50	6.30	5.50	1.05
10	Revenue	9,841,825	34,368,894	23,206,388	6,689,205
	Profit	(3,717,175)	20,809,894	9,647,388	6,689,205
12.5	Production	4.70	9.80	7.28	1.69

	Revenue	13,216,165	53,462,724	30,716,819	9,780,747
	Profit	(342,835)	39,903,724	17,157,819	9,780,747
	Production	6.80	9.10	7.64	0.84
15	Revenue	19,121,260	49,643,958	32,235,783	7,717,685
	Profit	5,562,260	36,084,958	18,676,783	7,717,685

The variation on production is smaller when the irrigation intensity is at 5, 7.5 10 and 15 cm (see table 3). Contrarily, the high variance on production were found at irrigation intensity of 0 cm, 1 cm ad 2.5 cm. Higher irrigation intensity affect smaller variation in paddy production. This evidence proofs water is a risk reducing input for paddy yield. which is come as no surprise because paddy is a water-sensitive crop.

Result is table 3 shows a possible profit loss at irrigation intensity of 0 cm, 7.5 cm, 10 cm and 12.5 cm. Contrarily, irrigation intensity at 1 cm, 2.5 cm and 15 cm never gain any profit loss. The highest minimum profit is found at irrigation intensity of 1 cm. Figure 1 shows a high variation in profit at irrigation intensity of 0 cm, 1, cm, 2.5 cm, and 12,5 cm. Farmer, who has high risk averse preference, might not going to apply irrigation intensity of 0 cm and 12.5 cm because of its potential profit loss and high profit variance. Farmer will experience possible profit loss if the paddy yield is less than 5.5 tones and paddy price is less than Rp. 2,800.

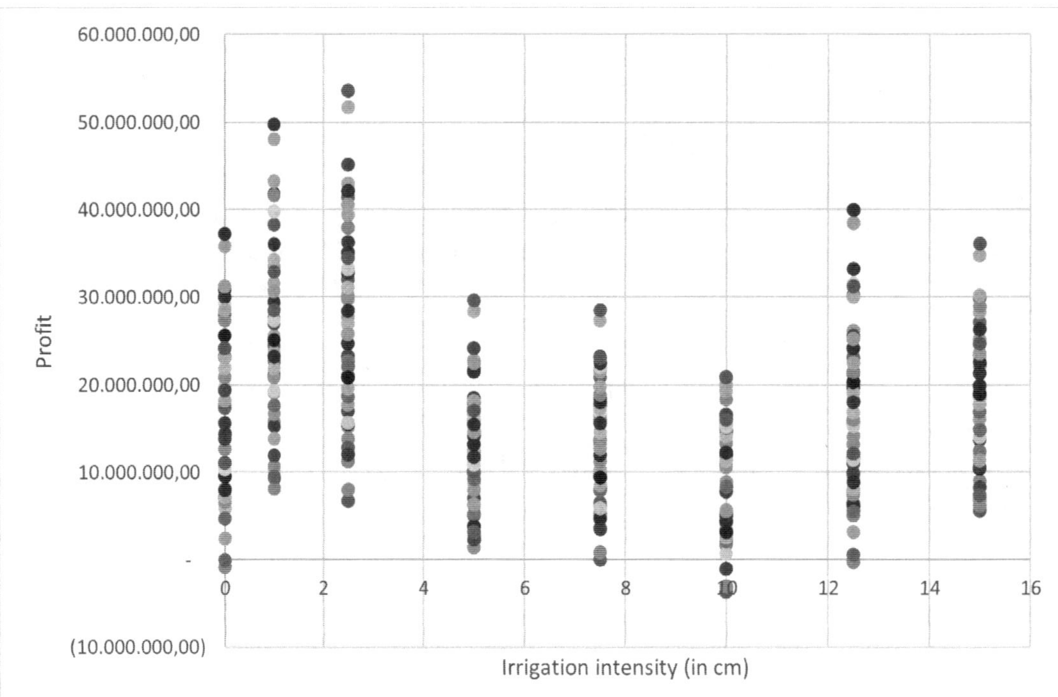

Figure 1. Profit distribution by irrigation intensity

Figure 1 shows an increasing trend of profit when farmer increase irrigation intensity from 0 cm to 2.5 cm. Contrarily, the trend of profit is decreasing when irrigation intensity is above 2.5 cm but below 10 cm. The variation in profit seems decreasing in conjunction with the increase in irrigation intensity. There is a trade-off between raising the profit with the decrease in profit variance.

SERF analysis shows the highest certainty equivalent for risk neutral farmer is at irrigation intensity of 2.5 cm (see Figure 2). This is because the highest average profit is found at irrigation intensity of 2.5 cm. The second-best choice is irrigation intensity at 1 cm and followed by 15 cm. The least irrigation intensity choice is at 10 cm.

The optimal irrigation choice is changing when farmer has risk averse preference. Figure 2 shows a slight change in risk averse preference, i.e. 0.1667, would change the optimal irrigation intensity at 1 cm. A change in irrigation intensity choices is because the smaller profit variation for irrigation intensity of 1 cm compared to 2.5 cm, even though the 2.5 cm irrigation intensity has the highest average profit.

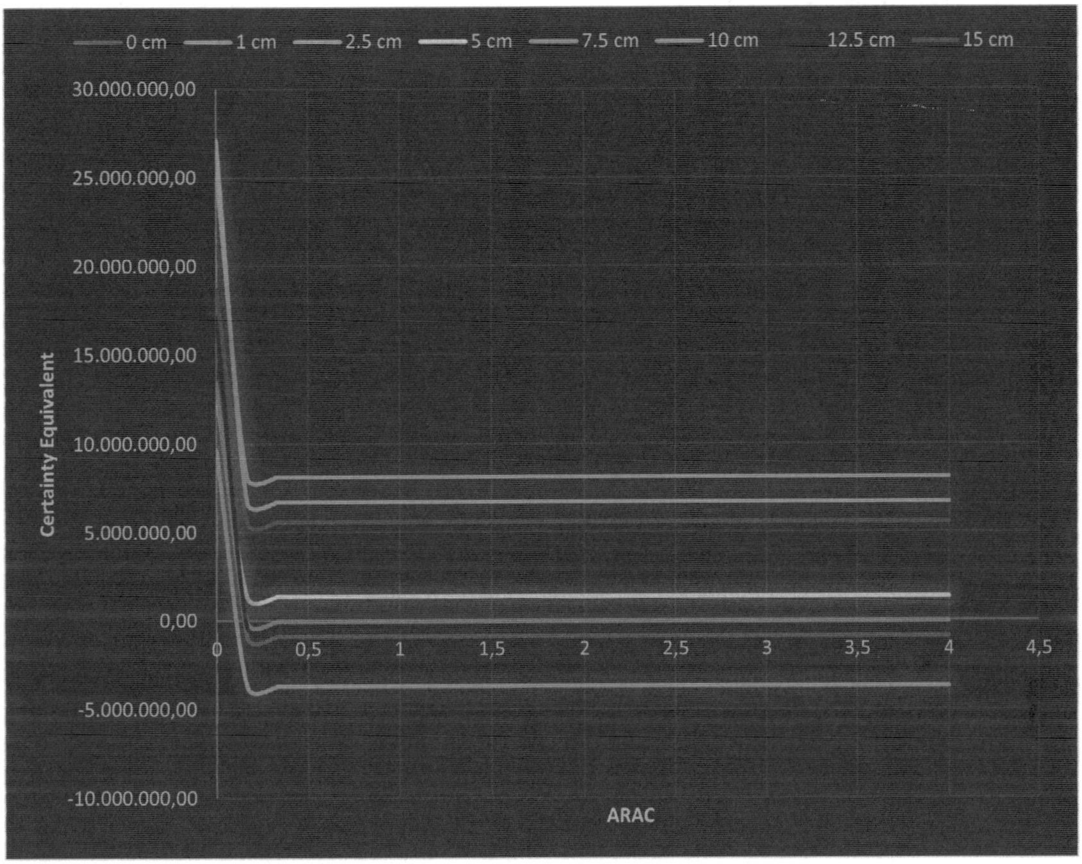

Figure 2. Certainty equivalent estimation with SERF model

The optimal choice of 1 cm irrigation intensity is not changing for farmer with very high risk-averse preference. This finding is similar with Wibowo et al [2] but the rate of change in ARAC that cause changing optimal irrigation choices from risk neutral behavior is not the same. Wibowo et al. [2] found ARAC 0.125 causes optimal irrigation intensity to change from 2.5 cm to 1 cm. Our study found a higher ARAC at 0.1667 to stimulate the changing in irrigation intensity choice. The difference is because our current model includes the price volatility that increase the risk. Thus, the increase in risk would affect higher risk averse preference to change the optimal choice from risk neutral behavior.

4. Conclusion

Limited water resources, high volatility of crop prices and variability in weather are the main issue in agricultural profitability. Climate change and high weather variability result in a high risk of failure in crop production. Risk in crop production and crop prices volatility would change agricultural profitability, thus changes the optimal decision of input use. The optimal decision of input use is not only affected by the profitability but also by risk preference of the farmer. Irrigation water is the main input for agricultural, thus choice of irrigation intensity may change because of the risk in profitability.

The highest production is found when farmer apply 2.5 cm irrigation intensity. The variation on production is smaller when the irrigation intensity is at 5 cm, 7.5cm, 10 cm and 15 cm. Higher irrigation intensity affect smaller variation in crop yield, which proofs water is a risk reducing input. The optimal choice for risk neutral farmer is at irrigation intensity of 2.5 cm. However, as farmer has more risk averse behavior, the optimal choices is changing to irrigation intensity of 1 cm. Our finding is similar with previous study [2] but the rate of change in ARAC that cause the change in optimal irrigation choices is not the same. Our study found higher ARAC that causes the change in optimal irrigation choices. The differences in finding is because our model include the price volatility that increase the risk in crop profitability. Thus, the increase in risk would affect higher risk averse preference to change the optimal choice from risk neutral behavior.

5. References

[1] Purwono and Purnamawati H 2008. Budidaya 8 Jenis Tanaman Pangan Unggul. Penerbit Swadaya, Depok

[2] Wibowo R P, and Rizaldi T 2019 Deficit irrigation for rice farming with production risk due to weather variability. In *IOP Conference Series: Earth and Environmental Science* **260**(1) p. 012021

[3] Wibowo R P, Hendricks N ., Kisekka I and Araya A. 2017 Using a crop simulation model to understand the impact of risk aversion on optimal irrigation management *Transactions of the ASABE* **60**(6), pp 2111-2122.

[4] Upendram, S., Wibowo, R., & Peterson, J. M. 2015 *Irrigation technology upgrade and water savings on the Kansas High Plains aquifer* (No. 1375-2016-109405).

[5] Peterson, J.M., and Y. Ding. 2005 Economic adjustments to groundwater depletion in the high plains: do water-saving irrigation systems save water? *American Journal of Agricultural Economics* **87**:147–159

[6] Foster T, N. Brozović N and Butler A P 2014 Modeling irrigation behavior in groundwater systems." *Water Resources Research* **50**:6370–6389.

[7] Chavas, J 2004. Risk analysis in theory and practice. Elsevier Academic Press.

[8] Bras R L, and Cordova J R 1981. Intraseasonal water allocation in deficit irrigation *Water Resources Research* **17**(4):866–874.

[9] BPS. 2019. https://www.bps.go.id/

Sustainability Assessment of Natural Rubber Primary Processing: A Review

Muhammad Haikal Sitepu[1*], Abdul Rahim Matondang[1], MeilitaTryana Sembiring[1]

[1]Industrial Engineering Department, Faculty of Engineering, Universitas Sumatera Utara, Medan, Indonesia

*Email: mhd_haikalkarana@usu.ac.id

Abstract. Natural Rubber is an important commodity for the world due to its valuable properties such as good elasticity and renewable. Natural rubber is used to produce various important products such as tire, sport equipment, and medical devices. With increasing demand for these products, the demand of natural rubber is projected to jump in the future. To fulfil this demand, the increasing of natural rubber production is unavoidable. However, to use natural rubber in producing those final products, it needs to flow through primary processing such as crumb rubber manufacturing process, crepe rubber processing and latex centrifuge processing. These processes become source of income for countries such as Indonesia, Malaysia and Thailand as well as citizen of those countries. Despite of these positive impacts, the environmental impacts of natural rubber primary processing such as greenhouse gas emission, waste, high water and energy consumption are identified. Thus, to maintain sustainability, monitoring of sustainability impacts from natural rubber primary processing is necessary. Sustainability assessment is a procedure to evaluate sustainability impacts of products, process and policies. This paper aims to review sustainability assessment process applied in Natural Rubber Primary Processing.

1. Introduction

Natural rubber is used to make various important products such as tire, engine components, sport equipment, medical and laboratory devices. An increasing demand of these products force the natural rubber industry to escalate their production. Natural rubber industry consists of various organizations with different activities that configure natural rubber industry supply network. This network consists of several stages including rubber plantation stage, primary natural rubber processing, secondary natural rubber processing stage, customer stage and recycle or disposal stage [1].

Natural rubber primary processing are implemented to change latex from rubber plantations into several work in process products such as crumb rubber, crepe rubber, high concentrated latex and rubber smoke sheet. Different processes are required to make these work in process products. For example, [2] explained the process to make crepe rubber. It started from transforming low concentrated latex from plantation into cubical shape using coagulation process. Then this cubical is rolled and dried to reduce the water contents. At this stage, the cubical form are transformed into blanket form. This is followed by cutting these blanket into standard size required by the suppliers. Figure 1 displays the natural rubber primary processing.

In general, rubber plantations and primary natural rubber processing are located in developing countries such as Indonesia, Thailand Sri Lanka and Malaysia while secondary processing are found in developed countries such as European Union, United States, Japan, South Korean and China. Natural rubber plantations and primary processing have been source of income for citizen in those developing countries. However, each country may have different combination of natural rubber primary processing. For example, Indonesia is dominated by crumb rubber manufacturing and rubber smoke sheet processors [1] while Sri Lanka is dominated by crepe manufacturing and rubber sheet processors [2]. This situation is influenced by the mastership of technology for processing latex by rubber plantations and the structure of industry that absorb the products from natural rubber primary processing. In Indonesia, majority rubber plantations apply coagulation process for their latex to form coagulated latex while in other countries such as Sri Lanka and Malaysia, latex is sold in form of low concentrated latex (fluid) and is not coagulated (see figure 1).

Despite the positive impacts of natural rubber primary processing, these processes have been responsible for some environmental impacts such as air and water pollution as well as high water consumption. With increasing concern for environmental and social impacts in natural rubber industry, stakeholders in this industry formulated sustainability standard. Sustainable Natural Rubber Initiative (SNR-i) is one standard formulated by International Rubber Study Group. One requirement in this standard is to push the primary processors to monitor the use of water, the wastewater and the emission resulting from the processes. Sustainability assessment is a process to evaluate integrated nature-society system in short or long term perspective in order to define which activities should or should not be taken to make society to be more sustainable [3,4]. This process can be used to evaluate sustainability impacts of products, processes and policies. This paper aims to review sustainability assessment processes that focused for assessing natural rubber primary processing.

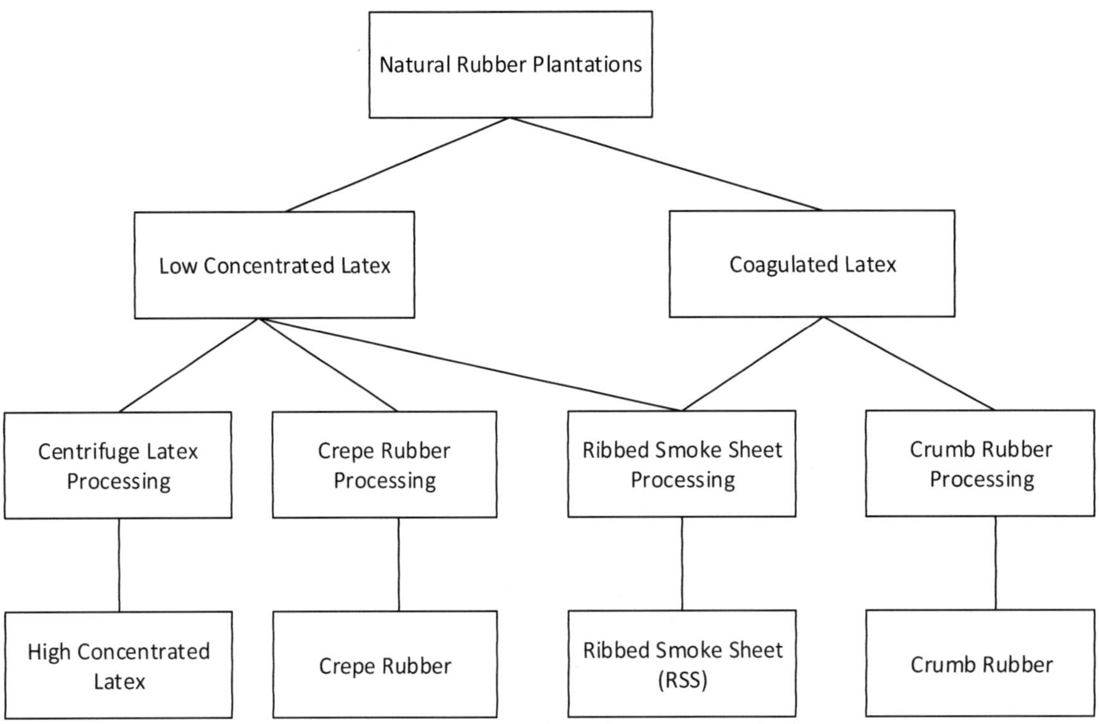

Figure 1. Natural Rubber Primary Processing

2. Research Methods

This research has objectives to review and to analyze sustainable assessment research in natural rubber primary processing and to highlight potential gaps in literature that require further investigation. To achieve these objectives, a review of literature is used as research method. Papers with focus on sustainability assessment in natural rubber primary processing from different Journals and Sources were reviewed.

2.1. Research Process

The research process consists of two main steps. First step is focused on searching and selecting journal papers. The keywords used for searching and selecting the papers were sustainability assessment and natural rubber industry. The searching was implemented on Scopus database. Further restriction related to publish year of paper between 2009 to 2019 was applied in the searching process.

This is followed by analysing those papers using thematic analysis. Four themes were applied including: objects of assessment, tools for assessment, indicators used in the assessment and presentation of assessment results. Object of assessment is first theme that used to present natural rubber primary processes that have been assessed in the literature. Tools for assessment are the second theme that used to map methods, models or tools used in assessment process. Third theme is indicators for assessment, which refers to criteria used in assessment process. Fourth theme is presentation of assessment result, which refers to how the assessment results are presented and which decisions are supported by the results. Figure 2 presents the research process.

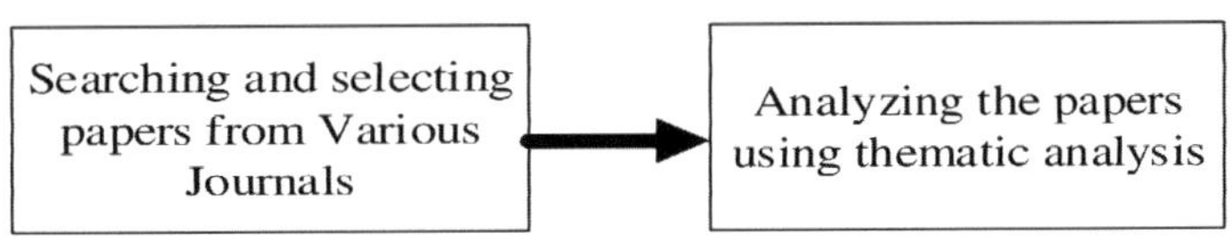

Figure 2. Research Process

3. Result and Discussion

3.1. Statistic of Reviewed Papers

The process of searching and paper selection resulted 11 related papers from various journals including Journal of Cleaner Production, Resources, Conservation and Recycling, Procedia Social and Behavioural Sciences, Kasetsart Journal of Social Sciences, Atmospheric Pollution Research and Environmental Development. From the reviewed papers, it has not been found the paper that focused to review sustainability in natural rubber primary processing, particularly paper with a focus to review sustainability assessment process. This paper fills this gap by reviewing the sustainability assessment process applied in natural rubber primary processing. Figure 1 displays distribution of reviewed papers based on journals.

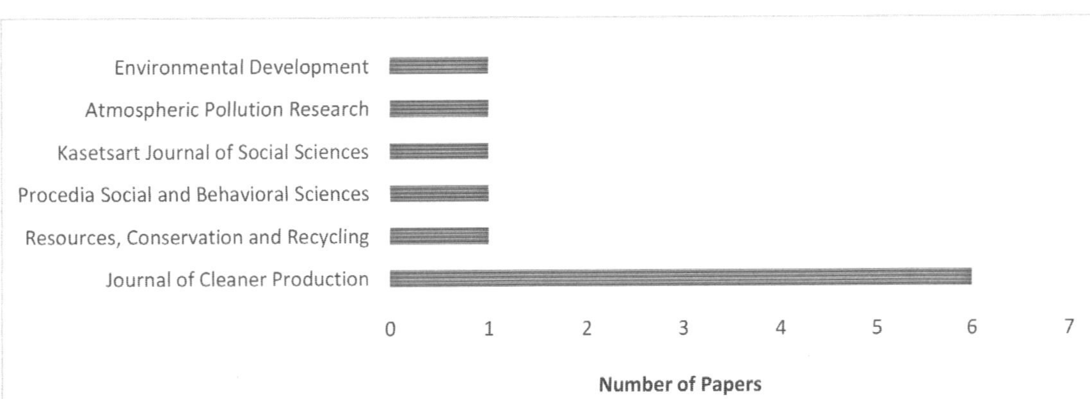

Figure 3. Distribution of Reviewed Papers Based on Journals

3.2. *Objects of Sustainability Assessment*

The main issue in the reviewed papers is to investigate environmental impacts from natural rubber primary processing. [5] investigated the environmental impacts from crepe manufacturing in Sri Lanka. They found that this process consumed high fresh water and released wastewater and greenhouse gas emission to environment. For more detail, water is highly consumed within wet processing including coagulation, maceration, and rolling. The similar result was found by [6] when investigated the environmental impacts of ribbed smoke sheet production in Indonesia. They found that coagulation and milling were the processes consumed the highest water. Furthermore, [7] found the burning (smoke) and milling station were sourced for greenhouse gas emission.

[8] investigated environmental impacts of centrifuge latex process. They found that this process uses chemicals, consumes energy, uses combustion and produces wastewater. These can cause environmental impacts such as global warming, acidification, eutrophication, human toxicity and photochemical oxidation. Furthermore, [9] and [10] compared greenhouse gas emission between centrifuge processor, crumb rubber processor and ribbed smoke processor in Thailand. They found that crumb rubber processor produced highest greenhouse gas emission compared to other processor. This was mainly sourced from the use of electricity and diesel within the process. [11] evaluated eco efficiency in crumb rubber manufacturing in Indonesia. They found that environmental impacts were sourced from the use of acid during coagulation process and the use of plastic on packaging process.

Table 1. Classification of Papers Based on the Object of Sustainability Assessment

Object of Sustainability Assessment	Reviewed Papers
Crepe Rubber Processing	[2] [5][12]
Crumb Rubber Processing	[11][9][10][13][12]
Rubber Smoke Sheet Processing	[6] [7][9][10][12]
Centrifuge Processing	[14][9][10][12]

3.3. *Tools and Indicators for Sustainability Assessment*

Since the focus of reviewed paper is the environmental impacts, material flow analysis and life cycle assessment are the sustainability assessment tools used by the most reviewed papers. Material flow analysis is good to identify the flow of materials from first process until the final process. This tool can be used to identify the consumption of latex, water, chemicals and the production of emission from each process. [5,6] used material flow analysis to investigate environmental impacts of crepe manufacturing and ribbed smoke sheet. Life cycle analysis was used by [8] to evaluate centrifuge latex process, by [11] to assess crumb rubber processing and by [7] to evaluate ribbed smoke sheet processing. Other approaches were demonstrated by [10] that used ecological footprint, by [12] that used mathematical models to predict greenhouse gas emission from activities in natural rubber primary processing and by [5] that used material flow cost analysis to evaluate the flow of cost within crepe rubber processing.

In term of indicators, majority of reviewed papers used environmental indicators including water consumption, energy consumption, chemicals use, total wastewater, greenhouse gas emission. Only one reviewed paper [5] was found to use economic indicators including cost of energy and materials. There is no social indicator has been used in assessment process in reviewed papers.

3.3.1. *Presentation of Sustainability Assessment Result*

Since majority of reviewed papers used material flow analysis and life cycle assessment to evaluate sustainability, the result of assessments were presented using measurement units. For example, [5] presented greenhouse gas emission using Kg e and wastewater using Metric Ton (MT). Furthermore, [10] displayed their result in form of index that is calculated using mathematical equation. For example, index for greenhouse gas emission was calculated by multiplying total greenhouse gas emission with area required to sequestering a ton CO_2 and equivalence factor of forest.

In term of supporting decisions, some reviewed papers demonstrated the use of assessment result to define the alternative for reducing the environmental impacts. For example, [5] used the assessment result to determine the options to reduce water, chemicals and energy consumption for crepe manufacturing. In similar way, [8] used the assessment result to determine the options for improvement in centrifuge latex processing. Furthermore, [6] demonstrated the use of assessment to define productivity improvement in ribbed smoke sheet processing.

4. Possibility for Future Research and Conclusion

4.1. *Possibility for Future Research*

Based on the analysis of reviewed papers several possibility for future research are identified:

- Due to limited research in reviewed papers that focused on social and economic impacts, further research are required to assess social and economic impacts from the natural rubber primary processing in developing countries.
- The opportunities is widely open to use different sustainability assessment tools such as multi criteria indicators, simulation and composite indicators.
- There is limited research to evaluate crumb rubber processing. Hence, further research is required to assess crumb rubber processing
- The use of assessment result to support decisions making are limited, hence there are opportunities for demonstrating the development of policies based on the sustainability assessment result.

4.2. *Conclusion*

Based on the results, some conclusions are obtained as follows.

- The review indicates that the dominant environmental impacts from natural rubber primary processing are high water consumption, greenhouse gas emission and wastewater.
- The review indicates that majority of reviewed papers are focused to assess environmental impacts of natural rubber primary processing. The evaluation of social and economic impacts are rare in reviewed papers.
- The review indicates that material flow analysis and life cycle assessment are dominant tools used in review papers to assess sustainability.

References

[1] Sitepu M H, McKay A and Holt R J 2016 Towards a Framework for Sustainable Development Planning in the Indonesian Natural Rubber Industry Supply Network *Procedia CIRP* **48** 164–9
[2] Dunuwila P, Rodrigo V H L and Goto N 2018 Sustainability of natural rubber processing can be improved: A case study with crepe rubber manufacturing in Sri Lanka *Resour. Conserv. Recycl.* **133** 417–27
[3] Ness B, Urbel-Piirsalu E, Anderberg S and Olsson L 2006 Categorising tools for sustainability assessment
[4] Sala S, Ciuffo B and Nijkamp P 2015 A systemic framework for sustainability assessment *Ecol. Econ.* **119** 314–25
[5] Dunuwila P, Rodrigo V H L and Goto N 2018 Financial and environmental sustainability in manufacturing of crepe rubber in terms of material flow analysis, material flow cost accounting and life cycle assessment *J. Clean. Prod.* **182** 587–99

[6] Marimin, Darmawan M A, Machfud, Islam Fajar Putra M P and Wiguna B 2014 Value chain analysis for green productivity improvement in the natural rubber supply chain: a case study *J. Clean. Prod.* **85** 201–11
[7] Phairuang W, Tekasakul P, Hata M, Tekasakul S, Chomanee J, Otani Y and Furuuchi M 2019 Estimation of air pollution from ribbed smoked sheet rubber in Thailand exports to Japan as a pre-product of tires *Atmos. Pollut. Res.* **10** 642–50
[8] Jawjit W, Pavasant P and Kroeze C 2015 Evaluating environmental performance of concentrated latex production in Thailand *J. Clean. Prod.* **98** 84–91
[9] Jawjit W, Kroeze C and Rattanapan S 2010 Greenhouse gas emissions from rubber industry in Thailand *J. Clean. Prod.* **18** 403–11
[10] Musikavong C and Gheewala S H 2017 Ecological footprint assessment towards eco-efficient oil palm and rubber plantations in Thailand *J. Clean. Prod.* **140** 581–9
[11] Maulina S, Sulaiman N M N and Mahmood N Z 2015 Enhancement of Eco-Efficiency Through Life Cycle Assessment in Crumb Rubber Processing *Procedia - Soc. Behav. Sci.* **195** 2475–84
[12] Vidanagama J and Lokupitiya E 2018 Energy usage and greenhouse gas emissions associated with tea and rubber manufacturing processes in Sri Lanka *Environ. Dev.* **26** 43–54
[13] Sharib S and Halog A 2017 Enhancing value chains by applying industrial symbiosis concept to the Rubber City in Kedah, Malaysia *J. Clean. Prod.* **141** 1095–108
[14] Jawjit W, Pavasant P and Kroeze C 2015 Evaluating environmental performance of concentrated latex production in Thailand *J. Clean. Prod.* **98** 84–91

IOP Conf. Series: Materials Science and Engineering **648** (2019) 012042 doi:10.1088/1757-899X/648/1/012042

Modeling of Multilayer MMIC Coplanar Waveguide Low Pass Filter for Cellular Application

Cakra D. Sedayu, Emerson P. Sinulingga*, and Syafruddin Hasan

Department of Electrical Engineering, Universitas Sumatera Utara, Medan, Indonesia

*Email : emerson.sinulingga@usu.ac.id

Abstract. In the early development of Monolithic Microwave Integrated Circuit (MMIC) technology, most of the components use Microstrip structure. In order to connect the active components to the ground in the Microstrip design is limited by the thickness of the substrate to avoid circuit performance degradation due to vertical coupling. Thus it can be said that the Microstrip is not the primary choice of design in operating components at microwave and millimeter wave frequencies. The Coplanar Waveguide (CPW) design technique was proposed as an alternative to the Microstrip structure for MMIC components. With CPW design technique, it generates components and circuits which are cost-effective and allow miniaturization of components in MMIC environment. In this paper, CPW Low Pass Filter (LPF) electromagnetic modeling has been demonstrated based on multilayer MMIC technology for the 2,100 MHz mobile application. The LPF CPW specification has been modeled with the substrate of GaAS (ε_r=12.9) and Polymide (ε_r=3.7). The thickness of GaAS substrate and Polyimide dielectric is 600 micron and 2.5 micron respectively for each layer.

1. Introduction

The Monolithic Microwave Integrated Circuit (MMIC) technology is widely used in modern telecommunications devices because the advantages it possessed, namely having small dimensions, can be reproduced on a large scale and work at frequencies of 300 MHz to 300 GHz. One component in the RF front-end diagram of the cellular telecommunications system of 2,100 MHz is the Low Pass Filter. This component serves to negate frequencies above 2.1 GHz [1 - 3]. One type of design technique for LPF is an elliptical filter or Cauer filter. This filter has a faster gain transition between the pass band and stop band so that it gets faster cut-off frequency. Research on LPF components in MMIC technology has been carried out to obtain optimal performance and increase reliability. However it was not done using elliptic filter design techniques [4] - [8]. This study discusses elliptical low pass filter modeling using MMW CPW nultilayer technology.

2. Multilayer CPW MMIC Low Pass Filter

Low pass filter is a filter that only passes signals with a frequency lower than the cut-off frequency and weakens the signal with a frequency higher than the cut-off frequency. Cut-off frequency (Fc) is the frequency at which the response filter leaves a band error (or - 3 dB points from the peak) [9]. In addition to the response from insertion loss that determines the cut-off frequency of a filter, it is necessary to look at the return loss response that determines the working frequency of a filter. A device can work well when it has Voltage Standing Wave Ratio (VSWR) below 2. Therefore the return loss of a device must have a minimum response of -10 dB at the desired work frequency [10]. In this research, the component that has been designed has specifications as shown in Table 1. These specifications are based on standardization from previous studies [10].

Table 1. Design Specification

Parameter	Specification
Operating Frequency	2.1 GHz with VSWR < 2
Cut-off Frequency	≥ 2.2 Ghz
Number of Layers	3 Layers (1 Substrate and 2 Dielectrics)
Substrate and Layers' Material	GaAS and Polymide
Dielectric Constants	GaAs (ε_r = 12,9) and Polymide (ε_r = 3,7)
Substrate and Layer Thickness	GaAS (600 microns), Polyimide (2.5 microns)
Conductor Thickness	0.8 microns
Return Loss (S_{21})	< -10 dB

2.1. *Schematic of Elliptic Planar Low Filter*

Designing schematic elliptical LPF is done using ADS software in the Design Guide Filter menu. In the Design Guide Filter menu the 2.4 GHz cut-off frequency is selected with the 6 GHz stop frequency, and the attenuation of 30 dB. This application only has a cut-off frequency setting, but not an operating frequency. Therefore tuning was required so that the LPF can work at a 2.1 GHz frequency. Figure 1 shows the results of the schematic elliptic LPF after the tuning.

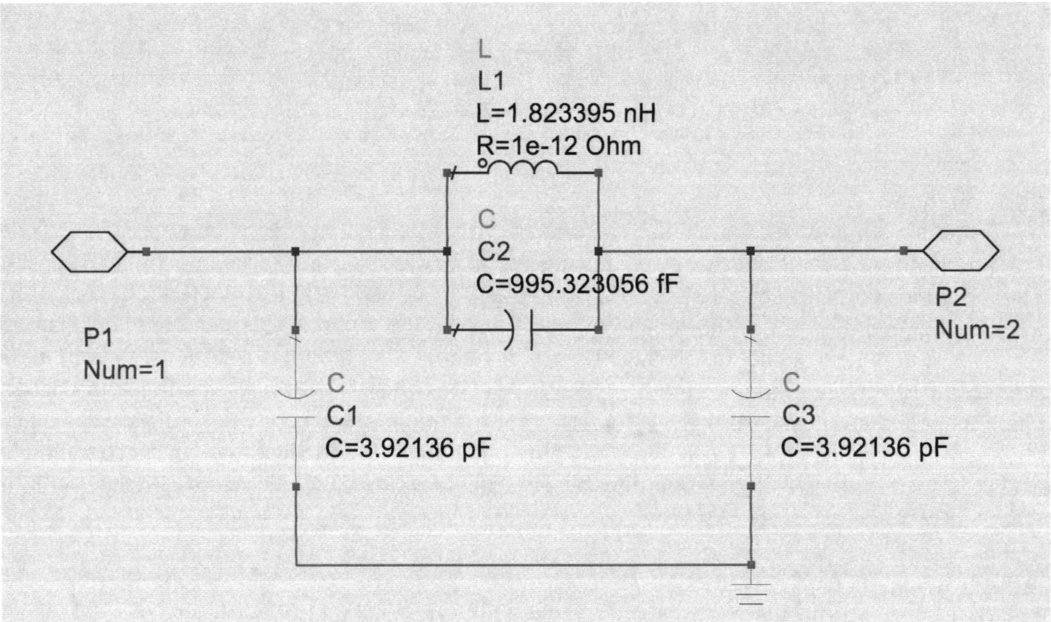

Figure 1. Schematic Circuit of Elliptic Low Pass Filter

2.2. *Coplanar Waveguide MMIC Low Pass Filter*

Based on the schematic design, there are four main components in an elliptic Coplanar Waveguide (CPW) LPF, namely three capacitors and one inductor. All components are transformed accordingly into MMIC LPF component as shown in Figure 2 and tuning is carried out so that the elliptic LPF component can work well at a frequency of 2.1 GHz and has a cut-off frequency above 2.2 GHz.

Figure 2 shows a layout design of elliptic CPW MMIC LPF. This component has a size of 1188x1029 μm^2.

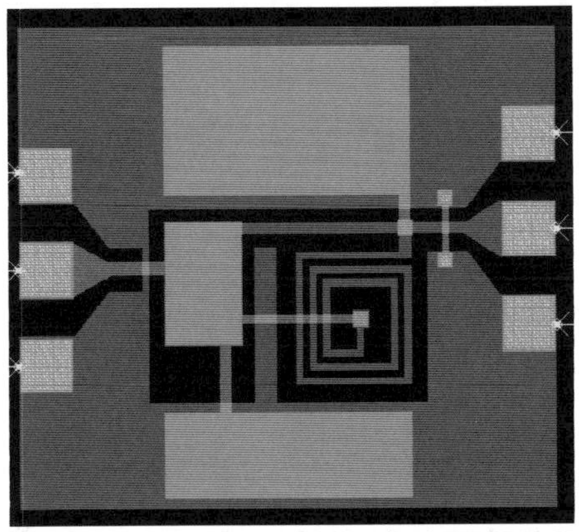

Figure 2. Layout Design of elliptic CPW MMIC LPF

2.3. Elliptic Multilayer CPW MMIC LPF
By using the same circuit pattern in the momentum component of the planar MMIC LPF, the transformation of each component in the multilayer CPW form is carried out. Tuning is also performed so that elliptical components of multilayer LPF can work well at a frequency of 2.1 GHz with a cut-off frequency above 2.2 GHz. Figure 3 shows a layout design of elliptic multilayer MMIC LPF with size of 973x721 μm^2.

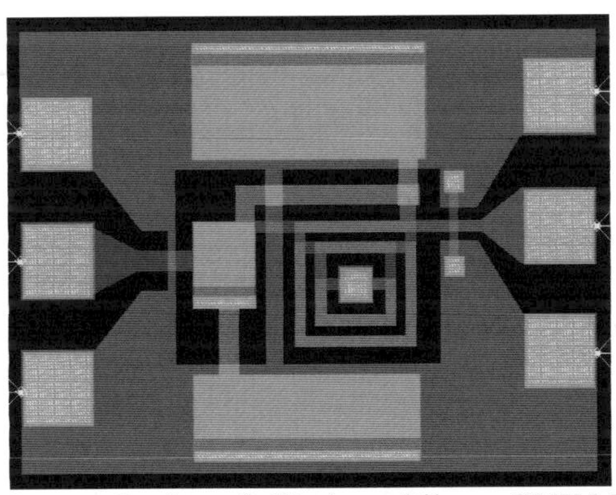

Figure 3. Layout Design of elliptic multilayer CPW MMIC LPF

The simulation process is carried out using stages of optimization based on 3D momentum references [11]. In this work steps are taken so that the simulation results have a better response and can approach the results of the measurements.

3. Results
Figure 4 shows a comparison of return loss and insertion loss of elliptical components of multilayer and planar MMIC LPFs respectively.

IOP Conf. Series: Materials Science and Engineering **648** (2019) 012042 doi:10.1088/1757-899X/648/1/012042

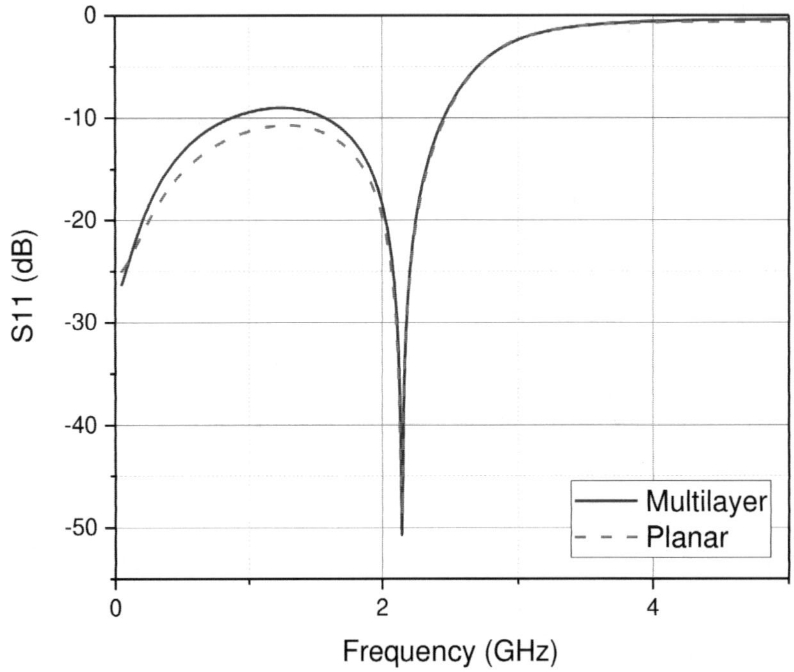

Figure 4. Optimized S_{11} Simulation Responses of multilayer dan planar CPW MMIC LPFs.

Elliptic planar MMIC LPF with optimal settings has a size of 1188x853μm^2 and elliptic MMIC multilayer LPF has a size of 973x695μm^2. It can be seen in Figure 4 that the return loss responses of planar and multilayer produces almost the same trend. Both graphs show frequency response that can work well at 2.1 GHz because it has a return loss value below -20 dB. This shows that with multilayer techniques can produce the same response with smaller component dimensions. The dimension size decreases by 49.85%. In addition, it can be seen also in Figure 4 that the return loss value decreased by 3.57% from -48.95 dB on planar LPF and -50.70 dB in multilayer LPF. In addition there was also a decline in the stop band by 43.02% from -21.589dB to -30.877dB. This proves that using multilayer techniques can produce a better response.

4. Conclusions

The use of multilayer techniques in LPFs results in smaller components with size reduction of 49.85% compared to planar LPF. This technique also produces a better return loss value of 3.57%, with a return loss of -48.95 dB on planar LPF and -50.70 dB in multilayer LPF. The simulation results of LPF components using multilayer techniques have been demonstrated with operating frequency of 2.1 GHz frequency and can be applied to cellular applications..

References

[1] Marsh S, 2006 Practical MMIC Design, *Artech House.*
[2] Pucel R A 1981 Design Considerations for Monolithic Microwave Circuits, *IEEE Transactions On Microwave Theory and Techniques,* vol. 29, pp.513-534.
[3] Robertson I D and Lucyszyn S 2001 RFIC and MMIC Design and Technology, *The Institution of Electrical Engineers*

[4] Budimir D, Robertson I D, Khalid A H and Rezazadeh A A 1997 Multilayer CPW lowpass filters on GaAs/Si MMICs *IEEE MTT/ED/AP/LEO Societies Joint Chapter United Kingdom and Republic of Ireland Section. 1997 Workshop on High Performance Electron Devices for Microwave and Optoelectronic Applications. EDMO* pp. 315-320.
[5] Fu K and Budimir D 1999 Electromagnetic modelling and design of multilayer coplanar waveguide lowpass filters *Symposium on High Performance Electron Devices for Microwave and Optoelectronic Applications. EDMO (Cat. No.99TH8401)*, London, UK, 1999, pp. 221-229.
[6] Alburaikan A 2016 Metamaterial Structure Inspired Miniature, Ph.D thesis, The University of Manchester, London
[7] Sulaiman A A and Awang Z 2004 On the optimization of the number of elements of MMIC low pass filter using EM simulation *2004 RF and Microwave Conference (IEEE Cat. No.04EX924)*, Selangor, Malaysia, 2004, pp. 56-59.
[8] Rao M S, Mahalakshmi I and Mourougayane K 2015 Elliptical function based differential band pass filter for IF signal processing and EMI applications *13th International Conference on Electromagnetic Interference and Compatibility (INCEMIC)*, pp. 138-141.
[9] Bowick C 1997 RF Circuit Design. Second Edition, United States.
[10] Chang P H, Chiang K, Lai J Y and Wang Y P 2009 Advanced integrated passive device (IPD) low pass filter designs on WLCSP *4th International Microsystems, Packaging, Assembly and Circuits Technology Conference*, pp. 132-135.
[11] Emerson P S, Rezazadeh A Aand Peter B K,2015 Modeling Technique and Measurements of CPW Compact GaAS Multilayer MMICs *IEEE.*

IOP Conf. Series: Materials Science and Engineering **648** (2019) 012043 doi:10.1088/1757-899X/648/1/012043

Smart Nation Initiative: Strategy and Implementation

F Fahmi[1,4*], H Pranoto[2,4], MS Lubis[3,4]

[1]Department of Electrical Engineering, Faculty of Engineering, Universitas Sumatera Utara, Indonesia
[2]Polytechnic states of Medan, Indonesia
[3]Department of Information Technology, Universitas Sumatera Utara, Indonesia
[4]APIC Sumatera Regional, Indonesia

*Correspondence Email: fahmimn@usu.ac.id

Abstract. Population growth due to urbanization raises a variety of typical urban problems such as decreasing the quality of public services, reducing the availability of residential land, congestion on highways, swelling levels of energy consumption, garbage accumulation, increasing crime rates, and other social problems. These problems are not facing only in cities, but also in regencies, provincial and even at the national level. APIC was formed to answer the need for an organization or collaboration that brings together various stakeholders to solve these problems. The collaboration of various stakeholders in one forum that can synergize all potentials by involving all potential to overcome the problems in a smart way. APIC views the need for an ecosystem that consists of human resources, governance, and adequate technology, which is needed to create smart villages, smart cities, smart regencies, smart provinces and at the end: smart nation. Implementing a smart solution starts with solving a small-distributed problem, but it is carried out appropriately and supported by all regional stakeholders thoroughly through synergy. These solutions are implemented in a strategic, gradual, and continuous manner. One of strategic way proposed was The APIC Innovation & Marketplace Hub that will open Supply & Demand of various Smart solutions (cities, public service providers, companies, etc.) that are tested and validated to be implemented. It required sufficient ecosystem to support the hub that is now undergoing and monitored through the Technology Readiness Level (TRL) to show the status of developing smart city.

1. Introduction

The rate of urbanization in Indonesia increases trough years very rapidly. It is estimated that by the end of 2035 (15 years from now) more than 66.6% of the population will inhabit big cities in Indonesia, in some city it is even getting higher up to 85% (BPS 2010).

Table 1. Urbanization trend

Province	2010	2015	2020	2025	2030	2035
Banten	67.0	67.7	69.9	73.7	78.8	84.9
Jawa Timur	47.6	51.1	54.7	58.6	62.6	66.7
Jawa Barat	65.7	72.9	78.7	83.1	86.6	89.3
Jawa Tengah	45.7	48.4	51.3	54.3	57.5	60.8
Bali	60.2	65.5	70.2	74.3	77.8	81.2
NTB	41.9	45.4	49.4	53.6	58.1	62.7
INDONESIA	**49.8**	**53.3**	**56.7**	**60.0**	**63.4**	**66.6**

IOP Conf. Series: Materials Science and Engineering **648** (2019) 012043 doi:10.1088/1757-899X/648/1/012043

Previous research showed city population growth due to urbanization raises a variety of typical urban problems[1] such as decreasing the quality of public services, reducing the availability of residential land, congestion on highways, swelling levels of energy consumption, garbage accumulation, increasing crime rates, and other social problems as depicted in figure 1.

Figure 1. Problems in the city[1]

These problems are not facing only in cities, but also in regencies, provincial and even at the national level. The big problems now are how to make the Village/Regency remain attractive to the community so as to prevent excessive urbanization. And also how can the potential in the Village/Regency can be maximized for the benefit of its citizens. In this way, the smart nation can be achieved and implemented.

The purpose of this study is to propose a strategic implementation to initiate the Indonesia smart nation, covering not only the smart city but also smart villages, smart regencies, and smart provinces. It is expected that the result of this study can be used to accelerate the initiative to build a smart nation.

2. Material and Methods

2.1 APIC (Association of Smart Indonesia Initiatives / Assosiasi Prakarsa Indonesia Cerdas)
The Association of Smart Indonesia Initiatives (APIC) was born from the embryo of the Smart Indonesia Initiatives Forum. This community holds the 2nd Smart City Forum on July 20-21 2016 in Surabaya. Located in the City of Heroes, 21 July 2016, APIC declared [2]. APIC was formed to answer the need for an organization or collaboration that brings together various stakeholders of this nation. It is encouraging the collaboration of various stakeholders in one forum that can synergize all potentials by involving cities in Indonesia. APIC views the need for an ecosystem that consists of human resources, governance, and adequate technology, which is needed to create smart villages, districts, cities, and provinces. Implementing a smart solution starts by solving a small problem, but it is carried out appropriately and supported by all regional stakeholders. These solutions are implemented in a strategic, gradual, and continuous manner.

The vision of APIC is to be an organization that initiates and realizes the development of Smart Indonesia. With some missions, e.g., facilitating stakeholder collaboration in the exchange of experience and knowledge to educate Indonesian life, facilitating the creation of smart solutions that are integrated into realizing smart Indonesia, accelerate the creative development process of Indonesia through technological, governance and human synergy[2].

2.2 Strategy and Implementation

In order to implement a smart nation, there are several strategies developed. First is to develop model recommendations and regulations on Smart Indonesia governance, including its components, such as; City, Regency, Village, Province, and National scale, through the related compartments. Garuda Smart Model (GSM) was proposed in order to standardize the platform used for smart city development[3].

The second strategy is together with relevant partners, to initiate and to assist the development of Smart Indonesia, within the scope and possible components, from many aspect and compartments. It includes smart healthcare, smart citizen, Smart safety security and mitigation, smart tourism and culture, smart regulation and law, smart Infrastructure, smart energy and environment, smart industry and economy, smart transport mobility and logistics, smart model standard architecture and technology[4].

Next strategy, the third strategy is to develop standardization certification platform recommendations and related components of Smart Indonesia. The latest version of Garuda Smart Model was issued in 2018 cover latest technology, including big data and artificial intelligence [5].

Strategy number 4 is to help the process of increasing the capacity of human resources from the users, developers, servants to the government.Human resources always are the primary concern in developing the smart city and therefore, should take emphasized in the implementation [6].

Last, it is essential to increase the component of the Domestic Content Level of the Indonesia Smart ecosystem as the fifth strategy. The development of smart nation should bring positive impact and growth of the industry in the country itself. Some solution provided with high domestic content can be found, for example, in the environment [7], energy used [8], wastewater [9,10], planting [11], water pollution [12] and also for river height and speed monitoring system[14].

3. Results and Discussion

3.1 APIC Innovation & Marketplace Hub

Innovation is the process of translating ideas or inventions into products that create the value or solution that customers will need. The idea of innovation results must be replicated effectively and efficiently and must meet individual needs. In business, innovation often occurs when ideas are implemented by companies to meet customer needs and expectations better.

The marketplace as a term is a location for product transactions where sellers and buyers meet somewhere. Seller will sell the product in a place that has been provided by the Provider with the marketplace concept. The products (platforms, frameworks, models, applications) sold in the marketplace will then be advertised by the Provider to get potential buyers. A large number of sellers and buyers who join the marketplace and transactions that occur between sellers and buyers determines the success of the provider with the marketplace concept [13].

There are gaps between the two sides. Problems faced by governments, public service providers, and communities are less information about what solutions are available and how to choose the right solution. In the other hand the industry lack of information about what necessarily needed by the government to solve their problems. This situation is well described through figure 2 below.

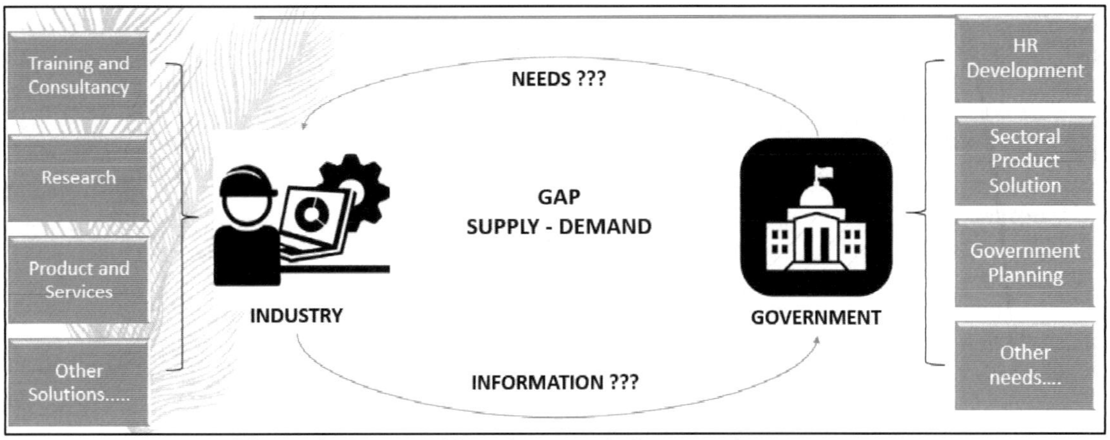

Figure 2. Gaps problems between industry and government

APIC Innovation & Marketplace Hub provides Supply & Demand Smart solutions (cities, public service providers, companies, etc.) that are utilized, tested, and validated to be implemented. APIC-Marketplace manages all stakeholders to be able to connect, exchange experiences, and show & provide the best solutions.

The marketplace required supports from all stakeholders : main and other stakeholders to create the marketplace ecosystem, as shown in figure 3. Main stakeholders include Cities, Public Service Providers, communities are key stakeholders to receive smart solutions, and meanwhile, the business people are the main suppliers of smart solutions. Other stakeholders include Consultant to identify needs and implement smart solutions to problems, Solution providers to Business people who provide smart solutions, Living Labs to environmental inspection / testing environment for smart solutions, Testing & Research institutes to independent institutions that validate smart solutions, Institution of standardization to an institution that sets standards for smart solutions, Developer Investor to invest in developing Smart City solutions and other smart regions, Foundation to Parties who contribute to the development of smart solution and Government, Company, Community to Parties who need smart solutions.

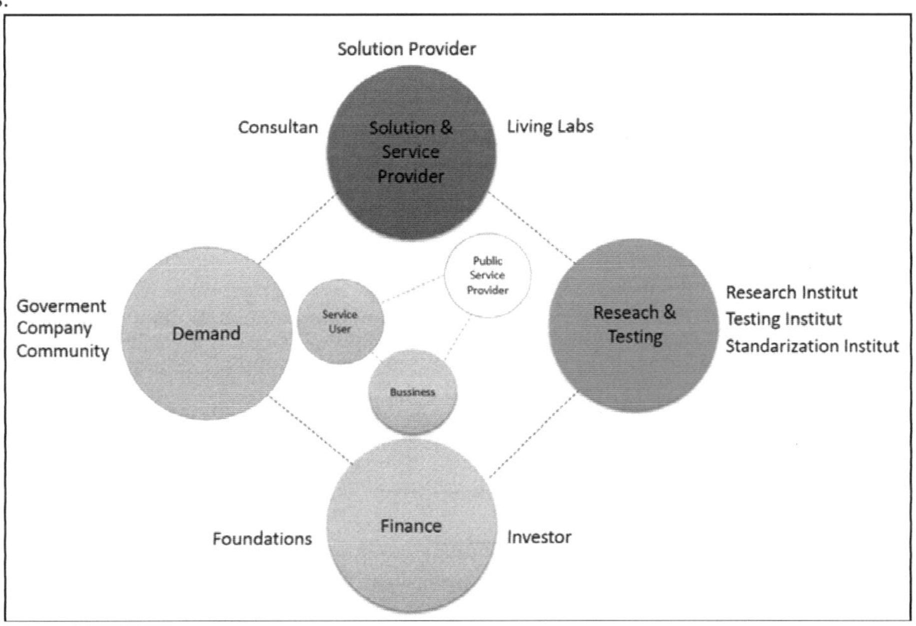

Figure 3. Marketplace ecosystem

APIC-Marketplace offers stakeholders a unique online and offline experience by bringing together the right stakeholders through an event and through the APIC Innovation Marketplace Hub for solutions that can be implemented. APIC Marketplace organizes the Offline APIC Marketplace at a specific time to connect and take advantage of experience and best practice solutions. Business side displays and provides their smart solutions that are independently tested and validated at APIC-Marketplace events. APIC-Marketplace adds value with Living Labs where Business side test, pioneer, and validate their business solutions.

3.2 Smart City Technology Readiness Level (TRL)

APIC-Marketplace uses the Technology Readiness Level (TRL) to show the status of developing Smart City solutions

The following Technology Readiness (TRL) levels are determined:

a. TRL 1 - basic principles observed
b. TRL 2 - a formulated technology concept
c. TRL 3 - experimental proof of concept
d. TRL 4 - technology validated in the lab – Prototype type A
e. TRL 5 - technology validated in the relevant environment (industry environment relevant in terms of possible technology)
f. TRL 6 - technology is shown in the relevant environment (industry environment relevant in terms of possible technology)
g. TRL 7 - demonstration of system prototypes in an operational environment type B
h. TRL 8 - complete and quality system
i. TRL 9 - a system that is actually proven in an operational environment (competitive manufacturing in terms of possible technology)

4. Conclusion

The existence of Smart Solution Marketplace is expected to be able to help the government, public service providers, communities to realize smart cities in Indonesia more effectively and productively while continuing to support the growth of the industries involved in it. Promotion of the work of APIC in the form of ideas or in the form of products that are ready to be implemented. In addition, a synergic relationship between all stakeholders is expected to be established so that the ideals of Indonesia Smart Nation can be achieved. Garuda Smart Model / Smart Nation Framework can be used to accelerate and streamline the progress of Indonesia smart nation.

Acknowledgments

Authors wishing to acknowledge to Mr. Suhono Supangkat as head of APIC Indonesia who provided data and information needed for this paper.

References

[1] Suhono S 2015 *Smart City Development in Indonesia and Asian-Africa Nations* (IEICE Global Plaza No.78) https://www.ieice.org/eng/activities/ieice_global_plaza/2015/no78.pdf
[2] APIC Indonesia http://www.apic.id/
[3] Tay KCK Supangkat SH Cornelius G and Arman AA 2018 *The SMART Initiative and the Garuda Smart City Framework for the Development of Smart Cities* (2018 International Conference on ICT for Smart Society (ICISS) pp 1-10) IEEE
[4] Chichernea V 2015 *Smart Cities Communities And Smart ICT Platform* (Journal of Information Systems & Operations Management 9(1))
[5] Sundari D Wang G Kaburuan ER and Fajar AN 2018 *Developing Garuda Smart City Model With Big Data Framework* (2018 Indonesian Association for Pattern Recognition International Conference (INAPR) pp 1-5) IEEE
[6] Thite M 2011 *Smart cities: implications of urban planning for human resource development.* (Human Resource Development International 14(5) pp623-631)

[7] Siregar B Nasution ABA and Fahmi F 2016 *Integrated pollution monitoring system for smart city* (2016 International Conference on ICT For Smart Society (ICISS) pp 49-52) IEEE
[8] Ariska MF Siregar B Andyani U and Fahmi F 2017 *Power meter monitoring for home appliances based on mobile data communication* (2017 International Conference on Smart Cities, Automation and Intelligent Computing Systems (ICON-SONICS) pp116-120) IEEE
[9] Siregar B Menen K Efendi S Andayani U and Fahmi F 2017 *Monitoring quality standard of waste water using wireless sensor network technology for smart environment* (2017 International Conference on ICT For Smart Society (ICISS) pp 1-6) IEEE
[10] Siregar B Fadhillah D Andayani U Pranoto H and Fahmi F 2017 *Simulation of waste transport monitoring based on garbage load capacity using load cell (*2017 International Conference on ICT For Smart Society (ICISS) pp 1-7) IEEE
[11] Siregar B Efendi S Pranoto H Ginting R Andayani U and Fahmi F 2017 *Remote monitoring system for hydroponic planting media.* (2017 International Conference on ICT For Smart Society (ICISS) pp 1-6) IEEE
[12] Rahmat R F, Athmanathan, Syahputra M F, Lydia M S 2017 *Real time monitoring system for water pollution in Lake Toba* (2016 International Conference on Informatics and Computing, ICCI 2016) pp 383-388 IEEE.
[13] Cosgrave E Arbuthnot K and Tryfonas T 2013 *Living labs, innovation districts and information marketplaces: A systems approach for smart cities* (Procedia Computer Science, 16, pp668-677)
[14] Nasution T H, Siagian E C, Tanjung K and Soeharwinto 2018 Design of river height and speed monitoring system by using Arduino *IOP Conference Series: Materials Science and Engineering* vol 308

IOP Conf. Series: Materials Science and Engineering **648** (2019) 012044 doi:10.1088/1757-899X/648/1/012044

Simulation and Control of Clean Water Supply on Campus Toilets Using Passive Infrared Receiver Sensor Technology and Flow Liquid Meter

U Andayani[1*], Erna B Nababan[1], Ilhammuddin Hasibuan[1], B Siregar[1], F Fahmi[2]

[1]Department of Information Technology, Universitas Sumatera Utara, Indonesia
[2]Department of Electrical Engineering, Universitas Sumatera Utara, Indonesia

*Correspondence Email: ulfi.andayani@usu.ac.id

Abstract. The use of tap water on campus toilets is often problematic due to careless user. The water faucet on campus toilets is often left open, and this leads to wasteful and inefficient use of water. The purpose of this research is as a solution to save and control water usage on campus toilets with automatic faucet so that water usage can be used only when they are necessary or needed. Control of water usage on campus toilets using fluid mechanics and physics kinematics method by using solenoid valve as a valve to open and close the flow of water and flow liquid meter to calculate the water discharge data through it. While, PIR sensor that serve to detect the presence of humans in the toilets so that the solenoid valve can open and let the water flows. While Arduino UNO as microcontroller and Wi-Fi shield used to deliver the water discharge data to application that had been built. The test results show that the method used can work with stable to control water supply on campus toilets with the distance no more than 3 m and large number of users will not affect the performance of the sensor or system built.

1. Introduction

Water is the source of life for every living being. Also, water is one of the main needs in our live. The high level of human needs of water is not proportional to the availability of water on earth, because of all the water that is in the earth, 97% is sea water, while the remaining 3% is fresh water and only 1% is available for human use.

The amount of wasteful use of water causes wasteful use of water. Until now the level of water demand is increasing along with the increasing rate of world population growth. So, it is not exaggeration if UNESCO predicts by 2020 the world will experience a global water crisis.

Monitoring of water usage in toilets is expected to reduce excessive water use. This monitoring is supported by Passive Infrared Receiver (PIR) sensor. Passive Infrared Receiver is an infrared-based sensor. However, it is not as common as most infrared sensors that consist of an IR LED and a phototransistor. PIR does not emit anything like IR LEDs. As the name implies 'passive', this sensor only responds to the energy of the passive infrared beam held by every object detected by it. Objects that can be detected by these sensors are usually the human body.

This is due to an IR filter that filters the wavelength of passive infrared light. IR Filter in a PIR sensor module can filter the wavelength of passive infrared light between 8 to 14 micrometres, so the wavelength produced from the human body ranging between 9 to 10 micrometres can be detected by the sensor.

Thus, when someone walks through the sensor, the sensor captures the emission of a passive infrared beam emitted by the human body which has a different temperature from the environment causing the pyroelectric material to react and generate an electric current due to the heat energy carried by the passive infrared beam. Then an amplifier circuit amplifies the current which is then compared by the comparator to produce the output.

Based on previous research this series of automatic water faucet is manufactured using microcontroller and the motion sensor is used to activate the automatic switch circuits which serve as input to the washer pump to control the valve to close or open. This will open automatically when the sensor is blocked. When the sensor is obstructed then the series will be active.

2. Previous Research

Several studies have been done to conserve water usage in toilets. In 2010, Gabriel conducted research for automatic faucet control in male toilets using PIR sensors. In his research, Gabriel utilized the PIR sensor as its control centre and object detection in the form of human body. The PIR sensor will give instructions to the relay to move the solenoid that acts as a valve. The results of analysis showed that the PIR sensor in this tool can detect the object within a maximum distance of 1.5 meters [1].

A second study in 2011, Alfarobi undertook research for the manufacture of an automatic faucet system for water savings in ablution using motion sensors. In his research, he made a system of automatic ablution faucet by placing the sensor at a certain point, so that when someone finishes, the faucet will automatically close [2].

In 2014, Subandi conducted a research for automation on water faucet to save water consumption. He used an ultrasonic sensor to detect the presence of a limb that is directed to the faucet and stop the flow of water if no limb is directed to the sensor. Ultrasonic sensor will convert physical quantities into mechanical quantities. He used an Atmega16 microcontroller as the system controller. Test results indicated that there was a difference in the amount of volume of water released using automatic faucet and ordinary faucet [3].

In 2014, Yano et al. conducted a research to measure water consumption and compare savings from tap water usage using Wireless Sensor Network. This research used RFbee Sensor to collect, send, and receive data. Data retrieved by RFbee sensors will be collected and then sent wirelessly to a computer connected directly to the RFbee sensor [4].

In 2017, Rahmat et al conducted a research showed that computation of science, especially fluid mechanics and kinematics physics is implemented on microcontroller. Arduino chip-based ATmega328P can be used to detect the location of leaks in pipes by using a data rate of water flow and the system is able to work stably to determining the location of the leak with a maximum distance of 2 meters and it can determine the location of the leak closest to the actual location of the leak with an average flow rate of 10 litters per minute [5].

3. Methodology

This research consists of several steps begins from data collection which is retrieved automatically through the sensor. For more information about the input and output process of the built system can be seen in Figure 1.

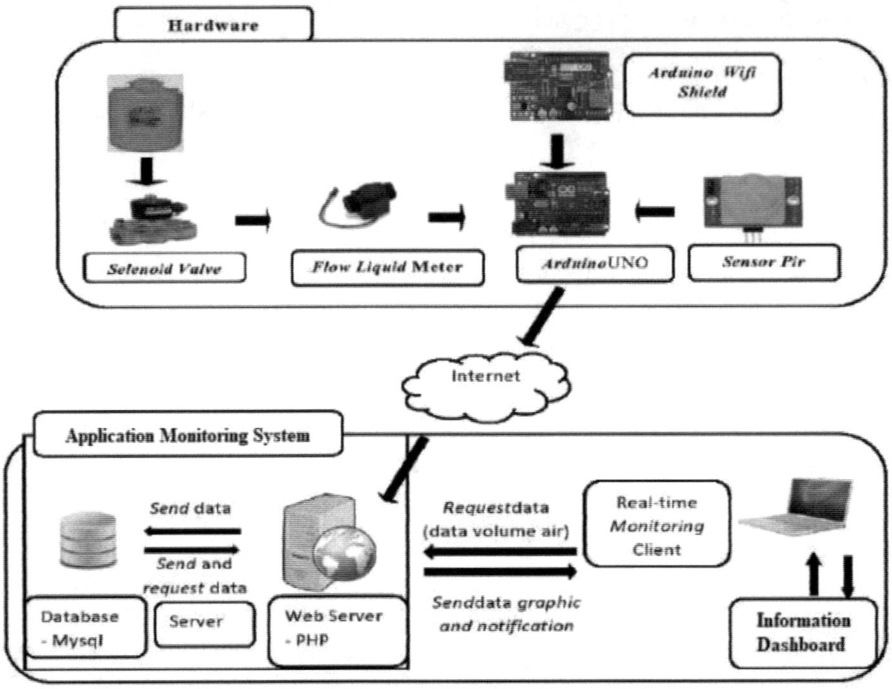

Figure 1. General architecture.

3.1 Hardware Sensor

This section shows the process of data retrieval by the sensor to then be sent to Arduino as well as data transmission by Arduino to the monitoring application system assisted by Wi-Fi shield.

This process starts from the water that flows on the pipe and goes through the solenoid valve. Then, it flows through the flow liquid meter sensor that previously installed in the pipe. The sensor will collect data from the water passing through it and will be sent to the Arduino. Flow liquid meter sensor will transmit data to Arduino via digital pin 2.

Then, Arduino calculates how much water discharge per second that passes through a flow liquid meter sensor. The water discharge per second that had been obtained is first stored in the server where the water discharge data is collected. Server will then directly send the data to the monitoring application system in real time. Collection of data from Arduino to server using Wi-Fi shield, to connect server with Arduino, Arduino will access the IP address of the server. After it is connected, Arduino will send water discharge data using POST method by accessing web page on server. Wi-Fi shield is mounted with stackable method on Arduino and will connect Arduino with server directly using RJ-45.

3.2 Application Monitoring System

Monitoring application system that will be built is a web-based application system using PHP.

3.3 Web Server

Web server used is a standalone web server. This web server will serve as a place of service and data processing between Arduino, database, and client.

Web server will receive the water discharge data sent by the server where the data is collected. This water discharge data will then be stored into the database and ready to be processed and be represented back to the client in the form of graphs. These graphs will be displayed within a certain time interval and will always be updated automatically every second if the Arduino sends the water discharge data to the system. The water discharge data will be processed and ready to be represented back to the client either when they access the web server or not.

3.4 Client

The client will access a web page on the web server to perform monitoring and only the specific client who has permission will be able to access this page. This page will display the amount of water discharge used per day or the amount of water discharge that flows on the flow liquid meter. Water discharge will be displayed in the form of graphs and tables. Water usage graphs will be changed when the water flows through the flow liquid meter and updated automatically in the form of graphs and tables.

3.5 Data Used

The data used in this research is data which is obtained directly from the sensor. Then the Arduino will receive data from the sensor. The data is number of turns of the mill inside the sensor caused by the flow of water through the liquid meter flow sensor. The number of turns of the mill is further processed so that water discharge data is obtained through the sensor every second. Then Arduino will send the data assisted by Wi-Fi shield into the server that had been provided. The data transmitted is in the form of volume water discharge data that had been used.

3.6 Design of Solenoid valve Sensor, Relay, PIR Sensor with Arduino

Arduino has several pins that serve as a place of data processing and power. On this system, pins are used for processing data sent from the sensor. The solenoid valve sensor will be connected to the digital pin 6 to receive data from the sensor. To open and close the flow of water, adaptor is used to assist by connecting the relay as a current limiter.

PIR is an infrared-based sensor. However, it is not as common as most infrared sensors that consist of an IR LED and a phototransistor. PIR does not emit anything like IR LEDs. As the name implies 'passive', this sensor only responds to the energy of the passive infrared beam held by every object detected by it. Objects that can be detected by these sensors are usually the human body. PIR sensor detects infrared waves that come from the body heat of living creatures including humans. These waves are emitted as humans move. The PIR sensor will connect to the Arduino via digital pin 8 can be seen in Figure 2.

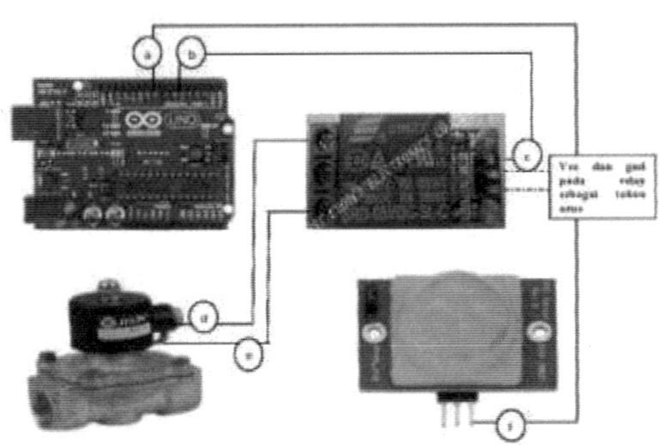

Figure. 2. Design of solenoid valve sensor, PIR sensor with Arduino.

3.7 Design of flow liquid meter with Arduino

Arduino has several pins that serve as a place of data processing and power. On this system, pins are used for processing data sent from the sensor. The flow liquid meter sensor will be connected to the digital pin 2 to receive data from the sensor, GND pin and 5V pin as power for the sensor. To design flow liquid meter with Arduino can be seen in Figure 3.

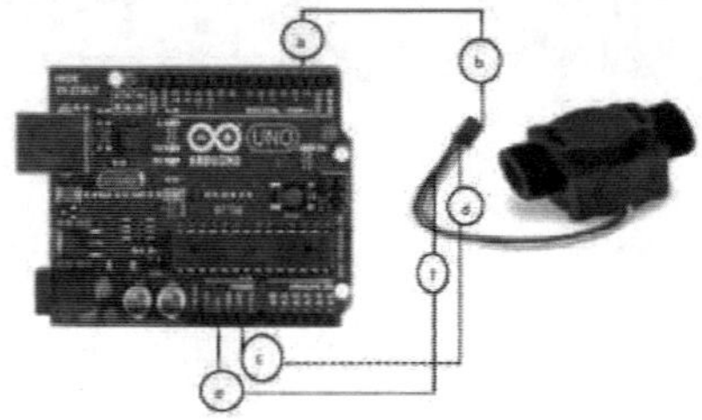

Figure 3. Flow liquid meter design with Arduino.

3.8 Design of Wi-Fi Shield and Arduino

Wi-Fi shield is an additional module used on the Arduino to connect to a server using a Wi-Fi network connection. This module will be mounted with stackable method on the Arduino as shown in Figure 4.

Figure 4. Design of Wi-Fi shield and Arduino.

4. Results and Discussion

System performance test is performed to determine the performance of the system in monitoring whether the system is running well or not. The test performed will display the graph of the amount of water used only when the tool works. The display of the graph will drop because of the interaction of the PIR sensor detects the presence of humans in the room, this will cause the water in the tube to be reduced. Graph can be seen in Figure 5.

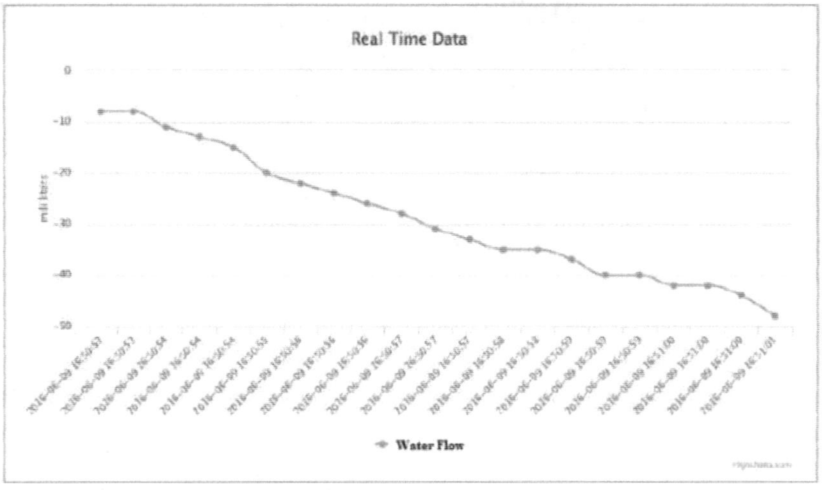

Figure 5. The display of water usage in graph.

The second phase test will display the test results based on the large number of toilet users. The results of this test will be displayed in tabular form based on testing that have been done. Based on the results of testing the number of toilet users, if there are no toilet users, then the system or tool built will not work. This is due to the PIR sensor does not accept or detect the presence of humans in the toilet. So, the water will not flow because the valve on the solenoid valve is not open. When one person enters the toilet, the built-in sensor or system will work, and that's because the PIR sensor detects the presence of a human in the toilet. Then the valve on the solenoid valve will open so that water can flow when it's needed. What is detected by the PIR sensor is the heat wave generated by the human body. Similarly, when 2, 3, and 4 people enter the toilet, the built system or tool can still work well. From the tests conducted based on the number of toilet users can be concluded that the tool can work well. This study shows that the large number of users will not affect the performance of the sensor or system built. The table can be seen in Table 1.

Table 1. PIR Sensor testing based on number of toilet users.

No	Number of toilet users	Work	Does Not Work
1	0	-	Does Not Work
2	1	Work	-
3	2	Work	-
4	3	Work	-
5	4	Work	-

The next test will display the test results ranging from distance that can still be detected by the sensor. The test results performed will be displayed in tabular form. Based on the tests that have been done, the maximum distance that can be detected by the PIR sensor can only work at a distance of 3 meters, but the PIR sensor test with a distance of 3 meters is not as accurate as a distance of 1 meter or 2 meters. Because the heat wave generated by the human body is too far to reach the PIR sensor. If the toilet user is more than 3 meters away, then the PIR sensor will not detect the presence of humans in the toilet and the automatic sensor will not work. It can be seen in Table 2.

Table 2. PIR sensor testing based on distance detectable by sensor.

Distance / Meter	Detected	Not Detected
1 m	Detected	-
2 m	Detected	-
3 m	Detected	-
4 m	-	Not Detected

5. Conclusion

The conclusions can be taken based on simulation and control of clean water supply are:

- The built system can reduce the level of human negligence in the use of water on campus toilets.
- The built system can improve the efficiency of water usage in campus toilets to reduce and anticipate the wasteful use of water.
- PIR sensor will not operate if blocked by objects or something thick, such as doors and walls of the toilet room.
- Testing Distance for PIR sensor no more than 3 m.

IOP Conf. Series: Materials Science and Engineering **648** (2019) 012044 doi:10.1088/1757-899X/648/1/012044

Acknowledgments
The Universitas Sumatera Utara supported this research. All lecturers and staff members of the Information Technology Department.

References

[1] Sakliressy E G, 2010 Kendali Keran Otomatis pada Toilet Pria dengan Sensor PIR (Passive Infrared, Skripsi Universitas Gunadarma, 2010.

[2] Alfarobi, 2011 Pembuatan Sistem Kran Otomatis Untuk Penghematan Air pada Tempat Wudhu dengan Sensor Gerak," Skripsi Universitas Gunadarma.

[3] Subandi, 2014 Sistem Aplikasi Kran Otomatis Untuk Penghematan Air Berbasis Mikrokontrol Atmega 16 *Jurnal Teknologi Technoscientia*, **6**(2) pp. 203-210,.

[4] Oliveira V C and Yano I H 2014 Wireless Sensor Networks for Measuring the Consumption Of Save Water Taps *American Journal of Applied Science*, **11**(6), pp. 899-905, 2014.

[5] Rahmat R F, Satria I S, Siregar B and Budiarto R 2017 Water Pipeline Monitoring and Leak Detection using Flow *International Conference on Electrical Engineering, Computer Science and Informatics (IAES)*.

The Effect of Clamshell Ash Substitution to the Mechanical Properties of Concrete

Rahmi Karolina[1], Ade Sri Rezeki[2], Syahrizal[3], M.AP. Handana[4]

[1,2,3,4] Department of Civil Engineering, Universitas Sumatera Utara, jl. Perpustakaan No. 1 Kampus USU Medan

Email : [1]rachmie_caroline@yahoo.co.id, [2]adesrirezeki@yahoo.com

Abstract. Concrete is the main material for construction that is widely used throughout the world. The more widespread use of concrete and the increasing scale of development also shows more and more concrete needs in the future. The era development of this rapid globalization era has resulted in the increasing number of used goods/waste whose existence can be a problem for life, one of which is the presence of clamshell ash. For this reason, many things have been done in order to recycle to overcome the problem of the existence of this waste. One of them is that clamshell ash can be used as a substitute for cement in a concrete mixture. This research uses clamshell ash and lime with variations of 0%, 5%, 10%, 15%, and 20% from the use of cement. The substitution of clamshell ash and lime affects the strength of the concrete. There are an increase in slump value with the substitution of clamshell ash by 8,9,11,13,14 and lime substitution by 8,8,5,10,12,13. There are a decrease in compressive strength on the substitution of clamshell ash by 89.18%, 74.09%, 67.87%, 64.92% of normal concrete and compressive strength on lime substitution by 69.84%, 58.53%, 57.05 %, 55.82% of normal concrete. There are a decrease in split tensile strength in the substitution of clamshell ash by 95.96%, 92.3%, 81.7%, 75.8% of normal concrete and split tensile strength on lime substitution by 87.93%, 81.33%, 65, 92%, 48.37% of normal concrete.

1. Introduction

Clamshell ash comes from the processing of clamshell waste which is cleaned and mashed up by using a grinding machine until it becomes ash.

Table 1. Chemical composition of clamshell ash and lime

Component	Level of clamshell ash (% Berat)	Level of lime (% Berat)
CaO	55,1038	34,6476
SiO_2	0,924	0,564
Fe_2O_3	0,0017	0,0242
MgO	0,9475	0,1865
Al_2O_3	1,2283	7,7558

2. Scope of problem

1. Concrete quality of f'c = 20 Mpa

2. The samples used are cylinders with a diameter of 15 cm and a height of 30 cm for compressive strength, split tensile strength and absorption test.
3. The samples used for the macrostructure test is 10 cm in diameter and with a height of 1 cm.
4. Tests.
    - Compressive strength
    - Split tensile strength
    - Absorption
    - Macrostructure
5. The substitution of clamshell ash and lime used in each samples are 5%, 10%, 15% and 20%.

3. Objectives

The objectives of this research are

1. To know the workability of fresh concrete that uses clamshell ash as a substitute for cement in a concrete mixture.
2. To know the mechanical behavior of concrete that uses clamshell ash and lime as a substitute for cement in a concrete mixture and compares it with normal concrete. Mechanical behavior studied includes: compressive strength, split tensile strength, absorption and macrostructure.

4. Research method

The method used in this research is an experimental test in the laboratory. The characteristics of the material used are as follows:

4.1. Clamshell ash

The clamshell ash used in this research comes from the processing of clamshell waste obtained from the waste of boiled clamshell stall located on Jalan Gagak Hitam (Ringroad). Clamshells are cleaned first by soaking it for 4 hours, then cleaned by brushing and then dried under the sun. After the shells are clean and dry, the shells are mashed by using a grinding machine until it becomes ash in Kapur Karya Tunggal Refinery, jalan Bunga Sakura I.

4.2. Samples

In this research, the samples are cylindrical with a diameter of 15 cm and a height of 30 cm. Tests are carried out after concrete reaches 28 days curing. The samples with the variation of clamshell ash and lime used can be seen in table 2

Table 2. Number of samples

Variation	Compressiv e Strength of 28 days curing	Split Tensile Strength of Umur 28 days curing	Absorption of 28 days curing	Number of samples
Normal Concrete	3	3	3	9
Concrete + 5% clamshell ash	3	3	3	9
Concrete + 5% lime	1	1	1	3
Concrete + 10% clamshell ash	3	3	3	9
Concrete + 10% lime	1	1	1	3
Concrete + 15% clamshell ash	3	3	3	9
Concrete + 15% lime	1	1	1	3
Concrete + 20% clamshell ash	3	3	3	9
Concrete + 20% lime	1	1	1	3
	Total of Samples			**57**

5. Research location

Concrete Technology and Engineering Materials Laboratory, Civil Engineering, Faculty of Engineering, University of Sumatra Utara for the compressive strength, split tensile strength and absorption test. Mechanical Technology Laboratory and Metallurgy Laboratory, Department of Mechanical Engineering, Faculty of Engineering, University of Sumatra Utara for the macrostructure test and Analytical Chemistry Laboratory, Faculty of Mathematics & Natural Sciences, University of Sumatra Utara for the chemical composition of clamshell ash and lime.

6. Results and discussion

6.1. Slump

The test results of slump value (ASTM C143-90 A) with the substitution of the clamshell ash and lime can be seen in graph 1 below:

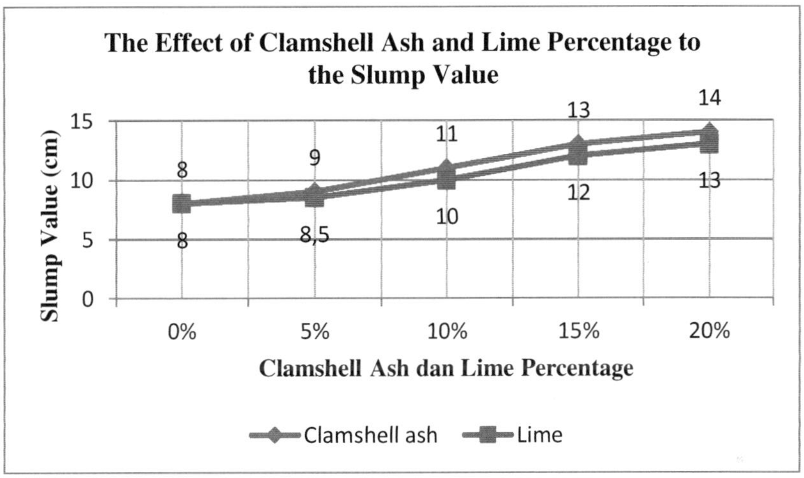

Figure 1. Graph of slump value to the substitution of clamshell ash and lime

6.2. Compressive strength

Concrete compressive strength test (ASTM C39-86) is carried out at 28 days of curing which is intended to obtain an overview of the development of concrete compressive strength by substituting clamshell ash and lime and the results are compared with normal concrete.

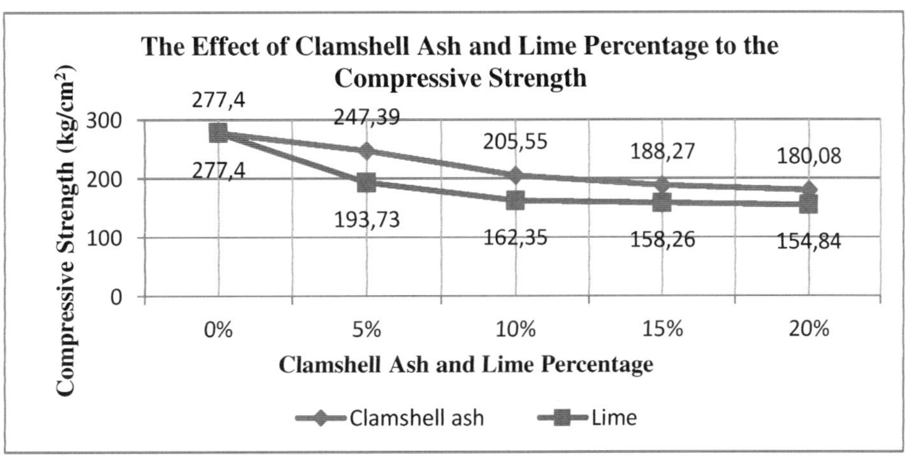

Figure 2. Graph of compressive strength test results

From the test results of concrete cylinders at 28 days curing, the results show that there is a decrease in strength at each level addition of clamshell ash and lime usage. So as to obtain a graph that decreases as the addition of shell and lime ash. A decrease in compressive strength occurs due to the content of silica compounds found in the clamshell ash and lime is very low. The calcium silicate compound is responsible for the mixed hardening process. Besides calcium silicate hydrate which is obtained from its active silica, tricalcium aluminate hydrate is also formed.

The strength of cement paste plays the most important role which is directly influenced by the quality of cement, water and its porosity. The small porosity will increase the strength of the cement paste and is greatly influenced by the water-cement ratio (w / c) in the mixture. The determining aggregate quality is its strength, its surface roughness and its gradation, besides being guaranteed to be free from dirt and reactive chemicals

6.3. Split tensile strength

Concrete split tensile strength test (ASTM C496–96) is carried out at 28 days curing which is intended to obtain an overview of the tensile stress of concrete with the substitution of clamshell ash and lime and the results are compared with normal concrete.

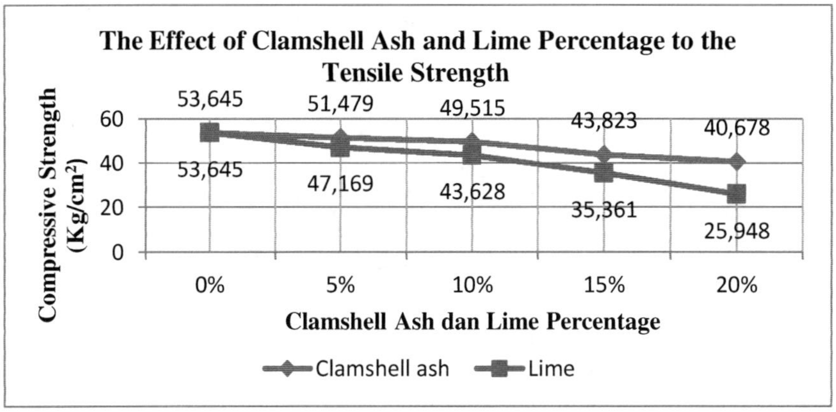

Figure 3. Graph of split tensile strength test results

6.4. Absorption

Absorption test (ASTM C642-97) is carried out at 28 days curing which is intended to obtain an overview of the absorption of concrete with the substitution of clamshell ash and lime and the results are compared with normal concrete.

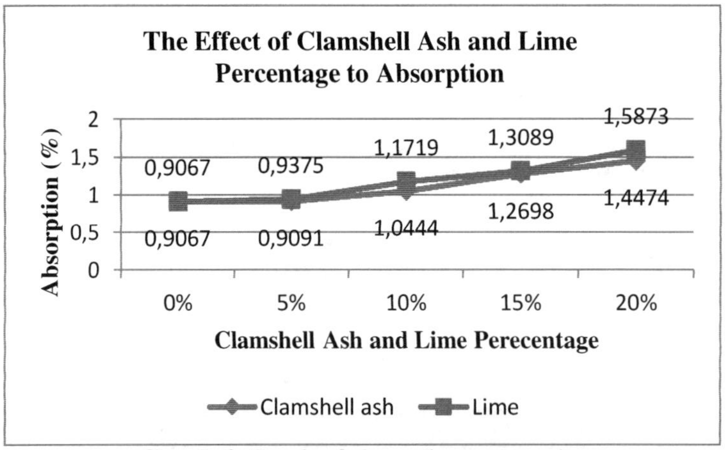

Graph 4. Graph of absorption test results

6.5. Macrostructure

Macrostructure observation (ASTM B-276) that is 100-200 times magnification by using the optical microscope type Rax Vision No.545491, MM-10A, 230V 50.

- Normal concrete

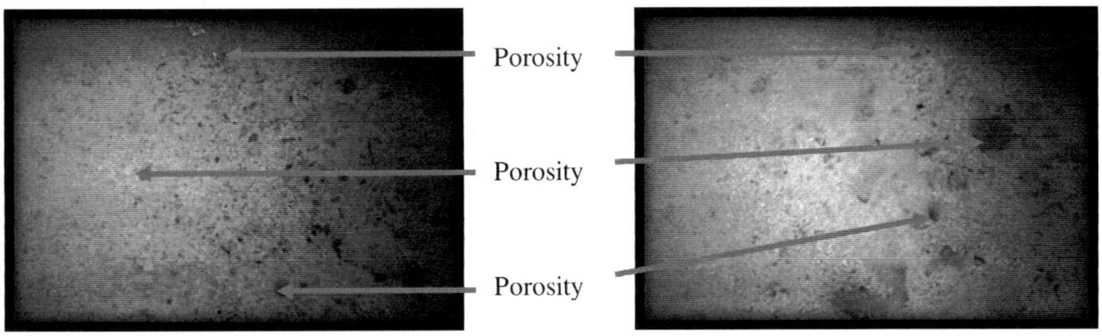

Figure 5. Macro photo of normal concrete at 100× and 200x magnification

- 20% clamshell ash substitution

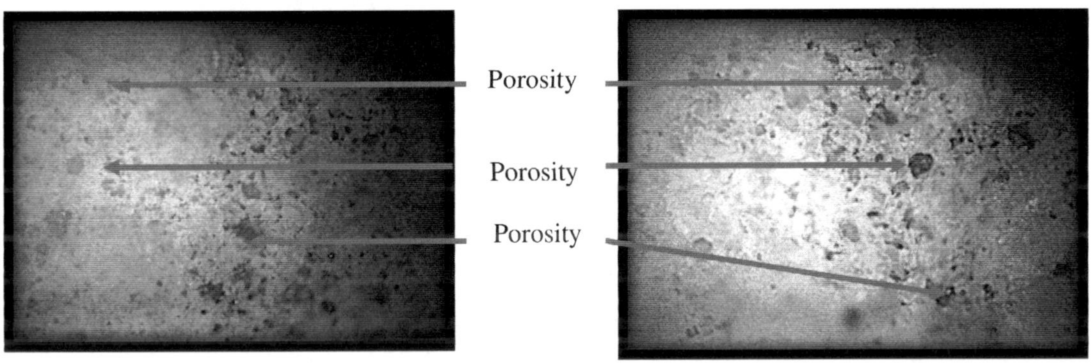

Figure 6. Macro photo of clamshell ash substitution at 100× and 200x magnification

- 20% lime substitution

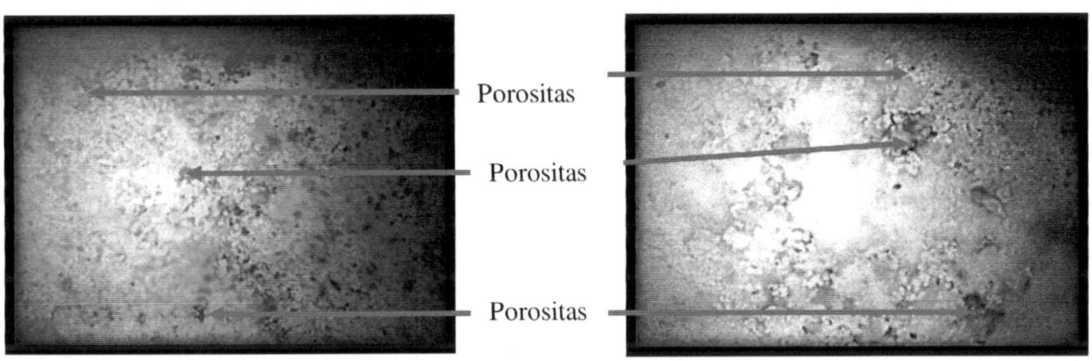

Figure 7. Macro photo of lime substitution at 100× and 200x magnification

The figures above show the results of macrostructure test on normal concrete, the substitution of clamshell ash and the substitution of lime with 100 × and 200x magnification. Visually, the samples can be seen directly and from the figures above, it can be seen that some defects in the porosity form where this will certainly result in a decrease in the mechanical properties of the concrete because it can be the source/start of cracking. The porosity value of normal concrete is less than the concrete mixture with the substitution of clamshell ash and lime.

The greater the percentage of clamshell ash and lime, the greater the porosity that occurs. So that there is a decrease in compressive strength, tensile strength and an increase in absorption value of concrete.

7. Conclusions

1. Substitution of clamshell ash and lime on concrete mixtures of 5%, 10%, 15% and 20% of cement use increases the slump value so that the workability of the concrete decreases.
2. The substitution of 5%, 10%, 15% and 20% clamshell ash of cement use in the concrete mixture decreases the compressive strength by 89.18%, 74.09%, 67.87%, 64.92% from each variation of normal concrete and split tensile strength by 95.96%, 92.3%, 81.7%, 75.8% from each variation of normal concrete.
3. 5%, 10%, 15% and 20% lime substitution of cement use in concrete mixture decreases the compressive strength value by 69.84%, 58.53%, 57.05%, 55.82% from each variation of normal concrete and split tensile strength by 87.93%, 81.33%, 65.92%, 48.37% from each variation of normal concrete.
4. Substitution of clamshell ash has greater compressive strength and split tensile strength than the substitution of lime. The greatest compressive strength in the substitution of clamshell ash is at a 5% percentage of 20.53Mpa and meet the planned concrete quality.
5. The greater the percentage of clamshell ash and lime mixture, the greater the porosity that occurs. So that there is a decrease in compressive strength, tensile strength and an increase in absorption value of concrete.

8. Suggestions

1. Further research is needed by using clamshell ash with different types of clamshells or other types of added material as a comparison material in planning concrete work.
2. Clamshell ash contains all the main chemical elements of cement even though in a lower percentage so that it can be used as a substitute for some cement if the appropriate technology is developed for proper utilization.
3. The combination of clamshell ash waste with other materials can also be considered in order to obtain better results.

9. Reference

[1] ASTM, Annual Books of ASTM Standards 1991 : *Concretes And Aggregates*, Vol.04.02 Construction, Philadelphia-USA: ASTM,1991,PA19103-1187.
[2] Isaac, Olufemi., & Manasseh, Joel. 2009. *Suitability of Periwinkle Shell as Partial Replacement for River Gravel in Concrete.* Department of Civil Engineering, University of Agriculture, Makurdi Benue State, Nigeria.
[3] Jakarta. Aji, Pujo & Rahmat Purwono. 2010. *Pengendalian Mutu Beton.* Itspress Surabaya.
[4] Kartini, Wahyu. 2007. *Penggunaan Serat Polypropylene Untuk Meningkatkan Kuat Tarik Belah Beton.* Jurusan Teknik Sipil, UPN Veteran : Jakarta Timur.
[5] Mulyono, Tri. 2003. *Teknologi Beton.* Penerbit ANDI Yogyakarta.
[6] Murdock, L.J., & Brook K.M. 1986. *Bahan dan Praktek Beton*. Penerbit: Erlangga, Jakarta.
[7] Nwofor, T.C., & Sule, S. 2012. *Stability of groundnut shell ash (GSA)/ordinary portland cement (OPC) concrete in Nigeria*. Department of Civil and Environmental Engineering, University of Port Harcourt, P.M.B 5323, Port Harcourt, Rivers State, Nigeria.
[8] Sagel, Olutoge, Festus A., Okeyinka, Oriyomi M., & Olaniyan, Olatunji S. 2012. *Assessment of The Suitability of Periwinkle Shell Ash (PSA) as Partial Replacement for Ordinary Portland*

Cement (OPC) in Concrete. Department of Civil Engineering, Laoke Akintola University of Technology, Ogbomoso, Nigeria.

[9] Sagel, R., Kole, P., & Kusuma, Gideon. 1993. *Pedoman Pengerjaan Beton.* Penerbit Erlangga : Jakarta.

[10] SK SNI 03-2847-2002. *Tata Cara Perhitungan Struktur Beton untuk Bangunan Gedung.* Badan Standar Nasional.

[11] SK SNI 03-2491-2002. *Metode Pengujian Kuat Tarik Belah Beton.* Badan Standar Nasional.

[12] Swamy, R.N. 1986. *Cement Replacement Materials.* Surrey University Press, London.

AUTHOR INDEX